Numerical Modeling of Superconducting Applications

Simulation of Electromagnetics, Thermal Stability, Thermo-Hydraulics and Mechanical Effects in Large-Scale Superconducting Devices

World Scientific Series in Applications of Superconductivity and Related Phenomena

ISSN 2424-8533

Series Editor: Guy Deutscher *(Tel-Aviv University, Israel)*

Published

Vol. 4 *Numerical Modeling of Superconducting Applications: Simulation of Electromagnetics, Thermal Stability, Thermo-Hydraulics and Mechanical Effects in Large-Scale Superconducting Devices*
edited by Bertrand Dutoit, Francesco Grilli and Frédéric Sirois

Vol. 3 *Superconducting Fault Current Limiter: Innovation for the Electric Grids*
edited by Pascal Tixador

Vol. 2 *MgB_2 Superconducting Wires: Basics and Applications*
edited by René Flükiger

Vol. 1 *Research, Fabrication and Applications of Bi-2223 HTS Wires*
edited by Kenichi Sato

Editor-in-Chief
Guy Deutscher

World Scientific Series in Applications of Superconductivity and Related Phenomena
VOL **4**

Numerical Modeling of Superconducting Applications

Simulation of Electromagnetics, Thermal Stability, Thermo-Hydraulics and Mechanical Effects in Large-Scale Superconducting Devices

Edited by

Bertrand Dutoit
École Polytechnique Fédérale de Lausanne, Switzerland

Francesco Grilli
Karlsruhe Institute of Technology, Germany

Frédéric Sirois
Polytechnique Montréal, Canada

NEW JERSEY • LONDON • SINGAPORE • BEIJING • SHANGHAI • HONG KONG • TAIPEI • CHENNAI • TOKYO

Published by

World Scientific Publishing Co. Pte. Ltd.
5 Toh Tuck Link, Singapore 596224
USA office: 27 Warren Street, Suite 401-402, Hackensack, NJ 07601
UK office: 57 Shelton Street, Covent Garden, London WC2H 9HE

British Library Cataloguing-in-Publication Data
A catalogue record for this book is available from the British Library.

World Scientific Series in Applications of Superconductivity and Related Phenomena — Vol. 4
NUMERICAL MODELING OF SUPERCONDUCTING APPLICATIONS
Simulation of Electromagnetics, Thermal Stability, Thermo-Hydraulics and Mechanical Effects in Large-Scale Superconducting Devices

ISBN 978-981-127-143-4 (hardcover)
ISBN 978-981-127-144-1 (ebook for institutions)
ISBN 978-981-127-145-8 (ebook for individuals)

For any available supplementary material, please visit
https://www.worldscientific.com/worldscibooks/10.1142/13282#t=suppl

Typeset by Stallion Press
Email: enquiries@stallionpress.com

Printed in Singapore

Contents

Introduction

Superconductivity is a particular electronic state that arises in certain materials when cooled below a critical temperature. This unique state results in zero electrical resistance for direct current (DC). For many of these materials, this property holds even in the presence of a high magnetic field (a few tesla up to nearly 100 tesla), which makes superconductors the only sensible option to build energy-efficient high-field magnets. In fact, magnets have been the main commercial application of superconductors, most of the time cooled to 4 K with expensive liquid helium. However, in 1986, a new class of superconductors called high-temperature superconductors (HTS) were discovered. Being superconductor at temperatures of 90 K or above, HTS can easily be cooled to around 65–77 K with liquid nitrogen, much cheaper than liquid helium. The suddenly cheaper cooling cost triggered the development of numerous additional applications, ranging from magnetic levitation to power devices (motors, cables, fault current limiters, transformers, etc.), as well as many others.

In the 1990s HTS wires quickly gained in maturity and demonstrations consisted mostly in replacing copper conductors by superconducting ones in order to assess the characteristics and benefits of HTS devices over conventional technology. However, it was quickly realized that in order to really take advantage of the properties of superconductors, most applications had to be significantly re-thought, and the need for modeling tools rapidly manifested itself. Indeed, good models allow, among other things, predicting the performance of conceptual devices and optimizing them before investing substantial resources to build prototypes. This saves considerable time and money and speeds up developments. In addition, good models reveal the effect of various design parameters on application performance, giving designers valuable insights to improve them and even think about radically new concepts.

The evolution of modeling tools for superconducting applications is correlated with a number of external factors. Firstly, before the 1990s, most models were based on analytical formulas developed for specific cases with relatively simple geometrical arrangements (slab, strip, ellipse, etc.). In addition, the electrical properties of low temperature superconductors were quite compatible with analytical approaches. The most well-known example is certainly the critical state model, also known as Bean's model (1964), which has driven the design of magnets and other applications for more than 40 years. Indeed, with the Bean model and the classical Maxwell equations, one can calculate losses due to flux variations, as well as current and magnetic field distributions in the device. In the same period, numerous analytical models for heat transfer, quench studies and mechanical analysis have also been developed or imported from other disciplines.

The advent of HTS materials engendered a first level of modeling complications. The rounded and very nonlinear E-J curve of HTS materials could not be approximated with the Bean model as well as before, rendering the Maxwell's equations very nonlinear and nearly impossible to solve analytically. In response to the consequent need for numerical simulation tools, researchers started rapidly to develop some. Progress has been exponential: at the end of the 1990s, only a handful of numerical simulation approaches had been published. Then, 25 years later, at the time of writing this book (2022), one can literally find thousands of scientific papers dealing with numerical simulations of HTS devices, most of them based on commercial finite element tools. A smaller but non-negligible portion of these papers come from researchers whose interests are related to the development of numerical methods themselves, and who proposed ways to speed-up computations by using smart mathematical formulations and/or assumptions to simplify the problem to be solved. As more and more applied superconductivity researchers got interested in this type of research, the modeling challenges specific to superconductors has even started to draw the attention of an increasing number of researchers who never touched superconductivity before. Finally, the steady increase of the computational speed of computers, the improvement of generic linear algebra solvers and the advent of sophisticated parallelization techniques that happened over the last 40 years had a huge impact on the complexity and scale of the problems that one can nowadays conceive to solve. Consequently, the development of HTS application has never progressed at such a fast pace.

This fast progress is encouraging, but it also poses the challenge of archiving in an organized way the knowledge in numerical modeling recently generated. Up to now, no comprehensive book that gathers this knowledge has been published, most likely because the task is titanic. The field is so wide and the borders so hard to define that it is difficult to determine where relevancy stops. Nevertheless, this multi-author book is a first attempt in this direction, even if we cannot hope to cover everything in an integrated manner. The choices of topics were made to reach a readership as wide as possible in the four main areas of numerical modeling used in applied superconductivity, namely electromagnetics (Chapter 1), quench and stability (Chapter 2), mechanical and structural analysis (Chapter 3) and thermal-hydraulics (Chapter 4). As a complementary source of information, it is worth mentioning the existence of the www.htsmodelling.com website, where various contributors share model files of HTS modeling.

In order to reach a readership as large as possible, the book has been published as Open Access. As editors, we are very grateful to the authors for the time spent in writing these chapters, as well as for their patience during the editing process that spanned over many years. We all hope that the book will serve its purpose to help modelers speed-up their work. We also hope that it will inspire others to push this effort further in the future. Enjoy reading!

Bertrand Dutoit, Francesco Grilli, Frédéric Sirois
Editors

Chapter 1

Electromagnetic Modeling of Superconductors

Enric Pardo

Institute of Electrical Engineering,
Slovak Academy of Sciences, Bratislava, Slovakia

Francesco Grilli

Karlsruhe Insititute of Technology, Karlsruhe, Germany

Superconductors enable many large-scale electric applications, both current and under research, with a high potential to cause important breakthroughs in human development. These are, for example, the reduction of emissions responsible for the climate crisis through energy generation (fusion and offshore wind turbines), electric transportation (electric and hybrid-electric airplanes or sea vessels), and energy-efficient electric networks (power-transmission cables and transformers). Superconductors also enable novel medical instruments, such as (high-field) magnetic resonance imaging (MRI) and accelerators for ion cancer therapy. Last but not least, superconducting magnets made it possible to conduct some of the largest experiments in fundamental research in the world, involving particle accelerators and detectors, such as the large hadron collider (LHC).

The abovementioned applications are just some examples, with the scope of superconducting applications being much wider.

The design of all these applications requires electromagnetic modeling. Although some simple studies could be done analytically, realistic configurations need computations by numerical modeling. Modeling can also suggest proof-of-concept ideas of novel devices or assist the analysis of electromagnetic characterization of superconducting materials (bulks, wires, tapes, or composite cables). Large-scale applications operate at currents or magnetic fields of characteristic frequencies ranging from mHz or below, such as in magnets, to kHz, such as in certain motors and generators for aviation, passing through at 50–60 Hz for electric grid devices. In any case, the frequency is low enough to neglect electromagnetic radiation, and hence, we can assume slowly varying magnetic fields.

This chapter presents the fundamentals of electromagnetic modeling for slowly changing magnetic fields or currents (quasimagnetostatics). The scope is aimed at researchers or master's and PhD students entering into the field, while keeping high rigor. Although this chapter is mainly a review, some parts are novel.

The outline of this chapter is as follows. Section 1.1 introduces the electromagnetic quantities and basic equations. Next, Section 1.2, we present analytical formulas that also serve to explain the main behavior of superconductors. In Section 1.3, we outline the physical and mathematical background of numerical methods in several formulations. Finally, Section 1.4 presents how to model power applications, including practical guidelines and the state of the art.

1.1. Introduction

In this section, we introduce the physical grounds of electromagnetics under slowly varying magnetic fields (quasimagnetostatics), and hence, we neglect electromagnetic radiation. Some sections go into detail and are rather rigorous in order to empower the knowledge of researchers entering the field.

1.1.1. *Maxwell equations in quasimagnetostatics*

This section outlines the Maxwell equations in quasimagnetostatics, which are the governing laws in electromagnetic modeling of large-scale superconducting applications. The complete Maxwell equations are [40, 57]

$$\nabla \cdot \mathbf{D} = q, \quad \text{Gauss' law,} \tag{1.1}$$

$$\nabla \cdot \mathbf{B} = 0, \quad \text{magnetic flux conservation law,} \tag{1.2}$$

$$\nabla \times \mathbf{E} = -\partial_t \mathbf{B}, \quad \text{Faraday's law,} \tag{1.3}$$

$$\nabla \times \mathbf{H} = \mathbf{J} + \partial_t \mathbf{D}, \quad \text{Ampere's law,} \tag{1.4}$$

where $\mathbf{E}$ is the electric field, $\mathbf{D}$ is the electric displacement, $\mathbf{B}$ is the magnetic flux density, $\mathbf{H}$ is the magnetic field, $\mathbf{J}$ is the current density, and q is the charge density. The symbol ∂_t denotes partial time derivative, and hence, $\partial_t \mathbf{B} = \partial \mathbf{B}/\partial t$. The displacement and magnetic field are defined as

$$\mathbf{D} \equiv \epsilon_0 \mathbf{E} + \mathbf{P}, \tag{1.5}$$

$$\mathbf{H} \equiv \mathbf{B}/\mu_0 - \mathbf{M}, \tag{1.6}$$

where $\mathbf{P}$ and $\mathbf{M}$ are the polarization, or the density of microscopic electric dipoles, and magnetization, or the density of microscopic magnetic dipoles, respectively. In order to solve Equations (1.1–1.4), it is necessary to use the constitutive relations of the material:

$$\mathbf{D} = \overline{\overline{\epsilon}}(\mathbf{E})\mathbf{E}, \tag{1.7}$$

$$\mathbf{B} = \overline{\overline{\mu}}(\mathbf{H})\mathbf{H}, \tag{1.8}$$

$$\mathbf{E} = \overline{\overline{\rho}}(\mathbf{J})\mathbf{J}, \tag{1.9}$$

where $\overline{\overline{\epsilon}}$, $\overline{\overline{\mu}}$, and $\overline{\overline{\rho}}$ are the permittivity, permeability, and resistivity, respectively, which are generally tensorial. In simpler words, this means that $\mathbf{D}$, $\mathbf{B}$, and $\mathbf{E}$ are not always in the same direction as $\mathbf{E}$, $\mathbf{H}$, and $\mathbf{J}$, respectively. The quasimagnetostatic assumption is to neglect the $\partial_t \mathbf{D}$ term in Equation (1.4), obtaining

$$\nabla \times \mathbf{H} = \mathbf{J}. \tag{1.10}$$

This is achieved for low enough frequencies. The influence of the displacement current, $\partial_t \mathbf{D}$, on the current distribution in closed current loops (or multi-turn coils) is negligible for conductor lengths, l, much shorter than the radiation wavelength, λ. This can be regarded as a rule of thumb for any situation since in magnetostatics, the current always forms a closed loop. Setting a stricter criterion than for typical antenna design, $l < \lambda/100$ instead of $l < \lambda/10$ [43, Section 5.1], the displacement current does not influence the current density for frequencies up to around 3 MHz and 3 kHz for conductor lengths of 1 m and 1 km, respectively. Even when the current density in the conductor is not influenced by the displacement current, there will still be a certain small radiation power loss due to the oscillating magnetic dipole moment. However, this contribution is typically negligible compared to the nonlinear Joule AC loss in superconductors.

An interesting consequence of (1.10) is the current conservation equation

$$\nabla \cdot \mathbf{J} = 0, \tag{1.11}$$

which appears because $\nabla \cdot \mathbf{J} = \nabla \cdot (\nabla \times \mathbf{H}) = 0$. Equation (1.11) means that the current density is neither created nor destroyed anywhere, forming closed loops.

1.1.1.1. *Faraday's integral law*

In electromagnetism, it is also often useful to take the integral Faraday's equation into account. This is found from (1.3), also called "differential Faraday's law," or just "Faraday's law." By integrating (1.3) on an imaginary surface, s, we obtain

$$\int_s \mathrm{d}\mathbf{s} \cdot (\nabla \times \mathbf{E}) = -\int_s \mathrm{d}\mathbf{s} \cdot \frac{\partial \mathbf{B}}{\partial t}, \tag{1.12}$$

where $\mathrm{d}\mathbf{s}$ is the surface differential, which is perpendicular to the surface. From Stokes' theorem of calculus, this equation turns into

$$\oint_{\partial s} \mathrm{d}\mathbf{l} \cdot \mathbf{E} = -\frac{\mathrm{d}\Phi}{\mathrm{d}t}, \tag{1.13}$$

where ∂s is the curve that encircles surface s, $\mathrm{d}\mathbf{l}$ is the length differential, and Φ is the magnetic flux crossing surface s defined as

$$\Phi \equiv \int_s \mathrm{d}\mathbf{s} \cdot \mathbf{B}. \tag{1.14}$$

To be precise, the integral Faraday's equation of (1.13) requires that surface s does not change over time.

1.1.2. *Macroscopic electromagnetic properties of superconductors*

Next, we present the constitutive material relations of (1.7)–(1.9) that are usually assumed for superconductors in power and magnet applications.

First, we consider the relation between $\mathbf{E}$ and $\mathbf{J}$. Only type-II superconductors with high pinning, usually named as hard superconductors, can carry sufficiently large currents for power and magnet applications. Although the electromagnetic properties are ultimately governed by vortex physics, their macroscopic behavior can be described by a relation between the macroscopic electric field $\mathbf{E}$ and the macroscopic current density $\mathbf{J}$. These quantities are averaged over volumes large enough in order to contain many vortices but small enough in order to be considered as "differential."

As a rule of thumb, the characteristic side of the differential volume should be much larger than the separation between vortices.

There are several physical models that explain the $\mathbf{E}(\mathbf{J})$ relation in superconductors. Originally, Anderson and Kim [5] obtained the following $\mathbf{E}(\mathbf{J})$ relation based on thermally activated flux creep:

$$\mathbf{E}(\mathbf{J}) = \frac{\mathbf{J}}{|\mathbf{J}|} E_c \exp\left[\frac{U_0}{kT}\left(\frac{|\mathbf{J}|}{J_c} - 1\right)\right], \tag{1.15}$$

where T is the temperature, U_0 is the activation energy of the vortex bundles, which depends on the material, $k = 1.38064852 \times 10^{-23}$ J/K is Boltzmann's constant, E_c is an arbitrary constant, and J_c is the current density that causes $|\mathbf{E}| = E_c$, which also depends on the material. Later, it was found that several mechanisms, such as collective flux creep or vortex glass, can be described by the general relation [17, 22]

$$\mathbf{E}(\mathbf{J}) = \frac{\mathbf{J}}{|\mathbf{J}|} E_c \exp\left[-\frac{U(|\mathbf{J}|)}{kT}\right], \tag{1.16}$$

with

$$U(|\mathbf{J}|) = U_0 \frac{(J_c/|\mathbf{J}|)^\alpha - 1}{\alpha}, \tag{1.17}$$

where α is a material constant. The case of $\alpha = -1$ corresponds to the thermally activated flux creep relation of (1.15). The limit of $\alpha \to 0$ results in $U(|\mathbf{J}|) = U_0 \ln(J_c/|\mathbf{J}|)$ and

$$\mathbf{E}(\mathbf{J}) = E_c \frac{\mathbf{J}}{|\mathbf{J}|}\left(\frac{|\mathbf{J}|}{J_c}\right)^n, \tag{1.18}$$

with $n = U_0/kT$. This power-law $\mathbf{E}(\mathbf{J})$ has been found in many experiments, appearing in most hard superconductors for $|\mathbf{J}|$ close to J_c. Equation (1.18) is the most commonly used relation for superconductor modeling. Figure 1.1 shows this power-law relation for several power-law exponents, n. From (1.18), we see that the resistivity tensor $\overline{\overline{\rho}}$ of the general constitutive equation (1.9) is just the scalar function

$$\rho(\mathbf{J}) = \frac{E_c}{J_c}\left(\frac{|\mathbf{J}|}{J_c}\right)^{n-1}. \tag{1.19}$$

As a consequence, the power law above assumes that the electric field follows the same direction as the current density, and hence, the resistivity is isotropic.

A simpler physical model is the critical-state model (CSM) [13], which states that *Any electromotive force, whatever small, induces a current*

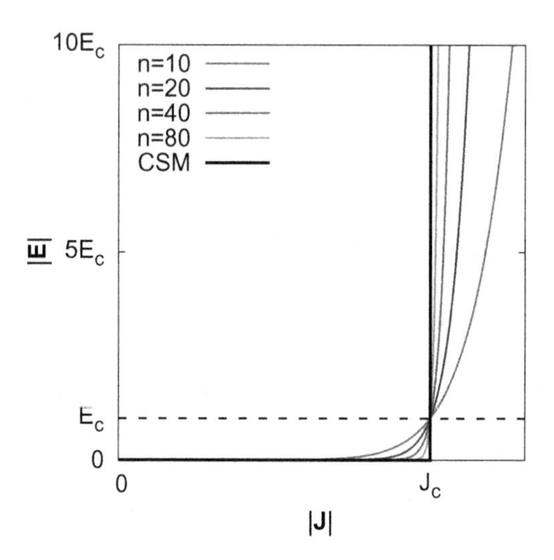

Figure 1.1. The nonlinear $\mathbf{E}(\mathbf{J})$ relation of a superconductor is usually well described by the power law in (1.18). On increasing the power-law exponent n, the power law approaches the multivalued $\mathbf{E}(\mathbf{J})$ relation of the CSM of (1.20).

density with magnitude equal to the critical current density J_c. This statement translates to the multivalued $\mathbf{E}(\mathbf{J})$ relation of Figure 1.1. The CSM, in its most general form, corresponds to the limit of $n \to \infty$ of the power-law relation of Equation (1.18), as shown in Figure 1.1. Low-temperature superconductors (LTS) often exhibit power-law exponents above 100, and REBCO high-temperature superconductors (HTS) may exhibit exponents as high as 40 in low magnetic fields. The simplicity of the CSM enables us to obtain invaluably useful analytical formulas and helps to quantitatively interpret the results obtained from more realistic $\mathbf{E}(\mathbf{J})$ relations. Mathematically, the CSM can be expressed as $\mathbf{J}(\mathbf{E}) = |\mathbf{J}|(|\mathbf{E}|)\mathbf{E}/|\mathbf{E}|$, with

$$|\mathbf{J}|(|\mathbf{E}|) = \begin{cases} J_c & \text{if } |\mathbf{E}| > 0, \\ \text{any } |\mathbf{J}| \text{ with } |\mathbf{J}| \leq J_c & \text{if } |\mathbf{E}| = 0. \end{cases} \tag{1.20}$$

As we discuss in Section 1.2.1, in many configurations, such as cylinders in an axial applied field, $|\mathbf{J}|$ always vanishes where $|\mathbf{E}| = 0$ for the whole history. Then, for those cases, $|\mathbf{J}|$ only takes values 0 or J_c. However, for general shapes such as three-dimensional (3D) rectangular prisms [70] or thin films [79], there appear regions with $|\mathbf{J}|$ grading between 0 and J_c.

The critical current density J_c, both in the CSM and the power law, and the power-law exponent n generally depend on the magnetic flux density

$\mathbf{B}$ and its orientation relative to the superconductor crystallographic axis. Then, the general material relation is of the form $\mathbf{E}(\mathbf{J},\mathbf{B})$. Both J_c and n typically decrease with $|\mathbf{B}|$. A common magnetic-field dependence of J_c for isotropic materials is the Kim model:

$$J_c(B) = \frac{J_0}{(1 + B/B_0)^m}, \tag{1.21}$$

where $B = |\mathbf{B}|$, $J_0 = J_c(B = 0)$, and B_0 and m are positive constants. In addition, the critical current density for a given $|\mathbf{B}|$ often increases when $\mathbf{B}$ becomes parallel (or close to parallel) to $\mathbf{J}$, causing anisotropic resistivity. Although this effect is of physical interest [11, 25, 46], it does not have an impact on most power applications.

In addition, it is usually assumed that the first critical magnetic field, H_{c1}, is negligible, as well as the reversible magnetization curve of the superconductor. Then, the relation between $\mathbf{B}$ and $\mathbf{H}$ of (1.8) is simply $\mathbf{B} = \mu_0\mathbf{H}$.

In quasimagnetostatics, $\mathbf{D}$ is not directly related to neither $\mathbf{B}$ nor $\mathbf{H}$, and hence, it is not necessary to take any relation between $\mathbf{D}$ and $\mathbf{E}$ into account.

1.1.3. *Vector and scalar potentials and their relation to the sources*

The electromagnetic problem can also be set as a function of the scalar and vector potentials, ϕ and $\mathbf{A}$, respectively. The vector and scalar potentials play a key role in integral and variational methods, in addition to the finite element methods (FEMs) directly solving these potentials. This section may be of interest to readers working or willing to work with numerical methods using $\mathbf{A}$ and ϕ as electromagnetic variables, but it may be skipped by others.

The origin of $\mathbf{A}$ and ϕ is the following. The fact that $\nabla \cdot \mathbf{B} = 0$ implies that $\mathbf{B}$ can be written as a function of a vector field $\mathbf{A}$:

$$\mathbf{B} = \nabla \times \mathbf{A} \tag{1.22}$$

because $\nabla \cdot (\nabla \times \mathbf{A}) = 0$ for any $\mathbf{A}$. Faraday's law implies that

$$\nabla \times (\mathbf{E} + \partial_t \mathbf{A}) = 0, \tag{1.23}$$

and hence, $\mathbf{E} + \partial_t\mathbf{A}$ can be written as a function of a scalar field ϕ as $\mathbf{E} + \partial_t\mathbf{A} = -\nabla\phi$ or

$$\mathbf{E} = -\partial_t\mathbf{A} - \nabla\phi \tag{1.24}$$

since $\nabla \times \nabla\phi = 0$ for any ϕ.

Next, consider the vector and scalar potentials, $\mathbf{A}'$ and ϕ', respectively, that follow

$$\mathbf{A}' = \mathbf{A} + \nabla f, \tag{1.25}$$

$$\phi' = \phi - \partial_t f, \tag{1.26}$$

where f is any scalar function. These vector and scalar potentials, $\mathbf{A}'$ and ϕ', describe the same $\mathbf{B}$ and $\mathbf{E}$ as $\mathbf{A}$ and ϕ, respectively. This conversion is called gauge transformation. In the following, we show that there is a gauge, called Coulomb's (or London's) gauge, where the scalar potential corresponds to the electrostatic potential, which is exclusively created by the electric charges.

Let us assume that there is a gauge of the vector potential $\mathbf{A}_c$ that follows

$$\nabla \cdot \mathbf{A}_c = 0. \tag{1.27}$$

In the last paragraph of this section, we see that there is always a vector potential that follows this condition. As a consequence of $\nabla \cdot \mathbf{A}_c = 0$, the divergence in Equation (1.24) becomes

$$\nabla \cdot \mathbf{E} = -\nabla^2 \phi_c, \tag{1.28}$$

where ϕ_c is the scalar potential in the considered gauge. Combining this equation with Gauss' law (1.1) results in

$$\nabla^2 \phi_c = -\frac{q}{\epsilon_0}, \tag{1.29}$$

which is the Laplace equation for the electrostatic potential, and hence, the scalar potential is the electrostatic potential. The solution to this equation can be found by the Green function method (Section 1.1.4), with the result being

$$\phi_c(\mathbf{r}) = \frac{1}{4\pi\epsilon_0} \int_v \mathrm{d}^3\mathbf{r}' \frac{q(\mathbf{r}')}{|\mathbf{r} - \mathbf{r}'|}, \tag{1.30}$$

where v is the volume of the region that contains the charges. For the vector potential, Ampere's law (1.10) in the void, where $\mathbf{B} = \mu_0 \mathbf{H}$, results in

$$\nabla^2 \mathbf{A}_c = -\mu_0 \mathbf{J}, \tag{1.31}$$

where we use the vector relation $\nabla \times \nabla \mathbf{A}_c = -\nabla^2 \mathbf{A}_c + \nabla(\nabla \cdot \mathbf{A}_c)$ and $\nabla \cdot \mathbf{A}_c = 0$. Equation (1.31) consists of three Laplace equations, one for each component of **A**, as in (1.29) but with different constants. By analogy, the solution is

$$\mathbf{A}_c(\mathbf{r}) = \frac{\mu_0}{4\pi} \int_v \mathrm{d}^3\mathbf{r}' \frac{\mathbf{J}(\mathbf{r}')}{|\mathbf{r}-\mathbf{r}'|}, \tag{1.32}$$

where v is the region where there are currents. This equation is strictly valid for the void. In the presence of magnetic materials, (1.32) represents the contribution from the currents.

Therefore, we have seen that for any $q(\mathbf{r})$ and $\mathbf{J}(\mathbf{r})$, there is always a solution to ϕ and $\mathbf{A}$ that follows the condition $\nabla \cdot \mathbf{A} = 0$. Moreover, for this case, the scalar potential is the electrostatic potential. Furthermore, in this gauge, the scalar and vector potentials can be calculated by direct integration from q and $\mathbf{J}$, respectively, by means of (1.30) and (1.32). Thus, q and $\mathbf{J}$ are the sources of the scalar and vector potentials, respectively.

1.1.3.1. *Long straight conductors (infinite)*

For very long (or "infinite") straight conductors transporting current or under a transverse applied field, it is useful to find $\mathbf{A}$ as a function of $\mathbf{J}$ in the cross-section. Taking z as the longitudinal direction, $\mathbf{J}(\mathbf{r}) = J(x,y)\mathbf{e}_z$, where $\mathbf{e}_z$ is the unit vector in the z direction. Hence, the vector potential in Coulomb's gauge follows $\mathbf{A}(\mathbf{r}) = A(x,y)\mathbf{e}_z$. For a thin straight wire with current I_w and length $2l$ much larger than its thickness, direct integration of (1.32) yields

$$A_{\text{wire}}(\mathbf{r}_2) = -\frac{\mu_0}{2\pi} I_w \ln\left[\frac{|\mathbf{r}_2 - \mathbf{r}_2'|}{l}\right], \tag{1.33}$$

where $\mathbf{r}_2$ is the vector position in the plane, $\mathbf{r}_2 = x\mathbf{e}_x + y\mathbf{e}_y$ in Cartesian coordinates, and $\mathbf{r}_2'$ is the central position of the conductor. The deduction of the above equation used the fact that $|\mathbf{r}_2 - \mathbf{r}_2'| \ll l$. The conductor half-length in (1.33) is usually dropped because it only adds a constant term that vanishes when evaluating $\nabla \times \mathbf{A}$. For very long conductors with any $J(\mathbf{r}_2)$, Equation (1.33) becomes

$$A(\mathbf{r}_2) = -\frac{\mu_0}{2\pi} \int_s \mathrm{d}^2\mathbf{r}_2' J(\mathbf{r}_2') \ln|\mathbf{r}_2 - \mathbf{r}_2'| + \frac{\mu_0}{2\pi} I \ln l, \tag{1.34}$$

where I is the net transport current and s is the region in the cross-sectional plane where there are currents. Again, the constant term with $\ln l$ can be omitted.

For infinitely long problems, the gauge transformation of (1.26) can be largely simplified into (1.35). We have seen that for Coulomb's gauge, $\mathbf{A}(\mathbf{r}) = A(x,y)\mathbf{e}_z$. If we also impose the infinitely long symmetry to the vector potential $\mathbf{A}'$ after the transformation, we get that $\mathbf{A}'(\mathbf{r}) = A'(x,y)\mathbf{e}_z$. As a consequence, the gauge fixing function f satisfies $\nabla f(\mathbf{r}) = [\partial_z f](x,y)\mathbf{e}_z$. Therefore, $\partial_x f = \partial_y f = 0$, and hence, f depends on neither x nor y. Because of the infinitely long symmetry, $\partial_z f$ does not depend on z, and hence, $f(z)$ needs to be linear with z. Then, $\partial_z f$ is uniform in the whole space. Thus, for infinitely long problems, the gauge transformation for the vector potential reduces to

$$A'(x,y) = A(x,y) + \partial_z f. \tag{1.35}$$

A single scalar, $\partial_z f$, determines the gauge at each instant of time, and hence, we can set the gauge by simply imposing a certain time dependence of the vector potential at an arbitrarily chosen point.

The longitudinal symmetry also causes the electrostatic scalar potential to be uniform within each conductor. This can be seen as follows. The infinitely long symmetry results in $\mathbf{J}(\mathbf{r}) = J(x,y)\mathbf{e}_z$. If the resistivity of the conductor is isotropic or the z direction corresponds to one of the main axes of the resistivity tensor, the electric field follows the z direction, and hence, $\mathbf{E}(\mathbf{r}) = E(x,y)\mathbf{e}_z$. From (1.24) and the symmetry of $\mathbf{A}$, the vector potential becomes $\nabla\phi(\mathbf{r}) = -[E(x,y) + \partial_t A(x,y)]\mathbf{e}_z$, and hence, $\nabla\phi(\mathbf{r}) = [\partial_z\phi](x,y)\mathbf{e}_z$. As we have seen for the gauge transformation, this condition results in $\partial_z\phi$ becoming uniform within each conductor. However, $\nabla\phi$ is not always uniform in the whole space. This can be seen in the following counterexample (Figure 1.2). If there are two very long conductors closed at the ends with two identical voltage sources at each end, there is a certain voltage drop between the conductors everywhere except at the central plane. Although $\nabla\phi$ follows the conductor's direction except very close to the ends, $\nabla\phi$ in the air in between generally has a transverse component. For this reason, $\partial_z\phi$ could be different at each conductor.

1.1.3.2. *Axial symmetry*

For problems with axial symmetry, such as cylinders and circular coils, the current density follows

$$\mathbf{J}(\mathbf{r}) = J(r,z)\mathbf{e}_\varphi, \tag{1.36}$$

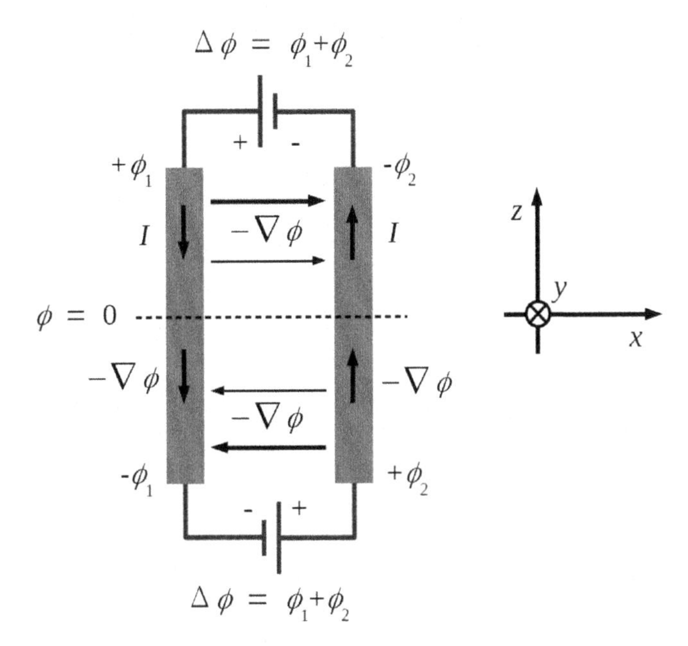

Figure 1.2. For very long conductors (usually referred as "infinite"), the modeling is done on the central xy plane, which is located at $z = 0$. Although the gradient of the scalar potential in Coulomb's gauge, $\nabla\phi$, is uniform at each conductor (solid blue rectangles), at $z \neq 0$, there is a transverse component of $\nabla\phi$ causing a transverse electric field in the air. The voltage sources at the ends are assumed identical, which are also generating a net current I at each conductor.

and hence, in Coulomb's gauge,

$$\mathbf{A}(\mathbf{r}) = A(r, z)\mathbf{e}_\varphi, \tag{1.37}$$

where r and z are the radial and axial coordinates, respectively, and $\mathbf{e}_\varphi$ is the unit vector in the angular direction. Now, a relevant case is the vector potential generated by a thin closed loop, with current I_l being (see Ref. [54, p. 112])

$$A_{\text{loop}}(r, z - z', r') = I_l \frac{\mu_0}{\pi k}\sqrt{\frac{r'}{r}}\left[\left(1 - \frac{k^2}{2}\right) K(k) - E(k)\right], \tag{1.38}$$

with

$$k^2 = \frac{4rr'}{(r + r')^2 + (z - z')^2}, \tag{1.39}$$

where $K(k)$ and $E(k)$ are the complete elliptic integrals of the first and second kind, respectively, and r' and z' are the coordinates of the loop. Then, for the whole body with axial symmetry,

$$A(r,z) = \int_{v_s} \mathrm{d}r'\mathrm{d}z' J(r',z') a_{\mathrm{loop}}(r, z-z', r'), \tag{1.40}$$

where $a_{\mathrm{loop}}(r, z-z', r') = A_{\mathrm{loop}}(r, z-z', r')/I_l$ and v_s is the region in the (r,z) plane where there are currents.

As for infinitely long problems, we can find important simplifications for the gauge transformation and the scalar potential at each conductor.

The gradient of the gauge transformation function in (1.26) satisfies $\nabla f(\mathbf{r}) = (1/r)\partial_\varphi f \mathbf{e}_\varphi$ with uniform $\partial_\varphi f$ in the whole space. This follows from the symmetry of $\mathbf{A}$ in Coulomb's gauge, $\mathbf{A}(\mathbf{r}) = A(r,z)\mathbf{e}_\varphi$, and the imposed symmetry of the transformed vector potential, $\mathbf{A}'(\mathbf{r}) = A'(r,z)\mathbf{e}_\varphi$. The simplification of the gauge transformation function implies that the gauge can be fixed by imposing the value of the vector potential at an arbitrary point for every time instant.

The gradient of the scalar potential within the conductors follows $\nabla\phi(\mathbf{r}) = (1/r)\partial_\varphi \phi \mathbf{e}_\varphi$, with uniform $\partial_\varphi\phi$, which could be different in each conductor. This follows from (1.24), (1.37), and $\mathbf{E} = E(r,z)\mathbf{e}_\varphi$, the last one being a consequence of (1.36) for isotropic resistivity tensors or those with a main axis in the angular direction. In the air, $\nabla\phi$ does not need to follow the angular direction since there could be a certain potential drop between conductors, as is the case of coils.

1.1.4. *Solution to the Laplace equation for electrostatics*

In this section, we find that the solution to the Laplace equation for electrostatics (1.29) is (1.30). This is used in the scalar and vector potential expressions as a function of their sources in Section 1.1.3. Readers not interested in the mathematical details may skip this section.

This can be seen by means of the Green function method, as follows. Any charge density $q(\mathbf{r})$ can be expressed as a combination of point charges as

$$q(\mathbf{r}) = \int_v \mathrm{d}^3\mathbf{r}' q(\mathbf{r}')\delta(\mathbf{r}-\mathbf{r}'), \tag{1.41}$$

where v is the region where there are charges and $\delta(\mathbf{r}-\mathbf{r}')$ is Dirac's delta, which follows $1 = \int_{\mathbb{R}^3} \mathrm{d}^3\mathbf{r}\delta(\mathbf{r}-\mathbf{r}')$ ($\mathbb{R}^3$ is the whole 3D space) and $\delta(\mathbf{r} -$

$\mathbf{r}') = 0$ except for $\mathbf{r} \to \mathbf{r}'$, where $\delta(\mathbf{r} - \mathbf{r}') \to \infty$. With $G(\mathbf{r}, \mathbf{r}')$ being the solution to

$$\nabla^2 G(\mathbf{r}, \mathbf{r}') = -\frac{\delta(\mathbf{r} - \mathbf{r}')}{\epsilon_0}, \tag{1.42}$$

the solution to $\phi_c(\mathbf{r})$ is

$$\phi_c(\mathbf{r}) = \int_v \mathrm{d}^3\mathbf{r} G(\mathbf{r}, \mathbf{r}') q(\mathbf{r}') \tag{1.43}$$

because from (1.41), (1.42), and (1.43), it is seen that

$$\begin{aligned} \nabla^2 \phi_c(\mathbf{r}) &= -\frac{q(\mathbf{r})}{\epsilon_0} = -\frac{1}{\epsilon_0} \int_v \mathrm{d}^3\mathbf{r}' q(\mathbf{r}') \delta(\mathbf{r} - \mathbf{r}') = \int_v \mathrm{d}^3\mathbf{r}' \nabla^2 G(\mathbf{r}, \mathbf{r}') q(\mathbf{r}') \\ &= \nabla^2 \int_v \mathrm{d}^3\mathbf{r}' G(\mathbf{r}, \mathbf{r}') q(\mathbf{r}'), \end{aligned} \tag{1.44}$$

and hence, Equation (1.43) follows, except the constant terms.

Taking into account that $\nabla^2 = \sum_i \partial^2/(\partial x_i)^2 = \sum_i \partial^2/[\partial(x_i - x_i')]^2 \equiv \nabla^2_{\mathbf{r}''}$, where x_i is one of the Cartesian coordinates, $x_i \in \{x, y, z\}$ and $\mathbf{r}'' = \mathbf{r} - \mathbf{r}'$, we can see that Equation (1.42) becomes

$$\nabla^2_{\mathbf{r}''} G(\mathbf{r}, \mathbf{r}') = -\frac{\delta(\mathbf{r}'')}{\epsilon_0}. \tag{1.45}$$

Since the right-hand side of (1.42) only depends on $\mathbf{r}''$, $\nabla_{\mathbf{r}''} G$ only depends on $\mathbf{r}''$. Taking into account that $\delta(\mathbf{r}'') = \delta(r'')$, where $r'' = |\mathbf{r}''|$, and using the fact that $\nabla^2_{\mathbf{r}''}$ is an isotropic operator (there is no privileged direction), G should be isotropic and hence depends only on r'': $G(r'')$.

From Gauss' theorem,

$$\int_{\partial v} \mathrm{d}\mathbf{s}'' \cdot \nabla G(r'') = -\frac{1}{\epsilon_0} \int_v \mathrm{d}^3\mathbf{r} \delta(\mathbf{r} - \mathbf{r}') = -\frac{1}{\epsilon_0}. \tag{1.46}$$

Taking v as a sphere, where ∂v is its surface, we obtain

$$\nabla_{\mathbf{r}''} G(r'') = -\frac{\mathbf{e}_{r''}}{4\pi\epsilon_0 r''^2}, \tag{1.47}$$

where $\mathbf{e}_{r''}$ is the unit vector in the $\mathbf{r}'' = \mathbf{r} - \mathbf{r}''$ direction. Integrating over r'' results in

$$G(\mathbf{r}, \mathbf{r}') = \frac{1}{4\pi\epsilon_0 |\mathbf{r} - \mathbf{r}'|}. \tag{1.48}$$

Combining (1.43) with (1.48) yields

$$\phi_c(\mathbf{r}) = \frac{1}{4\pi\epsilon_0} \int_v \mathrm{d}^3\mathbf{r}' \frac{q(\mathbf{r}')}{|\mathbf{r} - \mathbf{r}'|}. \tag{1.49}$$

1.1.5. *Integral relation between* B *and* J

There is also an integral relation between **B** and **J** that can be easily found from the vector potential in Coulomb's gauge from Section 1.1.3 by means of $\mathbf{B} = \nabla \times \mathbf{A}$.

From (1.32), we obtain the well-known Biot–Savart law for general 3D shapes [40]:

$$\mathbf{B}(\mathbf{r}) = \frac{\mu_0}{4\pi} \int_v \mathrm{d}^3\mathbf{r}' \frac{\mathbf{J}(\mathbf{r}') \times (\mathbf{r} - \mathbf{r}')}{|\mathbf{r} - \mathbf{r}'|^3}. \tag{1.50}$$

The equivalent law for infinitely long problems in the z direction can be obtained from (1.34):

$$\mathbf{B}(\mathbf{r}_2) = \frac{\mu_0}{2\pi} \int_s \mathrm{d}^2\mathbf{r}_2' \frac{J(\mathbf{r}_2')\mathbf{e}_z \times (\mathbf{r}_2 - \mathbf{r}_2')}{|\mathbf{r}_2 - \mathbf{r}_2'|^2}, \tag{1.51}$$

where $\mathbf{J} = J\mathbf{e}_z$ and $\mathbf{r}_2 = x\mathbf{e}_x + y\mathbf{e}_y$.

Finally, for axisymmetric configurations, $\mathbf{B} = \nabla \times \mathbf{A}$ from $\mathbf{A}$ in (1.40) results in

$$\mathbf{B}(r, z) = \int_{v_s} \mathrm{d}r'\mathrm{d}z' J(r', z')\mathbf{b}_{\mathrm{loop}}(r, z - z', r'), \tag{1.52}$$

where $\mathrm{d}r'\mathrm{d}z' J(r', z')\mathbf{b}_{\mathrm{loop}}(r, z - z', r')$ is the magnetic flux density created by a circular loop of radius r' located at $z = z'$ and current $\mathrm{d}r'\mathrm{d}z' J(r', z')$. Here, $\mathbf{b}_{\mathrm{loop}}$ has both r and z components, $\mathbf{b}_{\mathrm{loop}} = b_{r,\mathrm{loop}}\mathbf{e}_r + b_{z,\mathrm{loop}}\mathbf{e}_z$, where

$$b_{r,\mathrm{loop}}(r, z - z', r') = \frac{\mu_0}{2\pi\sqrt{(r' + r)^2 + (z - z')^2}} \left[K(k) + E(k) \frac{r'^2 - r^2 - (z - z')^2}{(r' - r)^2 + (z - z')^2} \right], \tag{1.53}$$

$$b_{r,\mathrm{loop}}(r, z - z', r') = \frac{-\mu_0(z - z')}{2\pi r\sqrt{(r' + r)^2 + (z - z')^2}} \left[K(k) - E(k) \frac{r'^2 + r^2 + (z - z')^2}{(r' - r)^2 + (z - z')^2} \right]. \tag{1.54}$$

In the expression above, k is given by (1.39), and $K(k)$ and $E(k)$ are the complete elliptic integrals of the first and second kind, respectively.

1.1.6. *Current potentials*

Modeling methods using the current potentials as state variables have recently become very popular. Thus, it is worth dedicating a few pages introducing these potentials and their physical meaning, which depend on the chosen gauge.

Thanks to the current conservation condition $\nabla \cdot \mathbf{J} = 0$ in quasimagnetostatics, the current density can be written as a function of the vector potential $\mathbf{T}$ as

$$\mathbf{J} = \nabla \times \mathbf{T}. \tag{1.55}$$

From this property and (1.10), we find that

$$\nabla \times (\mathbf{H} - \mathbf{T}) = 0, \tag{1.56}$$

and hence, $\mathbf{H} - \mathbf{T}$ is a conservative field. Thus, $\mathbf{H} - \mathbf{T}$ can be written as the gradient of a current scalar potential Ω:

$$\mathbf{H} = \mathbf{T} - \nabla\Omega. \tag{1.57}$$

Similar to the vector and scalar potentials in Section 1.1.3, the current vector and scalar potentials have gauge freedom. That is, the current vector and scalar potentials $\mathbf{T}'$ and Ω', related to $\mathbf{T}$ and Ω as

$$\begin{aligned} \mathbf{T}' &= \mathbf{T} + \nabla\phi \\ \Omega' &= \Omega + \phi, \end{aligned} \tag{1.58}$$

generate the same $\mathbf{J}$ and $\mathbf{H}$ as $\mathbf{T}$ and Ω, respectively. The physical interpretation of $\mathbf{T}$ and Ω depend on the gauge.

1.1.6.1. *Divergence-free gauge of* $\mathbf{T}$

For a divergence-free gauge, defined as

$$\nabla \cdot \mathbf{T} = 0 \tag{1.59}$$

everywhere, the divergence of $\mathbf{H}$ becomes $\nabla \cdot \mathbf{H} = -\nabla^2\Omega$. From $\mathbf{B} = \mu_0(\mathbf{H} + \mathbf{M})$, we obtain

$$\nabla^2\Omega = -q_m, \quad \text{with} \quad q_m \equiv -\nabla \cdot \mathbf{M}, \tag{1.60}$$

where q_m is called the magnetic pole density. Thus, Ω becomes the magnetic potential, which is analogous to the electrostatic potential (compare (1.60) with (1.29)), where q_m plays the same role of q/ϵ_0. The condition $\nabla \cdot \mathbf{T} = 0$ also implies that $\mathbf{T}$ can be written as another vector potential $\mathbf{G}$ so that

$$\mathbf{T} = \nabla \times \mathbf{G}. \tag{1.61}$$

This potential $\mathbf{G}$ is also subjected to gauge invariance. Choosing the gauge

$$\nabla \cdot \mathbf{G} = 0 \tag{1.62}$$

from $\nabla \times \mathbf{T} = \mathbf{J}$, we obtain

$$\nabla^2 \mathbf{G} = -\mathbf{J}, \tag{1.63}$$

which is analogous to Equation (1.31). Therefore, the solution to the above equation is

$$\mathbf{G}(\mathbf{r}) = \frac{1}{4\pi} \int_v \mathrm{d}^3\mathbf{r}' \frac{\mathbf{J}(\mathbf{r}')}{|\mathbf{r} - \mathbf{r}'|}, \tag{1.64}$$

which is proportional to the vector potential generated by the current density, $\mathbf{A}_J$, as $\mathbf{G} = \mathbf{A}_J/\mu_0$. Then, the current vector potential $\mathbf{T} = \nabla \times \mathbf{G}$ corresponds to the magnetic flux density generated by $\mathbf{J}$, $\mathbf{B}_J$, as

$$\mathbf{T} = \mathbf{B}_J/\mu_0. \tag{1.65}$$

Therefore, for the divergence-free gauge, $\mu_0\mathbf{T}$ and Ω correspond to the magnetic flux density generated by the currents and the magnetic potential generated by the magnetic materials, respectively.

The current vector and scalar potentials for any gauge can be written as

$$\begin{aligned} \mathbf{T} &= \mathbf{T}_{\mathrm{df}} + \nabla\phi, \\ \Omega &= \Omega_{\mathrm{df}} + \phi, \end{aligned} \tag{1.66}$$

where $\mathbf{T}_{\mathrm{df}}$ and Ω_{df} are the $\mathbf{T}$ and Ω from the divergence-free gauge, respectively.

1.1.6.2. *Magnetic-field gauge*

An alternative gauge is that defined by $\phi = -\Omega_{\rm df}$ in (1.66), and hence,

$$\mathbf{T} = \mathbf{T}_{\rm df} - \nabla\Omega_{\rm df},$$
$$\Omega = 0. \tag{1.67}$$

From (1.57), $\mathbf{T}$ becomes the magnetic field:

$$\mathbf{T} = \mathbf{H}. \tag{1.68}$$

1.1.6.3. *Current potential as magnetization*

For systems with no net current and no magnetic materials, the current vector potential can be interpreted as an effective magnetization, and hence, $\mathbf{J} = \nabla \times \mathbf{T}$ is taken as a magnetization current as done by Pardo and Kapolka [69]. For this case,

$$\mathbf{B} = \mu_0(\mathbf{H} + \mathbf{T}), \tag{1.69}$$

$$\nabla \times \mathbf{H} = 0. \tag{1.70}$$

As a consequence of the second equation, there exists a scalar potential Ω that follows

$$\mathbf{H} = -\nabla\Omega. \tag{1.71}$$

Again, we obtain $\nabla \cdot \mathbf{H} = -\nabla^2\Omega$. From (1.69), this results in

$$\nabla^2\Omega = -q_T \quad \text{with} \quad q_T \equiv -\nabla \cdot \mathbf{T}. \tag{1.72}$$

Consistently, Ω is the magnetic potential as a consequence of the effective magnetization pole density q_T. The advantage of this interpretation of $\mathbf{T}$ is that it vanishes outside the conducting (or superconducting) sample. Since $\mathbf{T}$ is taken now as a magnetization, there is no gauge invariance. In general, $\nabla \cdot \mathbf{T}$ does not vanish everywhere because $-\nabla \cdot \mathbf{T}$ corresponds to the effective magnetic pole density, although in some cases, it may occur that $\nabla \cdot \mathbf{T} \neq 0$ at the sample surface only. Actually, this interpretation of $\mathbf{T}$ can also be applied when transport currents and magnetic materials are present [69], although this extension is outside the scope of this book.

1.1.7. *Calculation of local dissipation and AC loss*

A quantity of practical importance for the design of large-scale applications is the power dissipation under changing or alternating magnetic fields, also called "AC loss."

In Section 1.1.7.1, we show that the local power loss dissipation per unit volume in both normal and superconducting materials is

$$p = \mathbf{J} \cdot \mathbf{E}. \tag{1.73}$$

Therefore, the evaluation of the AC loss for both kinds of materials is the same once $\mathbf{J}$ is known:

$$p = \mathbf{J} \cdot \mathbf{E}(\mathbf{J}), \tag{1.74}$$

where $\mathbf{E}(\mathbf{J})$ is the constitutive relation of the material. For a normal conductor, we have $\mathbf{E}(\mathbf{J}) = \rho\mathbf{J}$, while for a superconductor, $\mathbf{E}(\mathbf{J})$ is a nonlinear function of $\mathbf{J}$ (see Section 1.1.2). For the CSM, where the $\mathbf{E}(\mathbf{J})$ relation is multiple-valued ($\mathbf{E}$ can take several values for $|\mathbf{J}| = J_c$), $\mathbf{E}$ should be evaluated in a different way, such as from the vector and scalar potentials ($\mathbf{E} = -\partial_t\mathbf{A} - \nabla\phi$). From (1.73), the total instantaneous power loss in the sample is

$$P = \int_v \mathrm{d}^3\mathbf{r}\ \mathbf{J} \cdot \mathbf{E}, \tag{1.75}$$

and the loss per cycle for periodic excitations (applied magnetic field or current) is

$$Q = \oint \mathrm{d}t \int_v \mathrm{d}^3\mathbf{r}\ \mathbf{J} \cdot \mathbf{E}. \tag{1.76}$$

The hysteresis AC loss of magnetic materials, where there are no free currents $\mathbf{J}$, should be evaluated using the equations presented in Section 1.1.7.2.

1.1.7.1. *Fundamental aspects of the local loss dissipation*

In electromagnetism, the power dissipation per unit volume created by any current density $\mathbf{J}$ in a certain electric field $\mathbf{E}$ is $p = \mathbf{J} \cdot \mathbf{E}$ (Equation (1.73) above). This can be seen from the work exerted on the charge carriers, as follows. The force on a single carrier of charge C is $\mathbf{F} = C\mathbf{E}$. For a given small time δt, this force causes a work

$$\delta W = C\delta\mathbf{r} \cdot \mathbf{E} = C\delta t\mathbf{v} \cdot \mathbf{E}. \tag{1.77}$$

If the charge C belongs to a differential volume $\mathrm{d}V$, then $C = q\mathrm{d}V$, where q is the charge density. Then, the work density δw is

$$\delta w = \delta t q\mathbf{v} \cdot \mathbf{E} = \delta t\mathbf{J} \cdot \mathbf{E}. \tag{1.78}$$

Therefore, the power density, $p = \delta w/\delta t$, exerted by any current density is given by (1.73). This electric field is created by external sources from the observed differential volume since the charge in that volume does not create any net electric field at the center of the volume because of symmetry. Then, this work is exerted by external sources from the observation point. If $p > 0$, then the sources are providing energy, and hence, they are "losing" energy. Since the work δw on the differential volume is exerted locally, this energy gain must occur locally. This energy is transferred to another physical system, typically the kinetic energy of the carriers or the microscopic structure of the material. If this local electromagnetic energy is consumed to increase the thermal energy, we can say that the electromagnetic energy is dissipated. This is the link between the electromagnetic system and the thermal system in electrothermal problems. For $\mathbf{J}$ distributions in finite volumes, the net work will still be exerted by external sources.

In normal conductors, the work is usually transferred to the ionic lattice via conducting carrier collisions to impurities, lattice disorders, or oscillations. The result of the mechanisms is that for a given $\mathbf{E}$, $\mathbf{J}$ does not change in time, except for an extremely short transient. Then, a constitutive $\mathbf{E}(\mathbf{J})$ relation can be formulated.

Since the ultimate charge carriers in superconductors are electrons (or, more precisely, Cooper pairs), the same reasoning to obtain Equation (1.73) also applies to superconductors. However, we can also obtain (1.73) from vortex physics, as shown in the following. Then, we can ignore the charge carriers and assume that the microscopic origin of all electromagnetic properties in type-II superconductors are vortices, their gradient, and their movement. The starting point is that the driving force per unit length on a vortex is $\mathbf{F}_d = \mathbf{J} \times \mathbf{\Phi}_0$, where $|\mathbf{\Phi}_0|$ is the vortex flux quantum, which is $h/2e$, with e being the charge of the electron and h being Planck's constant. The direction of $\mathbf{\Phi}_0$ follows the direction of the vortex. Then, the rate of work per unit volume, p, on the vortices is

$$p = (\mathbf{J} \times \mathbf{\Phi}_0 n) \cdot \mathbf{v} = (\mathbf{\Phi}_0 n \times \mathbf{v}) \cdot \mathbf{J}, \tag{1.79}$$

where n and $\mathbf{v}$ are the vortex density and their velocity, respectively. From electromagnetic analysis, it can be shown that the vortices moving with a speed $\mathbf{v}$ create an electric field [36]:

$$\mathbf{E} = \mathbf{\Phi}_0 n \times \mathbf{v}. \tag{1.80}$$

The field $\mathbf{E}$ and $\mathbf{J}$ above are averaged over a small volume containing at least one vortex. Finally, by inserting Equation (1.80) into (1.79), we obtain $p - \mathbf{J} \cdot \mathbf{E}$.

1.1.7.2. *Hysteresis loss of magnetic materials*

In this section, we show that the hysteresis loss in a magnetic material under cyclic applied magnetic fields is (1.82) in general and (1.88) for soft ferromagnetic materials.

From Maxwell equations, it can be seen that the change in the free energy, $\delta\hat{F}$, at constant temperature of a sample with local magnetization $\mathbf{M}$ is [54, p. 116]

$$\delta\hat{F} = -\mu_0 \int_v \mathrm{d}^3\mathbf{r}\mathbf{M} \cdot \delta\mathbf{H}_a - \mu_0 \int_{\mathbb{R}^3} \mathrm{d}^3\mathbf{r}\mathbf{H}_a \cdot \delta\mathbf{H}_a, \tag{1.81}$$

where $\mathbf{H}_a$ is the applied magnetic field and $\delta\mathbf{H}_a$ is a variation in this field. If $\delta\hat{F} > 0$, then the external sources of $\mathbf{H}_a$ provide energy to the system. Then, the loss per cycle due to the cyclic applied fields is

$$Q = \oint \delta\hat{F} = -\mu_0 \int_v \mathrm{d}^3\mathbf{r} \oint \mathrm{d}\mathbf{H}_a \cdot \mathbf{M}, \tag{1.82}$$

where we use the fact that the cycle integral of the second term in (1.81) vanishes. This equation is valid for any applied magnetic field, which could be nonuniform and its direction could change during the cycle. If this loss per cycle is positive, it means that the source of the applied field is providing energy to the system, and hence, there is dissipation.

Equation (1.82) can also be written as a function of the total magnetic field $\mathbf{H}$ as

$$Q = -\mu_0 \int_v \mathrm{d}^3\mathbf{r} \oint \mathrm{d}\mathbf{H} \cdot \mathbf{M}. \tag{1.83}$$

The reason is that $\mathbf{H} = \mathbf{H}_M + \mathbf{H}_a$, with $\mathbf{H}_M$ being the magnetic field created by $\mathbf{M}$, and the cyclic integral $\int_v \mathrm{d}^3\mathbf{r} \oint \mathrm{d}\mathbf{H}_M \cdot \mathbf{M}$ vanishes. The latter can be seen as follows. First, the magnetic field created by a point dipole of magnetic moment $\mathbf{m}$ at position $\mathbf{r}'$ is [40]

$$\mathbf{H}_{\text{dipole}}(\mathbf{r}) = \overline{\overline{g}}(\mathbf{r} - \mathbf{r}')\mathbf{m}, \tag{1.84}$$

with the tensor $\overline{\overline{g}}(\mathbf{r} - \mathbf{r}')$ having components

$$g_{ij} = \frac{1}{4\pi r''^5}\left(3r''_i r''_j - \delta_{ij} r''^2_i\right), \tag{1.85}$$

where δ_{ij} are the identity matrix coefficients, $\mathbf{r}'' = \mathbf{r} - \mathbf{r}'$, $r'' = |\mathbf{r}''|$, and r''_i are the components of $\mathbf{r}''$. Since $\mathbf{M}(\mathbf{r})$ is the magnetic dipole density, the magnetic field created by $\mathbf{M}$, $\mathbf{H}_M$, is

$$\mathbf{H}_M(\mathbf{r}) = \int_v \mathrm{d}^3\mathbf{r}' \overline{\overline{g}}(\mathbf{r} - \mathbf{r}')\mathbf{M}(\mathbf{r}'). \tag{1.86}$$

Then,

$$\begin{aligned}
&\int_v \mathrm{d}^3\mathbf{r} \oint \mathrm{d}\mathbf{H}_M \cdot \mathbf{M} = \oint \mathrm{d}t \int_v \mathrm{d}^3\mathbf{r}\mathbf{M} \cdot \frac{\partial \mathbf{H}_M}{\partial t} = \\
&\oint \mathrm{d}t \int_v \mathrm{d}^3\mathbf{r} \int_v \mathrm{d}^3\mathbf{r}' \mathbf{M}(\mathbf{r})\overline{\overline{g}}(\mathbf{r} - \mathbf{r}')\frac{\partial \mathbf{M}(\mathbf{r}')}{\partial t} = \\
&\frac{1}{2}\oint \mathrm{d}t \int_v \mathrm{d}^3\mathbf{r} \int_v \mathrm{d}^3\mathbf{r}' \frac{\partial}{\partial t}[\mathbf{M}(\mathbf{r})\overline{\overline{g}}(\mathbf{r} - \mathbf{r}')\mathbf{M}(\mathbf{r}')] = 0.
\end{aligned} \tag{1.87}$$

The last expression vanishes due to the circle integral in time and the global time derivative of the integrand.

For isotropic, soft ferromagnetic materials, $\mathbf{M}$ is always parallel to $\mathbf{H}$, and the magnetization loops are very narrow. Then, the material response can be approximated by a single-valued $M(H)$ relation and the hysteresis loss density per cycle, Q_v, depends only on the magnetic field amplitude, H_m, and the bias magnetic field, H_{bias}. Then, the circle integral of (1.83) becomes $Q_v(H_m, H_{\text{bias}})$, and the total loss per cycle is

$$Q \approx -\mu_0 \int_v Q_v(H_m, H_{\text{bias}})\mathrm{d}^3\mathbf{r}. \tag{1.88}$$

In the above equation, the magnetic fields H_m and H_{bias} are total magnetic fields, and hence, they include the contribution of the magnetic material.

1.1.7.3. *Conductors and superconductors under uniform applied fields*

When a normal conductor or superconductor is submitted to only a uniform applied magnetic fields, $\mathbf{H}_a$, it can be seen from classical electrodynamics that Equation (1.76) yields

$$Q = -\mu_0 \oint \mathbf{m} \cdot \frac{\partial \mathbf{H}_a}{\partial t} \mathrm{d}t = -\mu_0 \oint \mathbf{m} \cdot \mathrm{d}\mathbf{H}_a, \tag{1.89}$$

where $\mathbf{m}$ is the magnetic moment, which is defined as

$$\mathbf{m} = \frac{1}{2} \int_v \mathrm{d}^3\mathbf{r}\ \mathbf{r} \times \mathbf{J}. \tag{1.90}$$

For very long samples, where the current loops close near the ends, the above equation results in

$$\mathbf{m} = l \int_s \mathrm{d}^2\mathbf{r}_2\ \mathbf{r}_2 \times \mathbf{J}, \tag{1.91}$$

where l is the sample length, s is the cross-sectional plane, and $\mathbf{r}_2$ is the position vector in that plane.

Equation (1.89) is valid for any periodic uniform applied vector field, including oscillating and rotating applied fields. This equation could also be obtained from (1.82) using the fact that $\mathbf{H}_a$ (and $\mathrm{d}\mathbf{H}_a$) is uniform.

1.2. Analytical Formulas and Main Electromagnetic Behavior

In this section, we deduce several analytical formulas to predict the electromagnetic behavior of superconductors. The main purpose of this section is to introduce the reader to the main features of the electromagnetic response of superconductors (bulks, wires, and tapes) and also superconducting conductors containing normal conducting parts, usually metals. The formulas presented in this section are also very useful for researchers working on numerical modeling in order to test or benchmark their numerical methods.

For composite wires and cables, and any superconducting object in general, we can distinguish between three types of electromagnetic response due to hysteresis currents, eddy currents, and coupling currents. In more detail:

- **Hysteresis currents**, also called superconducting currents, are those that form closed loops exclusively within superconducting regions.
- **Eddy currents** are those that close entirely within normal conducting parts.
- **Coupling currents** are currents that flow mostly on superconducting regions, usually filaments in multi-filamentary wires or tapes, but they

need to cross a normal conducting part in order to close the loop. Thus, they join or "couple" different superconducting regions.

1.2.1. *Hysteresis currents*

The Bean–London CSM, which we introduced in Section 1.1.2, allows us to find key analytical formulas of the elecromagnetic behavior of simple shapes. According to the CSM, the induced current density is independent of the frequency of the applied field or transport current, and hence, the superconductor shows hysteresis. Actually, the response also does not depend on the waveform of these inputs as long as they monotonically increase from their minimum to their maximum and monotonically decrease from maximum to minimum. The weak frequency dependence on power-law $\mathbf{E}(\mathbf{J})$ relations can be estimated with the scaling laws described in Section 1.2.1.7.

1.2.1.1. *Infinite cylinder under axial applied magnetic field*

As done by Bean [12], let us consider an infinitely long cylinder of radius R in zero-field cool situation, where no current density is present (see Figure 1.3). After applying an increasing magnetic field $\mathbf{H}_a$ in the axial direction z, $\mathbf{H}_a = H_a\mathbf{e}_z$, any closed circular circuit coaxial of the cylinder axis experiences an electromotive force, due to Faraday's law, as

$$\oint_{\partial s} \mathrm{d}\mathbf{l} \cdot \mathbf{E}_a = -\mu_0 \int_s \mathrm{d}\mathbf{s} \cdot \partial_t \mathbf{H}_a,$$

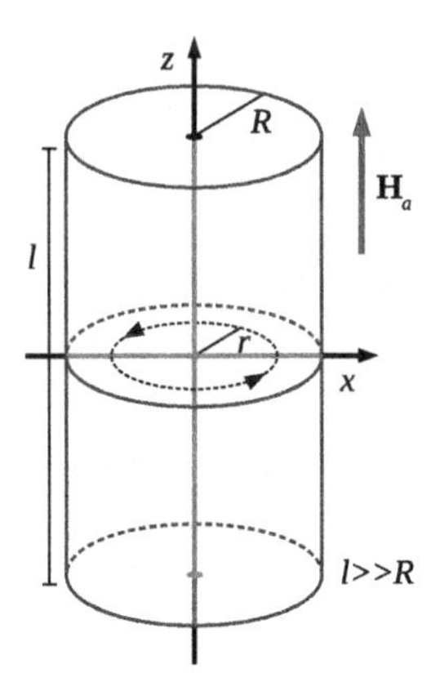

Figure 1.3. The CSM enables us to find analytical formulas of the response of a very long superconducting cylinder submitted to varying applied magnetic fields in the axial direction.

where $\mathbf{E}_a$ is the electric field due to the changing applied magnetic field or "applied electric field." Thanks to cylindrical symmetry, $\mathbf{E} = E\mathbf{e}_\varphi$ and $\mathbf{J} = J\mathbf{e}_\varphi$, with $\mathbf{e}_\varphi$ being the unit vector in the angular direction. The above integral relation results in

$$E_a = \frac{-\mu_0}{2} r \partial_t H_a.$$

Then, the rise in H_a causes a negative nonzero E_a. Following the CSM $\mathbf{E}(\mathbf{J})$ relation in (1.20), this negative applied electric field causes a current density $\mathbf{J} = -J_c\mathbf{e}_\varphi$ in a certain region. However, the cylinder is not completely filled with $J = -J_c$ everywhere since the newly induced current density partially shields the applied magnetic field. Thanks to the infinitely long geometry, a current layer from the surface to a depth d creates a uniform magnetic field of value $H_J = -J_c d$ at $r < R - d$. The current penetration depth d is such that $H = 0$ at $0 < r < R - d$, where r is the radial coordinate, because any change in H in the superconductor volume would create an electric field and hence current density (Figures 1.4(a) and 1.4(c)). We see now two key features of the CSM:

- For zero-field cool, the regions with zero current density has no magnetic field in the whole history after zero-field cool.
- Although the CSM allows intermediate values of $|\mathbf{J}|$ between 0 and J_c, these do not appear thanks to the circular and infinitely long geometry. Thus, $|\mathbf{J}|$ is either 0 or J_c.

Since $H = 0$ at $0 < r < R - d$, the magnetic field created by J follows $H_J = -H_a$, and hence, $d = H_a/J_c$. By further increasing H_a, the current further penetrates until the sample is saturated at

$$H_p = J_c R, \tag{1.92}$$

which is called penetration field. Then, the current density is

$$J = \begin{cases} -J_c & \text{if } R - d \le r \le R, \\ 0 & \text{otherwise}, \end{cases} \tag{1.93}$$

with

$$d = \begin{cases} H_a/J_c & \text{if } H_a \le H_p = J_c R, \\ R & \text{otherwise}. \end{cases} \tag{1.94}$$

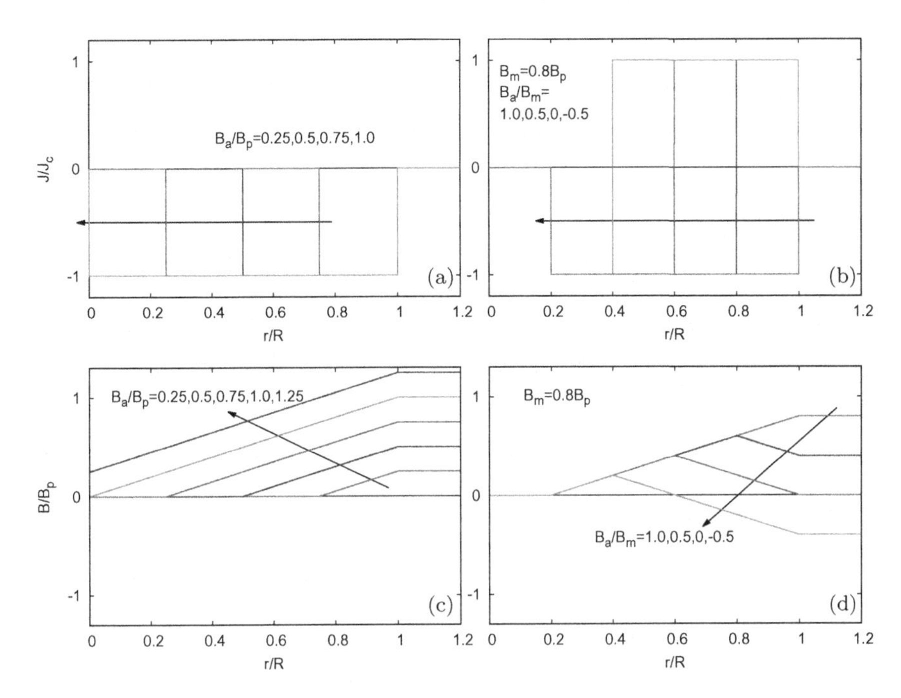

Figure 1.4. ((a) and (c)) At the initial curve, the current density, J, and the magnetic flux density, $B = \mu_0 H$, in a long cylinder of radius R monotonically penetrates from the surface inwards. (b) Later, when decreasing the applied flux density from the maximum, B_m, current density with the opposite sign penetrates from the surface inwards. (d) During this process, B is unchanged beyond a certain depth from the surface. Here, $B_p = \mu_0 H_p = \mu_0 J_c R$.

By Ampere's law, $J = -\mathrm{d}H/\mathrm{d}r$, and hence, H can be found by integration of J over r, resulting in

$$H = \begin{cases} H_a & \text{if } r \geq R, \\ J_c(r - R) + H_a & \text{if } R - d \leq r < R, \\ 0 & \text{otherwise.} \end{cases} \tag{1.95}$$

Next, we reduce the applied magnetic field from the maximum $H_a = H_m$ to $H_a = -H_m$ (reverse curve). Now, the electromotive force is opposite, and hence, the critical-state statement and Lenz law tell us that a region with $J = +J_c$ will appear. For the infinite cylinder, a layer with $J = +J_c$ of thickness d_r is induced from the surface, beyond which the magnetic field is unchanged or "frozen" and so is the current density. *Then, at the reverse curve, the regions enclosed by the region with reversed critical current has a frozen magnetic field and current density from the peak of the magnetic field.* The thickness of this layer is $d_r = (H_m - H_a)/2J_c$ because the local change in J is $2J_c$, instead of only J_c at the initial curve. Thanks to this, with $H_a = 0$, there is a certain remnant field in the superconductor. The maximum remnant field is achieved at $H_m \geq 2H_p = 2J_cR$. After further reducing H_a down to $-H_m$, the reverse critical region further penetrates until it fully removes the magnetic history. *Then, at the end of the reverse curve, the magnetic field and current density are the same as at the end of the initial curve but with the opposite sign.* The current density and magnetic field in the reverse curve are

$$J = \begin{cases} +J_c & \text{if } R - d_r < r \leq R, \\ -J_c & \text{if } R - d_m \leq r < R - d_r, \\ 0 & \text{otherwise,} \end{cases} \tag{1.96}$$

with

$$d_m = \begin{cases} H_m/J_c & \text{if } H_a \leq H_p = J_cR, \\ R & \text{otherwise,} \end{cases} \tag{1.97}$$

$$d_r = \begin{cases} (H_m - H_a)/2J_c & \text{if } H_m - H_a \leq 2H_p = 2J_cR, \\ R & \text{otherwise,} \end{cases} \tag{1.98}$$

which results in a magnetic field

$$H = \begin{cases} H_a & \text{if } r \geq R, \\ J_c(R - r) + H_a & \text{if } R - d_r < r < R, \\ J_c(r - R) + H_a & \text{if } R - d_m < r \leq R - d_r, \\ 0 & \text{otherwise.} \end{cases} \tag{1.99}$$

The magnetization process, as a result of increasing the magnetic field again from $-H_m$ to H_m (returning curve), is symmetric with respect to the reverse curve, following

$$J_\uparrow(\mathbf{r}, H_a) = -J_\downarrow(\mathbf{r}, -H_a), \tag{1.100}$$

$$H_\uparrow(\mathbf{r}, H_a) = -H_\downarrow(\mathbf{r}, -H_a), \tag{1.101}$$

where $J_\uparrow$ and $J_\downarrow$ denote the current density at the returning (increasing applied magnetic field) and reverse curve (decreasing applied magnetic field), respectively. In the above equation, we express the current density as a function of the vector position $\mathbf{r}$ because this expression is valid for a wide range of sample shapes. From the current profiles in Figure 1.4, it can be seen that the current density at the reverse and returning curves can be obtained by superposition of those at the initial curve as

$$J_{\downarrow}(\mathbf{r}, H_a) = J_i(\mathbf{r}, H_m) - 2J_i\left(\mathbf{r}, \frac{H_m - H_a}{2}\right), \tag{1.102}$$

$$J_{\uparrow}(\mathbf{r}, H_a) = -J_i(\mathbf{r}, H_m) + 2J_i\left(\mathbf{r}, \frac{H_m + H_a}{2}\right), \tag{1.103}$$

where J_i is the current density in the initial stage. Summarizing, the main features of the magnetization process in the CSM are as follows:

- Although the CSM allows current densities of magnitude below J_c, the response is such that $|J| = J_c$, wherever current density is present.
- For zero-field cool, the regions with zero current density has no magnetic field in the whole history since zero-field cool.
- At the reverse curve, the regions enclosed by the region with reversed critical current has a frozen magnetic field and current density from the peak of the magnetic field.
- At the end of the reverse curve, the magnetic field and current density are the same as at the end of the initial curve but with the opposite sign.
- The current density at the reverse and returning curves can be obtained from the initial one.

Once $J(\mathbf{r}, H_a)$ is known, the magnetic moment can be calculated from (1.90). From the sample symmetry, $\mathbf{m} = m\mathbf{e}_z$, where $\mathbf{e}_z$ is the unit vector in the z direction. The volume-averaged magnetization is $M = m/V$, where V is the sample volume. Applying (1.96) to (1.90), the magnetization of the initial curve is

$$M_i(H_a) = \begin{cases} \frac{J_c R}{3}\left[\left(1 - \frac{H_a}{H_p}\right)^3 - 1\right] & \text{if } H_a \leq H_p = J_c R, \\ -\frac{J_c R}{3} \equiv -M_s & \text{if } H_a \geq H_p = J_c R, \end{cases} \tag{1.104}$$

where M_s corresponds to the saturation magnetization (see Figure 1.5). For the infinite cylinder, the magnetization exactly saturates at the penetration field $H_a = H_p = J_c R$. The relation between the initial and reverse curves in (1.103) also applies to the magnetization since for any arbitrary current densities $\mathbf{J}_1(\mathbf{r})$ and $\mathbf{J}_2(\mathbf{r})$, it follows that $\mathbf{M} = (1/2V)\int \mathrm{d}^3\mathbf{r}\ \mathbf{r} \times [\mathbf{J}_1(\mathbf{r}) + \mathbf{J}_2(\mathbf{r})] = \mathbf{M}_1 + \mathbf{M}_2$, where $\mathbf{M}_1$ and $\mathbf{M}_2$ are the magnetizations for $\mathbf{J}_1(\mathbf{r})$ and $\mathbf{J}_2(\mathbf{r})$, respectively. For the parts of the loop where the sample is saturated,

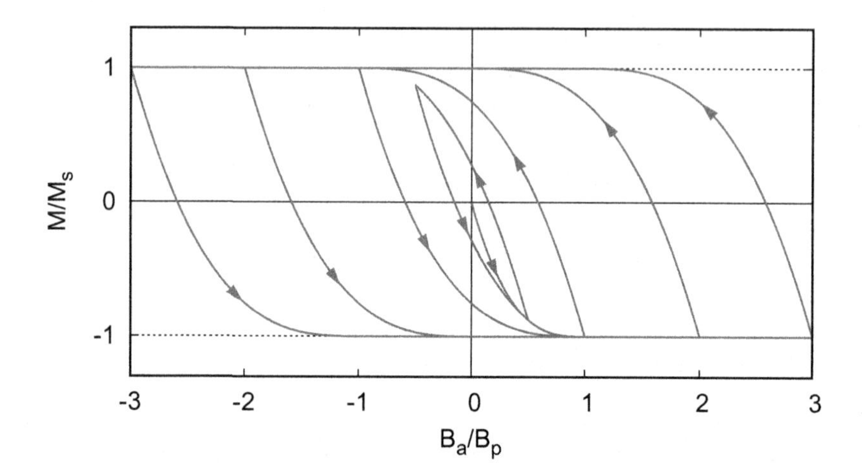

Figure 1.5. When submitted to periodic applied magnetic flux densities, B_a, a superconductor in the CSM presents hysteresis loops after the initial rise in B_a. The above results are for very long cylinders with the B_a amplitudes of $B_m/B_p = 0.5, 1, 2$, and 3, where $B_p = \mu_0 H_p = \mu_0 J_c R$ and the saturation magnetization is $M_s = J_c R/3$.

the width of the loop is $\Delta M = 3J_c R/3$, which enables experimentalists to obtain J_c from magnetization loops as

$$J_c = \frac{3\Delta M}{2J_c R}. \tag{1.105}$$

Knowing that the loss per cycle and unit volume is $Q_v = 2\mu_0 \int_{-H_m}^{H_m} \mathrm{d}H_a M_\downarrow(H_a)$ and that $M_\downarrow(H_a) = M_i(H_m) - 2M_i[(H_m - H_a)/2]$ and using (1.104), we obtain the AC loss [30]:

$$\frac{Q_v}{\mu_0 \pi H_p^2} = \begin{cases} h^3(4-2h)/(3\pi) & \text{if } h =\leq 1, \\ (4h-2)/(3\pi) & \text{if } h =\geq 1, \end{cases} \tag{1.106}$$

where $H_p \equiv J_c R$ and $h \equiv H_m/H_p$. The normalized AC loss above is dimensionless and depends only on the reduced applied magnetic field amplitude $h = H_m/J_c R$. The limits for low and high applied field amplitudes are of high practical importance since many other configurations show the same qualitative behavior:

$$Q_v = \frac{4\mu_0}{3J_c R} H_m^3 \propto H_m^3 \ \text{ if } H_m \ll H_p, \tag{1.107}$$

$$Q_v = \frac{4\mu_0 J_c R}{3} H_m \propto H_m \ \text{ if } H_m \gg H_p, \tag{1.108}$$

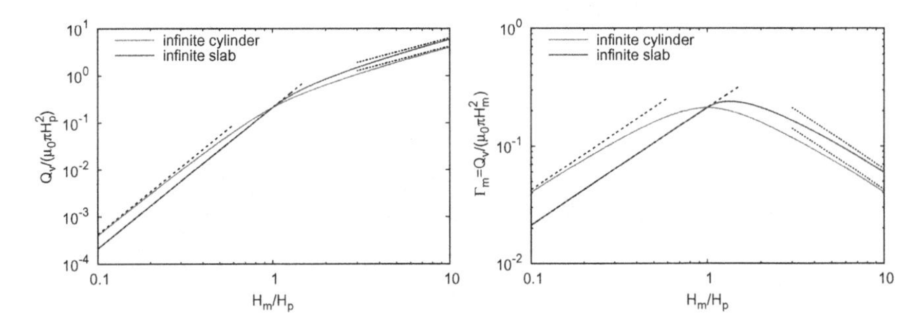

Figure 1.6. The AC loss of infinite superconducting cylinders and slabs in parallel applied magnetic fields monotonically increases with the applied field amplitude (left). The dimensionless, normalized AC loss $Q_v/(\mu_0\pi H_p^2)$ therein is independent of J_c and the dimensions of the body. The dimensionless loss factor in the right-hand plot, $\Gamma_m = Q_v/(\mu_0\pi H_m^2)$, is also independent of J_c and the sample size but enables a more detailed analysis. This factor causes a peak at $H_m/H_p = 1$ and $H_m/H_p = 4/3$ for the cylinder and slab, respectively, namely $H_p = J_cR$ and $H_p = J_ca$, respectively, for each shape.

with $H_p = J_cR$, as in (1.106). Alternatively, we can also normalize the AC loss with the amplitude of the applied field, H_m^2, instead of H_p^2 so that

$$\Gamma_m \equiv \frac{Q_v}{\mu_0\pi H_m^2}, \tag{1.109}$$

which corresponds to the imaginary part of the AC susceptibility [24]. This quantity allows us to analyze, in more depth, the details of the AC loss curve. In particular, the peak of the normalized loss of (1.106), H_{pk}, occurs at $H_{\text{pk}} = H_p$, where $H_p = J_cR$ is the penetration field (Figure 1.6). This enables us to obtain J_c from measurements under applied alternating magnetic fields.

1.2.1.2. *Infinite slab under parallel applied field*

Another case of practical importance is a slab with an applied field parallel to one infinite direction. For this idealized shape, the current is in the y direction, as shown in Figure 1.7. Following the same reasoning as for the cylinder, we find the same expression for J and H in the initial curve as Equations (1.93) and (1.94) but replacing R by the half-width of the slab, a. However, the magnetization is now

$$M_i(H_a) = \begin{cases} -H_a + \frac{H_a^2}{2J_ca} & \text{if } H_a \le H_p, \\ -J_ca/2 \equiv -M_s & \text{if } H_a \ge H_p, \end{cases} \tag{1.110}$$

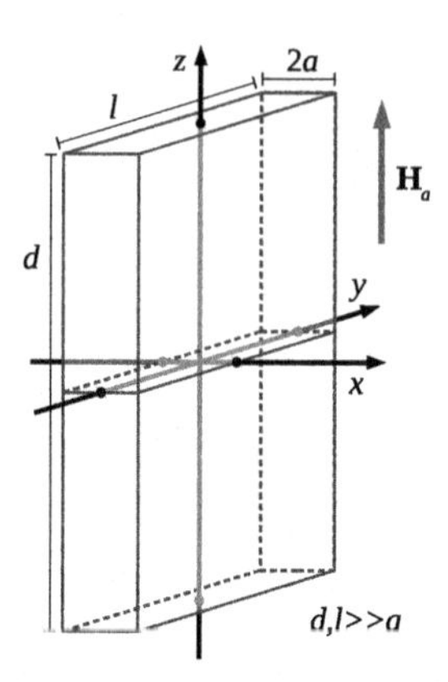

Figure 1.7. The response of a superconducting slab in the CSM to applied magnetic fields, $\mathbf{H}_a$, along one of the large dimensions also follows the analytical formulas.

with $H_p = J_c a$. Obtaining this magnetization requires (1.91) since the current loops close at infinity. Since the saturation magnetization now is $M_s = J_c a/2$, the current density is related to the width of the saturated part of the loop as

$$J_c = \Delta M/a. \tag{1.111}$$

From the magnetization loops obtained from this initial magnetization, the normalized AC loss is

$$q_v \equiv \frac{Q_v}{\mu_0 \pi H_p^2} = \begin{cases} 2h^3/(3\pi) & \text{if } h \leq 1, \\ (6h-4)/(3\pi) & \text{if } h \geq 1, \\ 2h/\pi & \text{if } h \gg 1, \end{cases} \tag{1.112}$$

where $h = H_m/H_p$. The loss factor is just $\Gamma_m = q_v/h^2$. In contrast to the cylinder, the peak of the loss factor is not at H_p but at $H_{\rm pk} = 4H_p/3 = 4J_c a/3$ (Figure 1.6).

1.2.1.3. *Circular wire with transport current*

The AC loss in a circular wire with alternating transport current $I(t) = I_m \sin(\omega t)$, such as that in Figure 1.8, cannot be obtained from a magnetization loop. Instead, we can use the local instantaneous loss $\mathbf{J} \cdot \mathbf{E}$. In this way, we obtain the instantaneous power loss by integrating over the volume and the loss per cycle by further integrating over one cycle.

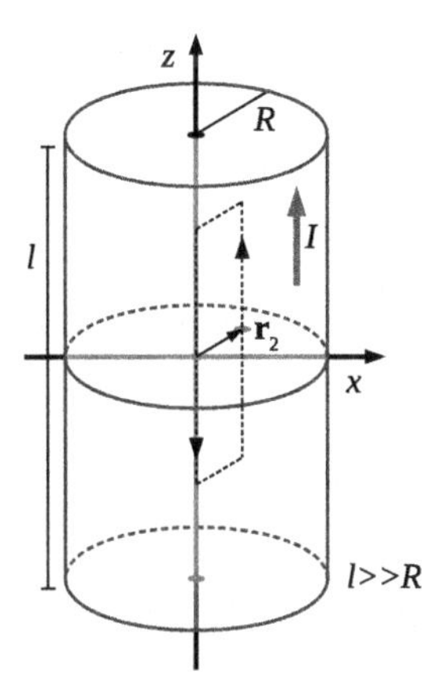

Figure 1.8. Long (or infinite) cylinder with transport current I. We use the closed rectangular loop to calculate the local electric field, which follows the z direction.

First, we find the current density by applying positive transport current after the zero-field cool situation. Due to the infinite geometry, $\mathbf{J}$ follows the transport current direction, which we choose as z:

$$\mathbf{J}(\mathbf{r}) = J(r)\mathbf{e}_z, \tag{1.113}$$

where r is the distance from the wire center. As a consequence of the CSM statement (Section 1.1.2), a current with density J_c will penetrate from the surface to the center in a cylindrical shell, as shown in Figure 1.10. This can be seen as circular vortices of magnetic flux that enter the wire surface in a gradient density determined by the Biot–Savart law, $\nabla \times \mathbf{B} = \mu_0 \mathbf{J}$, and $J = J_c$. The current density is then

$$J(r) = \begin{cases} 0 & \text{if } r < b, \\ J_c & \text{if } a \geq r \geq b, \end{cases} \tag{1.114}$$

where a and b are the wire and current-free zone radius, respectively, with b related to the current as

$$b = a\sqrt{1 - i}, \quad \text{with} \quad i = I/I_c. \tag{1.115}$$

In the above equation, the critical current is $I_c = \pi a^2 J_c$. As for long cylinders under axial applied magnetic fields (Section 1.2.1.1), the solution to the current density is such that $|J|$ is either J_c or 0, although the CSM allows any $|J| \leq J_c$. Again, the reason for the absence of intermediate $|J|$ between 0 and J_c is the high symmetry of the system. This can be seen

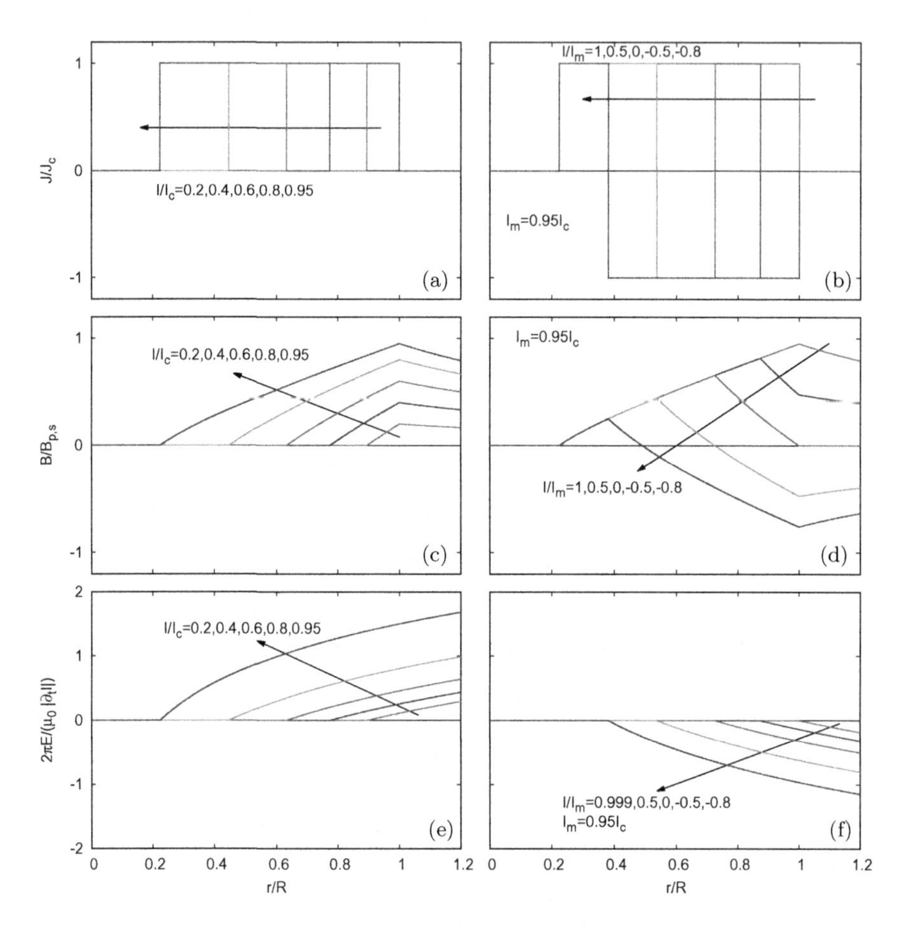

Figure 1.9. ((a) and (b)) The normalized current density, ((c) and (d)) magnetic flux density, and ((e) and (f)) electric field of a circular wire in the CSM under transport current are universal and hence independent of the critical current density (J_c), wire radius (R), frequency (f), and current waveform. The left-hand plots are for the initial curve (I increasing from 0 to I_m), while the right-hand plots are for the decreasing curve that follows (I decreasing from I_m to $-I_m$). Here, $B_{p,s}$ is the magnetic flux density on the wire surface at saturation, $B_{p,s} = \mu_0 J_c R/2$, and $\partial_t I$ is the time derivative of the current.

as follows. With the current increasing from 0 to $I > 0$, there will appear regions with $J > 0$. If the current density is induced as a cylindrical shell of $J > 0$ penetrating a certain distance from the surface, the magnetic field remains zero at the core with $J = 0$, with the changing magnetic fields only appearing in the region where $J > 0$. Since these changing magnetic fields cause a nonzero electric field, $E > 0$, the CSM statement tells us that

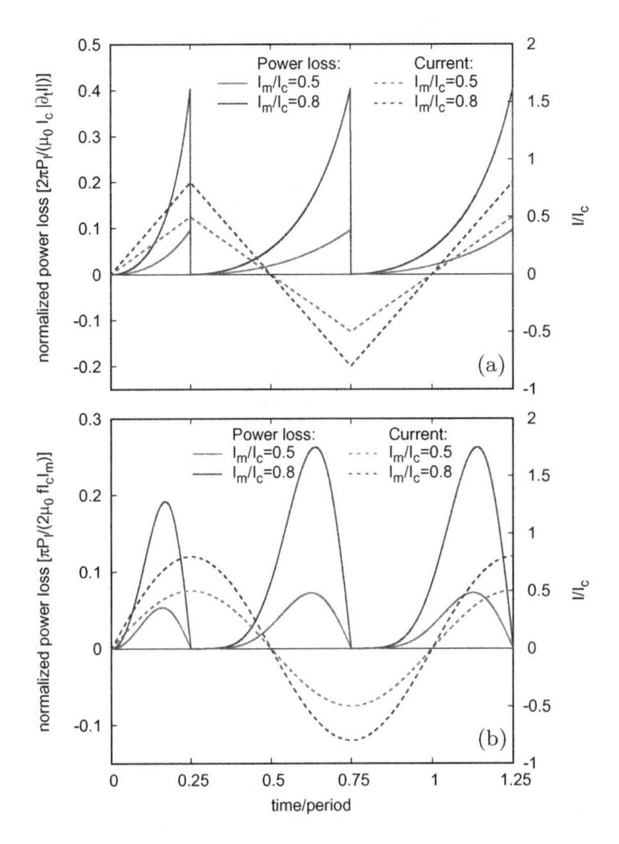

Figure 1.10. The dimensionless power loss in a circular wire in the CSM depends on the current waveform but not on J_c, wire dimensions, or frequency. The normalized power loss above is calculated using (1.119), (1.126), and (1.128). In (1.128), $|\partial_t I|_{\mathrm{av}}$ is the constant ramp rate of the triangular waveform (a) and $|\partial_t I|_{\mathrm{av}} = 4fI_m$ for the sinusoidal current, with $I(t) = I_m \sin(2\pi f t)$.

$J = J_c$ at the shell. Indeed, that is the only current distribution that is consistent with the CSM. If we imagine a region with finite J but below $|J_c|$ (at the core, for example), there will appear $E > 0$ due to Faraday's law, but that contradicts the CSM statement because $|J|$ can be lower than J_c only when the electric field completely vanishes.

The electric field can be found from the magnetic field by means of Faraday's law. Since for our case, $\mathbf{E}(\mathbf{r}) = E(r)\mathbf{e}_z$ and $\mathbf{B}(\mathbf{r}) = B(r)\mathbf{e}_\varphi$, with φ being the angular coordinate, Faraday's law reduces to $\partial_r E = \partial_t B$. By applying Ampere's law to the obtained current density, the magnetic flux density is

$$B(r) = \begin{cases} 0 & \text{if } r \leq b, \\ \mu_0 J_c[r - a^2(1-i)/r]/2 & \text{if } a \geq r \geq b, \\ \mu_0 i J_c a^2/(2r) & \text{if } r \geq a. \end{cases} \tag{1.116}$$

From Faraday's law and taking into account that $E(r = 0) = 0$ because $J(r = 0) = 0$,

$$E(r) = \int_0^r \mathrm{d}r' \partial_t B(r') = \int_0^r \mathrm{d}r' \partial_i B(r') \partial_t i. \tag{1.117}$$

Then, the electric field results in

$$E(r) = \begin{cases} 0 & \text{if } r \leq b, \\ (\mu_0 J_c a^2/2) \partial_t i \ln(r/b) & \text{if } a \geq r \geq b. \end{cases} \tag{1.118}$$

The integral of $J(r)E(r)$ over the cross-section is the instantaneous loss per unit length, resulting in

$$P_l = -\frac{\mu_0}{4\pi} I_c^2 \partial_t i [i + \ln(1-i)]. \tag{1.119}$$

An interesting limit is for low normalized currents, for which

$$P_l = \frac{\mu_0}{8\pi} I_c^2 i^2 \partial_t i \quad \text{if } i \ll 1. \tag{1.120}$$

At the beginning of the initial curve, the power loss increases with the square of the current.

In order to calculate the loss per cycle, we need to obtain the power loss not only at the initial curve but at the whole cycle. After the initial curve, the current decreases. Now, a negative current density $-J_c$ appears at the surface and penetrates inwards. Again, this can be regarded as circular vortices leaving the wire surface, creating a magnetic field gradient of opposite sign from that at the initial curve (Figure 1.9(d)). At the peak of the AC cycle, we recover the same situation as the end of the initial curve but with negative J. On increasing the current again, we repeat the same process but with all current density with the opposite sign. As for the case of the infinite cylinder in an applied magnetic field, the decrease in the current density of the decreasing current, $J_{\downarrow}$, and that of the subsequent increasing current, $J_\uparrow$, are obtained from the initial curve J_i:

$$J_\downarrow(i) = J_i(i_m) - 2J_i\left(\frac{i_m - i}{2}\right),$$

$$J_\uparrow(i) = -J_i(i_m) + 2J_i\left(\frac{i_m + i}{2}\right), \tag{1.121}$$

where i_m is the normalized current at the peak, $i_m = I_m/I_c$. Then, the current density for the decreasing curve is

$$J_\downarrow(r) = \begin{cases} 0 & \text{if } b_m > r, \\ J_c & \text{if } b_\downarrow > r \geq b_m, \\ -J_c & \text{if } a \geq r \geq b_\downarrow, \end{cases} \tag{1.122}$$

with

$$b_m = a\sqrt{1 - i_m},\ b_\downarrow = a\sqrt{1 - (i_m - i)/2}. \tag{1.123}$$

From this current density, we obtain the magnetic field:

$$B_\downarrow(r) = \begin{cases} 0 & \text{if } b_m \geq r, \\ \mu_0 J_c[r - a^2(1 - i_m)/r]/2 & \text{if } b_\downarrow \geq r \geq b_m, \\ \mu_0 J_c[-r + a^2(1 + i)/r]/2 & \text{if } a \geq r \geq b_\downarrow, \\ \mu_0 i J_c a^2/(2r) & \text{if } r \geq a. \end{cases} \tag{1.124}$$

Again, the electric field is obtained from (1.117) as

$$E_\downarrow(r) = \begin{cases} 0 & \text{if } b_\downarrow \geq r, \\ (\mu_0 J_c a^2 \partial_t i/2)\ln(r/b_\downarrow) & \text{if } a \geq r \geq b_\downarrow. \end{cases} \tag{1.125}$$

Integrating the instantaneous local loss $J(r)E(r)$ over the wire cross-section, the instantaneous power loss per unit wire length is

$$P_{l\downarrow} = \frac{\mu_0}{4\pi} I_c^2 \partial_t i \left[\frac{i_m - i}{2} + \ln\left(1 - \frac{i_m - i}{2}\right)\right]. \tag{1.126}$$

At the beginning of the decreasing curve, $i_m - i \ll 1$, the power loss is

$$P_{l\downarrow} = -\frac{\mu_0}{32\pi} I_c^2 (i_m - i)^2 \partial_t i \quad \text{if } i_m - i \ll 1. \tag{1.127}$$

The power loss at the decreasing curve increases quadratically with the change in current, $i_m - i$, starting from zero loss. Note that the power loss is positive since $\partial_t i$ is negative. In order to obtain universal results, independent of J_c, wire dimensions, and frequency, it is convenient to normalize the AC loss into the following dimensionless quantity:

$$p_l \equiv \frac{2\pi P_l}{\mu_0 I_c |\partial_t I|_{\text{av}}}, \tag{1.128}$$

with $|\partial_t I|_{\text{av}}$ being the average of the modulus of $\partial_t I$ in a full cycle. The typical time dependence for alternating currents of constant $|\partial_t I|$ (triangular waveform) is depicted in Figure 1.10(a), which shows a monotonous power-loss increase for every half-cycle and a discontinuous sharp decrease at the end. For sinusoidal currents, $I(t) = I_m \sin(2\pi f t)$, the average $|\partial_t I|$ is $|\partial_t I|_{\text{av}} = 4 f I_m$. The normalized power loss, now being $p_l = \pi P_l/(2\mu_0 I_c I_m f)$, shows a peak at each half-cycle and vanishes at the peak of the current (Figure 1.10(b)). Although the instantaneous power loss depends on the current waveform, this is not the case for the loss per cycle.

At the decreasing curve, all the above quantities can be obtained by the symmetry of the AC cycle, and hence,

$$\begin{aligned} J_\downarrow(i) &= -J_\uparrow(-i), \\ B_\downarrow(i) &= -B_\uparrow(-i), \\ E_\downarrow(i) &= -E_\uparrow(-i), \\ P_{l\downarrow}(i) &= P_{l\uparrow}(-i). \end{aligned} \tag{1.129}$$

The loss per cycle and unit wire length is the time integral in a whole period of P_l as

$$Q_l = \int_{t_0}^{t_0+T} \mathrm{d}t P_l(t) = 2\int_{i_m}^{-i_m} \mathrm{d}i \frac{1}{\partial_t i} P_{l\downarrow}(i), \tag{1.130}$$

where t_0 is any time after the end of the initial curve. After integration, the resulting Q_l is

$$Q_l = \frac{\mu_0 I_c^2}{2\pi}[i_m(2-i_m) + 2(1-i_m)\ln(1-i_m)]. \tag{1.131}$$

The normalized loss, defined as $q \equiv 2\pi Q_l/(\mu_0 I_c^2)$, only depends on the normalized current amplitude, i_m:

$$q = [i_m(2-i_m) + 2(1-i_m)\ln(1-i_m)]. \tag{1.132}$$

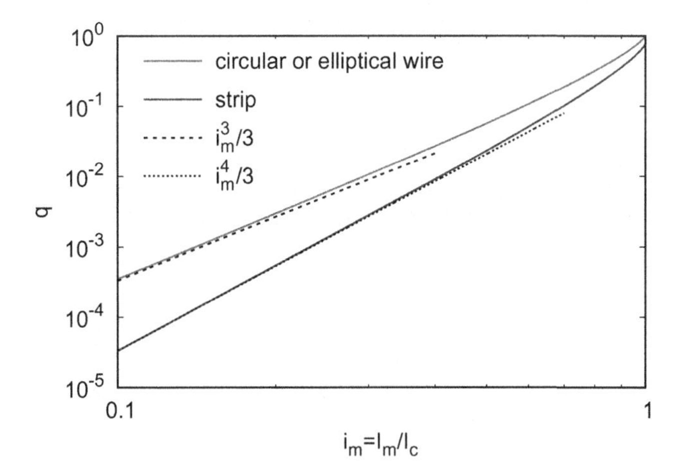

Figure 1.11. The nondimensional normalized transport AC loss, $q \equiv 2\pi Q_l/(\mu_0 I_c^2)$, of a circular or elliptical wire and a strip, obtained using (1.132) and (1.165), respectively.

For very low amplitudes, $i_m \ll 1$, and the limit of $i_m \to 1$, the normalized loss approaches

$$\begin{aligned} q &\approx i_m^3/3 \quad \text{for } i_m \ll 1, \\ q &= 1 \qquad\quad \text{for } i_m \to 1, \end{aligned} \tag{1.133}$$

The main behavior of the AC loss can be seen in Figure 1.11. As we explain in the following (Section 1.2.1.4), the AC loss for a wire with elliptical cross-section is the same as for a circular cross-section of the same I_c. In that figure, we have already added the AC loss of a strip, Equation (1.165), for completeness.

Alternatively, if you are not interested in the instantaneous power loss but only in the loss per cycle and unit length, you could calculate this quantity from the magnetic flux density at the peak. This can be seen as follows. For the CSM and $I < I_c$, there always exists a region with $J = 0$, and hence, $E = 0$ there. For a cylinder, $E = 0$ always at $\mathbf{r}_2 = 0$. From the integral form of Faraday's law (1.13), $\mathbf{E}$ at any point $\mathbf{r}_2$, with $\mathbf{E} = E(\mathbf{r}_2)\mathbf{e}_z$, follows

$$E(\mathbf{r}_2) = \partial_t \Phi(\mathbf{r}_2), \tag{1.134}$$

with

$$\Phi(\mathbf{r}_2) \equiv \int_0^{\mathbf{r}_2} (\mathrm{d}\mathbf{l} \times \mathbf{B}) \cdot \mathbf{e}_z, \tag{1.135}$$

where $\mathrm{d}\mathbf{l}\times\mathbf{B}$ is always in the z direction and $\Phi(\mathbf{r}_2)$ is the magnetic flux per unit length crossing the line between 0 and $\mathbf{r}_2$. Then, the loss per cycle is

$$Q=\int_S \mathrm{d}^2\mathbf{r}_2 \oint \mathrm{d}tJ\partial_t\Phi = 2\int_S \mathrm{d}^2\mathbf{r}_2 \int_{t(i=i_m)}^{t(i=-i_m)} \mathrm{d}tJ\partial_t\Phi_\downarrow, \tag{1.136}$$

where, at the last equality, we integrate only at the decreasing curve, and hence, we write the subindex "↓" in Φ. At this curve, $\partial_t\Phi_\downarrow$ is nonzero only at the reverse current penetration zone, where $J=-J_c$. Then,

$$\begin{aligned} Q &= -2J_c\int_S \mathrm{d}^2\mathbf{r}_2 \int_{t(i=i_m)}^{t(i=-i_m)} \mathrm{d}t\partial_t\Phi_\downarrow \\ &= -2J_c\int_S \mathrm{d}^2\mathbf{r}_2 \int_{i_m}^{-i_m} \mathrm{d}i\partial_i\Phi_\downarrow \\ &= -2J_c\int_S \mathrm{d}^2\mathbf{r}_2[\Phi_\downarrow(-i_m)-\Phi_\downarrow(i_m)]. \end{aligned} \tag{1.137}$$

Since at the reverse curve $J_\downarrow$ is related to J as (1.121), $\Phi_\downarrow$ follows the same relation:

$$\Phi_\downarrow(i)=\Phi_i(i_m)-2\Phi_i\left(\frac{i_m-i}{2}\right). \tag{1.138}$$

Then, from (1.137), we obtain

$$Q=4J_c\int_S \mathrm{d}^2\mathbf{r}_2\Phi_i(i_m). \tag{1.139}$$

Alternatively, we can also obtain a relation with the vector potential, $\mathbf{A}=A(\mathbf{r}_2)\mathbf{e}_z$, instead of Φ. Thanks to $\mathbf{B}=\nabla\times\mathbf{A}$, the integral of Φ in (1.135) becomes

$$\Phi(\mathbf{r}_2)=A_0-A(\mathbf{r}_2), \tag{1.140}$$

where $A_0=A(\mathbf{r}_2=0)$. Then, (1.139) takes the form

$$Q=4J_c\int_S \mathrm{d}^2\mathbf{r}_2[A_0-A(\mathbf{r}_2)]. \tag{1.141}$$

The loss per cycle calculated by these equations is the same as (1.131), as reported by Norris [62]. Actually, this equation can be extended to any infinitely long configuration where the current fronts penetrate monotonically from the surface inwards for any half cycle [36, 65].

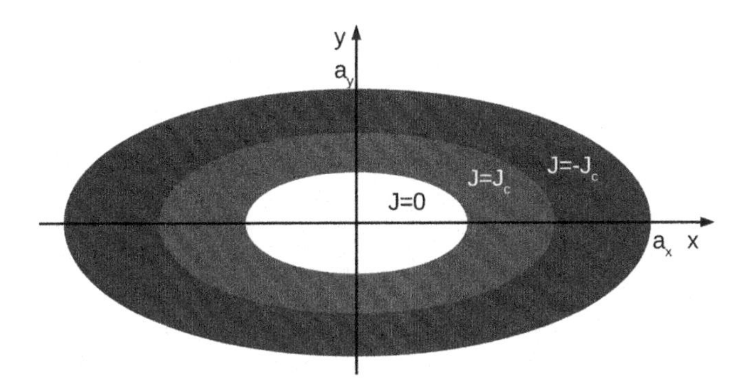

Figure 1.12. An elliptical wire in the CSM under alternating transport current exhibits current fronts as concentric ellipses of the same eccentricity as the whole wire.

1.2.1.4. *Elliptical wire with transport current*

Although the deduction is not straightforward, the magnetic flux density inside an elliptical wire with uniform current density J and of semi-axes a_x and a_y, whose eccentricity is $s \equiv a_y/a_x$, follows the simple relation [16, 62]

$$\mathbf{B}(x,y) = \frac{\mu_0 J}{1+s}\left(-y\mathbf{e}_x + sx\mathbf{e}_y\right), \tag{1.142}$$

although the expression for $\mathbf{B}$ outside the wire is more complex [16, 62]. Since the magnetic flux density within the ellipse depends on the eccentricity but not on the overall size, the flux density vanishes in a current-free core of the same eccentricity (Figure 1.12). Therefore, the current density in the initial curve will penetrate following elliptical current fronts of the same eccentricity as the whole wire, and hence the current density, which follows the z direction, will be (see Figure 1.12)

$$J(x,y) = \begin{cases} 0 & \text{if } x^2 + (y/s)^2 < b_x^2, \\ J_c & \text{if } b_x \leq x^2 + (y/s)^2 \leq a_x, \end{cases} \tag{1.143}$$

where b_x is the horizontal semi-axis of the current-free core. This quantity is related to the normalized current, $i = I/I_c$, as

$$b_x = a_x\sqrt{1-i}. \tag{1.144}$$

The current density at the reverse and returning curves can be easily obtained using (1.121).

Moreover, it can also be found that equation (1.131) for a circular wire also applies to circular wires of any eccentricity [62].

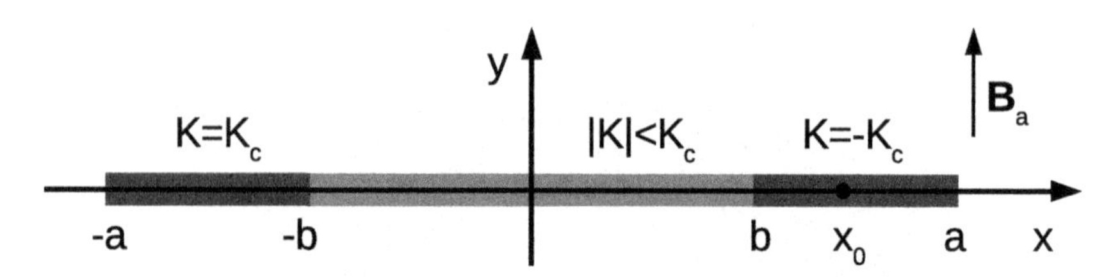

Figure 1.13. A thin strip with thickness d much smaller than its width, $2a$, in the CSM has regions with sheet current density, K, where $|K| = K_c \equiv J_c d$, and a central region with $|K| < K_c$ when submitted to a perpendicular applied magnetic flux density B_a.

1.2.1.5. *Thin strip under applied magnetic field*

The current density of a thin film in a perpendicular applied magnetic field (Figure 1.13) can also be analytically obtained by plane-to-plane conformal mapping [62, 105].

Let us first take the case of perfect shielding, corresponding to the limit of extremely large J_c. As shown by Huebener *et al.* [38] and Zeldov *et al.* [105], the sheet current density, K, (or current density per unit width) on a thin strip under perfect shielding is

$$K_{\text{shield},B_a}(x) = -\frac{2B_a}{\mu_0}\frac{x}{\sqrt{a^2 - x^2}}, \tag{1.145}$$

where a is the half-width of the strip, and the total flux density on the plane of the strip, having only the y component, is

$$B_y(x) = \begin{cases} \frac{B_a|x|}{\sqrt{x^2-a^2}} & \text{if } |x| > a, \\ 0 & \text{if } |x| \leq a. \end{cases} \tag{1.146}$$

As we can see, B_y becomes infinite at $x = a$. In practice, the finite thickness of the strip, d, limits B_y to a finite value. The finite J_c will also limit the sheet current density to finite values at the edge.

Next, let us consider the case of finite J_c, where there appear two segments of $|K| = J_c d$ penetrating toward the edge up to a distance b from the tape center that is, by now, unknown. Then, $K = +J_c d$ on the left segment and $K = -J_c d$ on the right (Figure 1.13). As a consequence of this current density, K at the inner region, where $|x| < b$, needs to also shield the flux density from the saturated segments. In order to calculate this, we first solve the shielding K reacting to a line current of intensity I_w at a certain $x = x_0 > b$ (Figure 1.13). Norris [62] shows that the sheet current density necessary to shield the flux density from this current is

$$K_{\text{shield},I_w}(x) = -\frac{I_w}{\pi(x_0 - x)}\sqrt{\frac{x_0^2 - b^2}{b^2 - x^2}}. \tag{1.147}$$

If we set $I_w = -J_c d \cdot dx_0$ at $b < x_0 \leq a$ and $I_w = +J_c d \cdot dx_0$ at $-b > x_0 \geq -a$ and integrate over x_0 on both sides, we obtain the K necessary to shield the magnetic flux density from the saturated regions as

$$K_{\text{shield},J_c}(x) = \frac{2J_c d \cdot x}{\pi\sqrt{b^2 - x^2}}\text{arcosh}\left(\frac{a}{b}\right) - \frac{2J_c d}{\pi}\arctan\left(\frac{x}{a}\sqrt{\frac{a^2 - b^2}{b^2 - x^2}}\right). \tag{1.148}$$

The total current density in $|x| < b$ also needs to shield the applied magnetic field, and hence, we should also add the contribution to (1.145) but replacing a by b there. Next, we find the relation between the half-width of the shielded region, b, and the applied flux density, B_a. To do this, we can note that both $K(x)$ in (1.145) and the first term in (1.148) are proportional to $x/\sqrt{b^2 - x^2}$, which causes them to diverge at $|x| = b$. However, the physical solution cannot diverge because the CSM assumption imposes the upper limit $|K| \leq J_c d$. Then, b is such that the terms with $x/\sqrt{b^2 - x^2}$ cancel each other, occurring for

$$b = \frac{a}{\cosh\left(\frac{B_a}{B_c}\right)}, \tag{1.149}$$

with $B_c \equiv \mu_0 J_c d/\pi$. Alternatively, we could find b by calculating B_y that K generates and imposing $B_y = 0$ at $|x| < b$, giving the same result. Then, the final sheet current density at the initial curve is

$$K(x) = \begin{cases} -\frac{2J_c d}{\pi}\arctan\left(\frac{x}{a}\sqrt{\frac{a^2-b^2}{b^2-x^2}}\right) & \text{if } |x| \leq b, \\ +J_c d & \text{if } -b \geq x \geq -a, \\ -J_c d & \text{if } b \leq x \leq a, \end{cases} \tag{1.150}$$

with b given by (1.149). The reader can see the current penetration process for the initial curve in Figure 1.14.

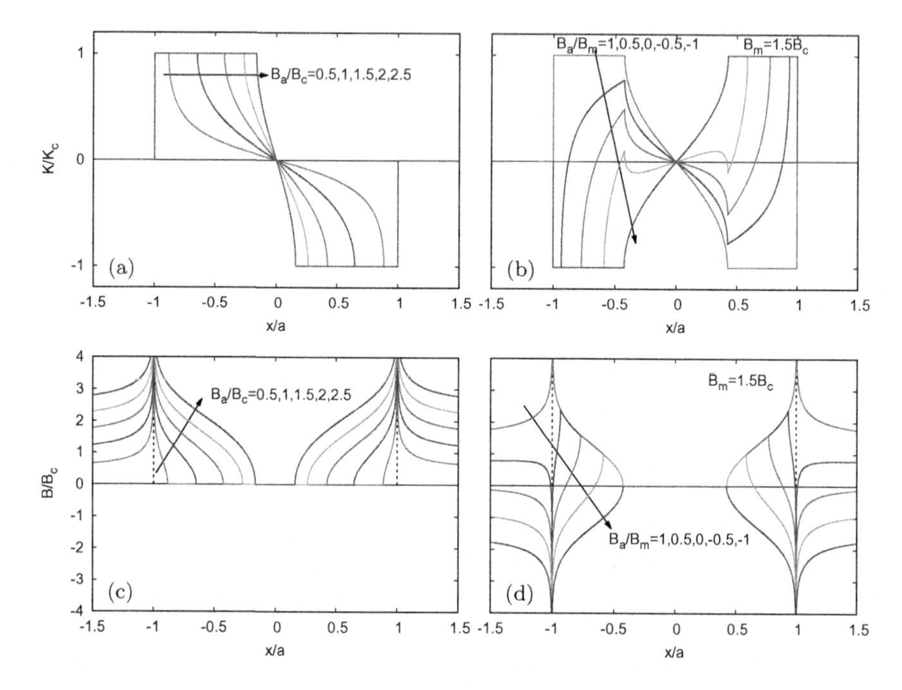

Figure 1.14. Top, the profiles of sheet current density, K, normalized to $K_c = J_c d$ for a thin strip with an applied perpendicular magnetic flux density B_a (a) at the initial curve and (b) at the decreasing curve, where $B_c = \mu_0 J_c d/\pi$. Bottom, the total magnetic flux density corresponding to the profiles on the top (c) for the initial curve and (d) the decreasing curve.

Using this $J(x)$, we obtain the magnetization, or magnetic moment per unit volume, from (1.91):

$$M = -M_s t, \tag{1.151}$$

with

$$M_s \equiv J_c a/2, \tag{1.152}$$

$$t \equiv \tanh\left(\frac{B_a}{B_c}\right). \tag{1.153}$$

The important results are that:

- the saturation magnetization at (at $B_a \gg B_c$) is the same as that of the slab.

- full saturation occurs only at $B_a \to \infty$, and hence, an ideal strip **never fully saturates.**

As a consequence of the second property, it is more convenient to define the penetration field as the peak of the AC loss factor, as detailed later in this section, which represents 98.57% of the full saturation. Alternatively, we can also define an arbitrary saturation ratio, r, so that $M = rM_s$. For example, for r =0.9, 0.95, 0.97, and 0.9857, we obtain penetration fields, B_p, of B_p/B_c =1.47, 1.83, 2.09, and 2.47, respectively.

Once $J(x)$ at the initial curve is known, $B(x)$ can be found by integrating the Biot–Savart law in two dimensions (2D) (1.51), resulting in

$$B(x) = \begin{cases} 0 & \text{if } |x| \le b, \\ B_c \text{artanh} \dfrac{a\sqrt{x^2 - b^2}}{|x|\sqrt{a^2 - b^2}} & \text{if } b \le |x| < a, \\ B_c \text{artanh} \dfrac{|x|\sqrt{a^2 - b^2}}{a\sqrt{x^2 - b^2}} & \text{if } |x| > a, \end{cases} \tag{1.154}$$

where b and t are given by (1.149) and (1.153), respectively. This flux density diverges at $|x| \to a$ (Figures 1.14(c) and 1.14(d)).

After reaching a maximum value of the applied flux density, B_m, the values of J, M, and B at the decreasing and increasing curves become directly related to the initial curve by (1.103), with J replaced by M or B as appropriate. Therefore, we can easily obtain these quantities for the whole magnetization loop.

From the magnetization loop, we can calculate the AC loss per cycle and unit length, Q_l, as $Q_l = 2ad \oint M(B_a)\mathrm{d}B_a$, resulting in

$$Q_l = 4a^2 d J_c B_m g(B_m/B_c), \qquad \text{with} \tag{1.155}$$

$$g(x) = (2/x)\ln \cosh x - \tanh x. \tag{1.156}$$

From this, we can define the dimensionless AC loss, q_l, as

$$q_l \equiv \frac{\mu_0 Q_l}{4\pi a^2 B_c^2} = (B_m/B_c)g(B_m/B_c), \tag{1.157}$$

which is independent of the strip J_c and its dimensions. Similarly, we can also define the dimensionless loss factor, Γ'_m, related to Q_l as

$$\Gamma'_m \equiv \frac{\mu_0 Q_l}{4\pi a^2 B_m^2} = (B_c/B_m)g(B_m/B_c), \tag{1.158}$$

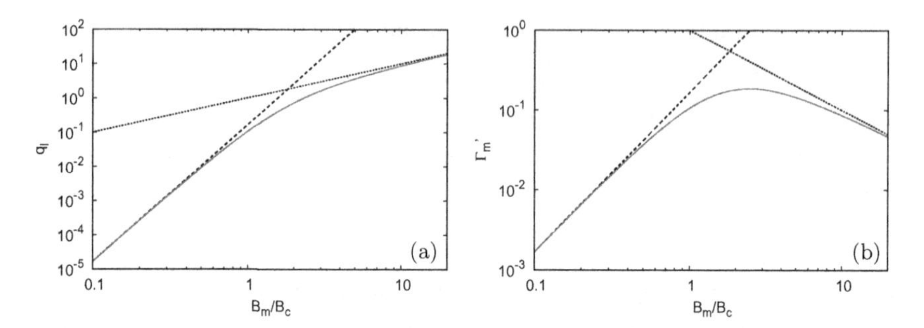

Figure 1.15. The normalized AC loss, $q_l \equiv \mu_0 Q_l/(4\pi a^2 B_c^2)$, for a strip with a perpendicular applied flux density of amplitude B_m monotonically increases with B_m, while the loss factor, $\Gamma'_m \equiv \mu_0 Q_l/(4\pi a^2 B_m^2)$, shows a peak. The dashed and dotted lines are the limits for the low and high B_m, respectively.

with the limits for the low and high applied AC flux density being

$$\begin{aligned} \Gamma'_m &= \tfrac{1}{6}\left(\tfrac{B_m}{B_c}\right)^2 && \text{if } B_m \ll B_c, \\ \Gamma'_m &= \tfrac{B_c}{B_m} && \text{if } B_m \gg B_c. \end{aligned} \tag{1.159}$$

Then, at low applied AC flux densities, the loss factor increases as B_m^2, and at high B_m, the loss factor decreases as $1/B_m$, with a peak in between (Figure 1.15(b)). From (1.158), we can find that the peak occurs at $B_{\rm pk} \approx 2.465B_c$ and has a value $\Gamma'_{m,\rm pk} \approx 0.1857$. Since $q_l = \Gamma'_m(B_m/B_a)^2$, the normalized AC loss (and the AC loss Q_l) does not show a peak, increasing monotonically with B_m (Figure 1.15(a)).

1.2.1.6. *Thin strip with transport current*

The current density, magnetic flux density, and AC loss of a thin film of width $2a$ and thickness d under an alternating transport current, I, can be obtained in a very similar way as for applied magnetic fields (Section 1.2.1.5). The CSM can only predict the case when $|I| < I_c$, with the critical current $I_c = 2adJ_c$.

Now, the starting point is that the sheet current density, K, for a perfect conductor (very high J_c or very low I limit) is, as found through conformal mapping by Swan [95] and Norris [62],

$$K(x) = \frac{I}{\pi\sqrt{a^2 - x^2}}, \tag{1.160}$$

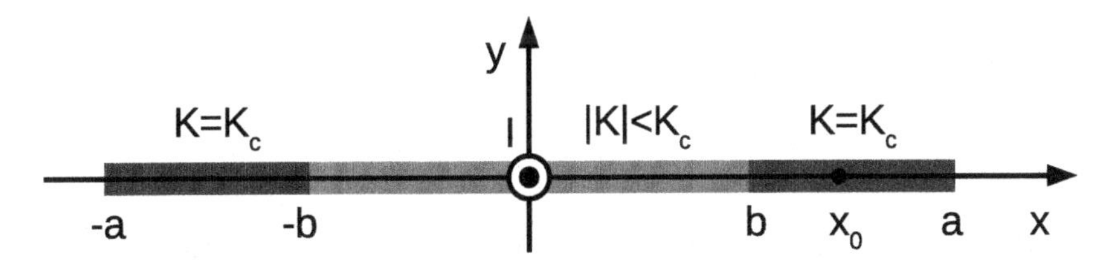

Figure 1.16. A thin strip of thickness d much smaller than its width, $d \ll 2a$, in the CSM under transport current has two regions with sheet current density $K = K_c$, where $K_c \equiv J_c d$, that enclose a central region with $K < K_c$.

and the magnetic flux density at the $y = 0$ line is $\mathbf{B} = B_y \mathbf{e}_y$, with

$$B_y(x) = \begin{cases} 0 & \text{if } |x| < a, \\ \dfrac{I}{2\pi} \dfrac{x}{|x|\sqrt{x^2 - a^2}} & \text{if } |x| > a. \end{cases} \tag{1.161}$$

When considering a finite J_c and increasing the current from 0 to $I > 0$, there will appear two segments with saturated sheet current density, $K = K_c \equiv J_c d$, penetrating from the right and left to a certain distance b from the center (Figure 1.16). However, in the central region, where $|x| < b$, there will still appear nonzero sheet current density with $|K| < K_c$. Since the current flowing in the saturated region is $I_{\text{sat}} = I_c(1 - b/a)$, the current flowing in the non-saturated region is $I_{\text{shield}} = I - I_c(1 - a/b)$. In the central region, where $|K| < K_c$, the CSM assumption tells us that both the electric field and B_y remain zero. The sheet current density necessary to shield B_y from the saturated regions can be calculated in the same way as for the applied field case. In general, the resulting net current in the shielded region, I_{shield,J_c}, is different from the I_{shield} above. Therefore, we need to add a current-carrying component proportional to (1.160) but with a replaced by b there. Finally, we obtain the physical value of b by imposing that $K(x)$ cannot diverge at $|x| = b$ since the CSM assumes $|K| \leq K_c$. The final solution is [62]

$$K(x) = \begin{cases} \dfrac{2J_c d}{\pi} \arctan \sqrt{\dfrac{a^2 - b^2}{b^2 - x^2}} & \text{if } |x| \leq b, \\ J_c d & \text{if } b \leq |x| \leq a, \end{cases} \tag{1.162}$$

with

$$b = a\sqrt{1 - i^2} \tag{1.163}$$

and $i \equiv I/I_c$.

Once $K(x)$ is known, the magnetic flux density can be found by integrating the Biot–Savart law, resulting in

$$B(x) = \begin{cases} 0 & \text{if } |x| \leq b, \\ \frac{B_c x}{|x|} \operatorname{artanh}\sqrt{\frac{x^2-b^2}{a^2-b^2}} & \text{if } b \leq |x| < a, \\ \frac{B_c x}{|x|} \operatorname{artanh}\sqrt{\frac{a^2-b^2}{x^2-b^2}} & \text{if } |x| > a, \end{cases} \tag{1.164}$$

where $B_c \equiv \mu_0 J_c d/\pi$.

From these formulas of the sheet current density and magnetic flux density of the initial curve, we can easily find those from the decreasing and increasing curves of the stationary AC cycle using (1.121). A typical AC behavior can be seen in Figure 1.17.

In order to calculate the AC loss per cycle and unit length, Q_l, we can use the fact that $E = 0$ at the center since $J = 0$ there, then calculate the magnetic flux per unit length $\Phi(x, y = 0)$, as defined in (1.135), and apply

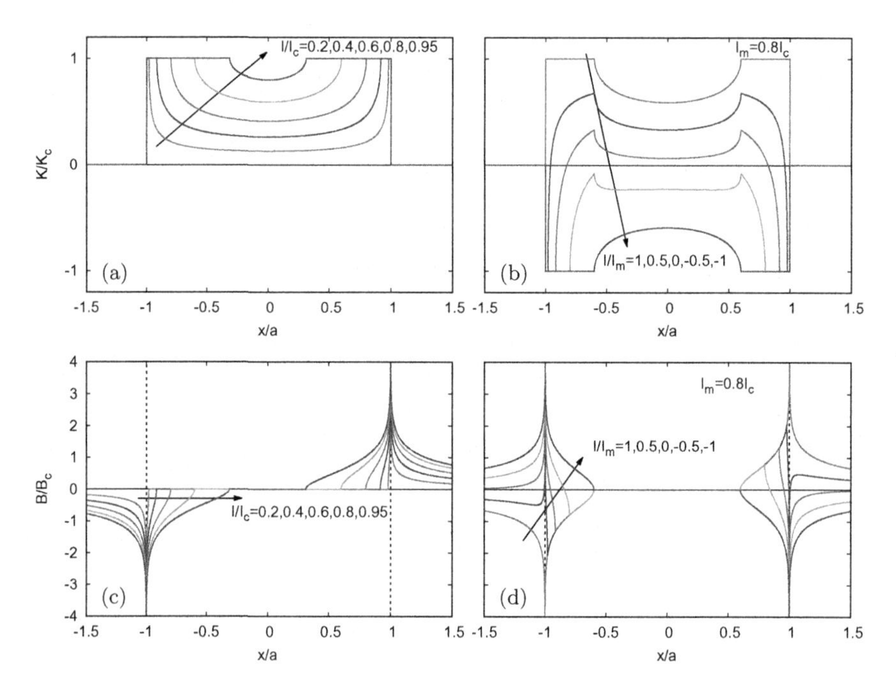

Figure 1.17. ((a) and (b)) Sheet current density in a thin film with alternating transport current and ((c) and (d)) the magnetic flux density that they generate. The left-hand plots, (a) and (c), are for the initial curve, and the right-hand plots, (b) and (d), are for the decreasing curve that follows.

(1.139). When doing so, Q_l results in [62]

$$q \equiv \frac{2\pi Q_l}{\mu_0 I_c^2} = 2[(1+i_m)\ln(1+i_m) + (1-i_m)\ln(1-i_m) - i_m^2], \quad (1.165)$$

where q is the nondimensional normalized loss as defined above. The main limits of the normalized loss are the following:

$$\begin{aligned} &q \approx i_m^4/3 && \text{if } i_m \ll 1, \\ &q = 4\ln 2 - 2 \approx 0.7726 && \text{if } i_m \to 1. \end{aligned} \quad (1.166)$$

It is important to note that for low current amplitudes, the AC loss increases with i_m^4 in contrast to only i_m^3 for elliptical wires.

1.2.1.7. *Universal scaling law for the power-law* $\mathbf{E}(\mathbf{J})$ *relation*

In this section, we show that for the power-law $\mathbf{E}(\mathbf{J})$ relation of (1.18), the vector potential, current density, and AC loss follow a universal scaling law, which is valid for any shape. We also find a relation with the CSM results, determining the frequency that the CSM predictions agree the best with those from the power law.

Next, we find the partial differential equation for the vector potential. For Coulomb's gauge, the vector potential follows Equation (1.31). Taking the isotropic power law from (1.18), the current follows

$$\mathbf{J} = \frac{J_c}{E_c}\left(\frac{|\mathbf{J}|}{J_c}\right)^{1-n}\mathbf{E}. \quad (1.167)$$

Then, Equation (1.31) becomes

$$\begin{aligned} \nabla^2\mathbf{A} &= -\mu_0\frac{J_c}{E_c}\left(\frac{|\mathbf{J}|}{J_c}\right)^{1-n}\mathbf{E}, \\ \nabla^2\mathbf{A} &= \mu_0\frac{J_c}{E_c}\left|\frac{\nabla^2\mathbf{A}}{\mu_0 J_c}\right|^{1-n}(\partial_t\mathbf{A} + \nabla\phi), \end{aligned} \quad (1.168)$$

where in the second equation, we used (1.24) for $\mathbf{E}$ and again (1.31) for $\mathbf{J}$. We can write all the quantities in dimensionless variables as follows:

$$\begin{aligned} \mathbf{r}' &\equiv \mathbf{r}/l, \\ \tau &\equiv t\omega, \\ \mathbf{a} &\equiv \mathbf{A}/\mu_0 l^2 J_c, \\ \varphi &\equiv \phi/\mu_0 l^3 \omega J_c, \end{aligned} \quad (1.169)$$

where l and ω are a certain characteristic length and angular frequency of the problem, respectively. For a tape submitted to a periodic applied magnetic field, for example, l and ω could be the tape width and the angular frequency associated to the applied field period, respectively. With this normalization, Equation (1.168) becomes

$$\nabla'^2 \mathbf{a} |\nabla'^2 \mathbf{a}|^{n-1} = k(\partial_\tau \mathbf{a} + \nabla' \varphi), \tag{1.170}$$

where k is the dimensionless parameter,

$$k = \frac{\mu_0 J_c l^2 \omega}{E_c}, \tag{1.171}$$

∇' is the nabla operator with respect to $\mathbf{r}'$ in (1.169), and ∂_τ is the partial derivative to the normalized time. From Equation (1.170), it is clear that $\mathbf{a}$ and φ only depend on parameter k. If we change k by a constant C so that the new k, $\tilde{k}$, is

$$\tilde{k} = Ck, \tag{1.172}$$

Equation (1.170) tells us that the new $\mathbf{a}$ and φ, $\tilde{\mathbf{a}}$ and $\tilde{\varphi}$, respectively, follow

$$\tilde{\mathbf{a}} = C^{\frac{1}{n-1}} \mathbf{a}, \tag{1.173}$$

$$\tilde{\varphi} = C^{\frac{1}{n-1}} \varphi. \tag{1.174}$$

In case the sample is submitted to an external applied magnetic field $\mathbf{H}_a$, which may be nonuniform, we should take into account the fact that the normalization of the vector potential in (1.169) defines a normalized magnetic field in the absence of magnetic materials:

$$\mathbf{h} \equiv \mathbf{H}/J_c l. \tag{1.175}$$

Since the normalized applied field follows $\mathbf{h}_a = \nabla' \times \mathbf{a}_a$, where $\mathbf{a}_a$ is the normalized vector potential for the applied field, scaling k as $\tilde{k} = Ck$ results in the scaling of $\mathbf{a}$ in (1.173) if $\mathbf{h}_a$ follows the scaling

$$\tilde{\mathbf{h}}_a = \mathbf{h}_a C^{\frac{1}{n-1}}. \tag{1.176}$$

For a cyclic applied field of given amplitude H_m, the dimensionless loss factor

$$\Gamma_m = \frac{Q_v}{\mu_0 \pi H_m^2}, \tag{1.177}$$

where Q_v is the loss per cycle and unit volume, which is related to $\mathbf{a}$, $\mathbf{h}_a$, and $\mathbf{h}_m$ as

$$\Gamma_m = \frac{1}{\mu_0 \pi H_m^2 V} \int_V \mathrm{d}^3\mathbf{r} \oint \mathbf{B} \cdot \mathrm{d}\mathbf{H}_a = \frac{1}{\pi h_m^2 V'} \int_{V'} \mathrm{d}^3\mathbf{r}' \oint (\nabla \times \mathbf{a}) \cdot \mathrm{d}\mathbf{h}_a, \tag{1.178}$$

where $V' = V/l^3$. The above equation shows us that scaling k as $\tilde{k} = Ck$ does not modify the loss factor, and hence,

$$\tilde{\Gamma}_m = \Gamma_m. \tag{1.179}$$

For sinusoidal applied fields, Γ_m corresponds to the imaginary AC susceptibility.

If the sample transports a current I, the relation between the normalized current with respect to the critical current, $i \equiv I/I_c$, and $\mathbf{a}$ is

$$i = \frac{I}{I_c} = \frac{1}{I_c} \int_S \mathrm{d}\mathbf{s} \cdot \mathbf{J} = \frac{-1}{\mu_0 I_c} \int_S \mathrm{d}\mathbf{s} \cdot \nabla^2 \mathbf{A} = \frac{-1}{S'} \int_{S'} \mathrm{d}\mathbf{s}' \cdot \nabla'^2 \mathbf{a}, \tag{1.180}$$

where S is the sample cross-section, $S' \equiv S/l^2$, and $\mathrm{d}\mathbf{s}' = \mathrm{d}\mathbf{s}/l^2$. Then, under the change $\tilde{k} = Ck$, the new i, $\tilde{i}$, is

$$\tilde{i} = iC^{\frac{1}{n-1}}. \tag{1.181}$$

The normalized loss $q = 2\pi Q_l/\mu_0 I_c^2$, where Q_l is the loss per cycle and conductor length, is related to $\mathbf{a}$ as

$$\begin{aligned} q &= \frac{2\pi}{\mu_0 I_c^2} \int_S \mathrm{d}s \oint \mathrm{d}t \mathbf{E} \cdot \mathbf{J} \\ &= \frac{2\pi}{\mu_0 I_c^2} \int_S \mathrm{d}s \oint \mathrm{d}t J_c E_c \left| \frac{\mathbf{J}}{J_c} \right|^{n+1} = \frac{2\pi J_c E_c}{\mu_0 I_c^2} \int_S \mathrm{d}s \oint \mathrm{d}t \left| \frac{\nabla^2 \mathbf{A}}{\mu_0 J_c} \right|^{n+1} \\ &= \frac{2\pi}{k S'^2} \int_{S'} \mathrm{d}s' \oint \mathrm{d}\tau \left| \nabla^2 \mathbf{a} \right|^{n+1}. \end{aligned} \tag{1.182}$$

Then, a new k, $\tilde{k} = Ck$, results in a new q:

$$\tilde{q} = C^{\frac{2}{n-1}} q. \tag{1.183}$$

For a certain current amplitude, I_m, the loss factor is defined as

$$\Gamma_t \equiv \frac{2\pi Q_l}{\mu_0 I_m^2} = \frac{q}{i_m^2}, \tag{1.184}$$

where $i_m \equiv I_m/I_c$. Then, the loss factor is independent of changes in k:

$$\tilde{\Gamma}_t = \Gamma_t. \tag{1.185}$$

The reasoning above is exactly the same if, in addition to the transport current, the sample is submitted to an applied magnetic field with the same period as the transport current, although the current and applied field do not need to be in phase.

Summarizing, the scaling laws for a power-law $\mathbf{E}(\mathbf{J})$ relation are

$$\begin{aligned}
\tilde{k} &= Ck, \\
\tilde{\mathbf{a}} &= C^{1/(n-1)}\mathbf{a}, \\
\tilde{\varphi} &= C^{1/(n-1)}\varphi, \\
\tilde{\mathbf{h}}_a &= C^{1/(n-1)}\mathbf{h}_a, \\
\tilde{i}_m &= C^{1/(n-1)}i_m, \\
\tilde{\Gamma}_m &= \Gamma_m, \\
\tilde{\Gamma}_t &= \Gamma_t.
\end{aligned} \tag{1.186}$$

Then, on increasing k in Equation (1.171) by increasing the frequency or J_c, the curves of the loss factor $\Gamma_m(h_m)$ and $\Gamma_t(i_m)$ shift to higher h_m and i_m, respectively (see Figure 1.18). Actually, we can construct k-independent (or frequency-independent or J_c-independent) curves by taking $\Gamma_m(h_m^*)$ and $\Gamma_t(i_m^*)$, where

$$\begin{aligned}
h_m^* &= h_m k^{-1/(n-1)} = \frac{H_m}{J_c l}\left(\frac{E_c}{\mu_0 J_c l^2 \omega}\right)^{1/(n-1)}, \\
i_m^* &= i_m k^{-1/(n-1)} = \frac{I_m}{I_c}\left(\frac{E_c}{\mu_0 J_c l^2 \omega}\right)^{1/(n-1)}.
\end{aligned} \tag{1.187}$$

The CSM corresponds to the limit $n \to \infty$, resulting in

$$\begin{aligned}
h_m^* &= \frac{H_m}{J_c l} = h_m, \\
i_m^* &= \frac{I_m}{I_c} = i_m,
\end{aligned} \tag{1.188}$$

and hence, h_m^* and i_m^* are equivalent to h_m and i_m, respectively, in the CSM. We name $\Gamma_m(h_m^*)$ and $\Gamma_t(i_m^*)$ as master curves, which only depend on the sample shape and the power-law n factor. For high n factors, the

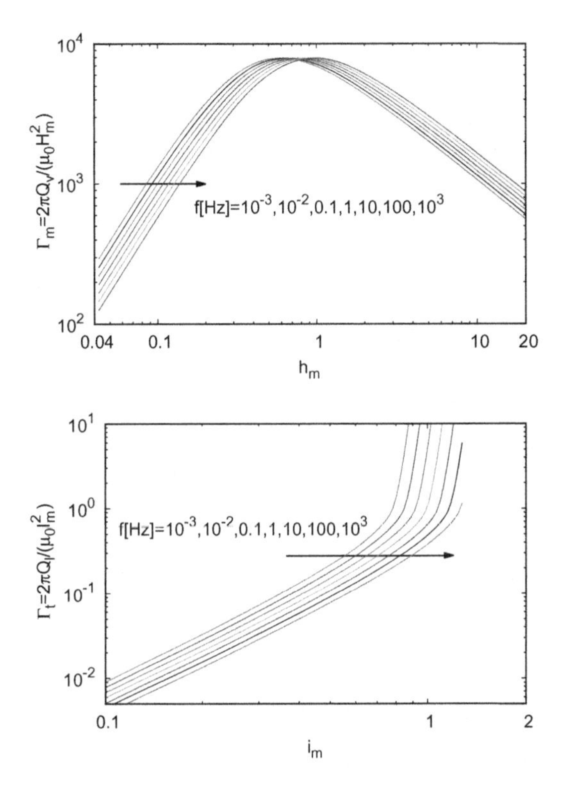

Figure 1.18. The magnetization and transport loss factors in a superconducting thin strip shift to the right with an increase in the frequency.

curve for the power law approaches that for the CSM. Then, the master curve for the CSM can be used to estimate the frequency and J_c dependence of superconductors with power-law $\mathbf{E}(\mathbf{J})$ relations. This can be done by taking h_m^* and i_m^* in (1.187) for the power law as input for the CSM curve:

$$\begin{aligned} \Gamma_{m\text{PL}}(H_m, \omega, J_c, n) &\approx \Gamma_{m\text{CSM}}[h_m^*(H_m, \omega, J_c, n)], \\ \Gamma_{t\text{PL}}(I_m, \omega, J_c, n) &\approx \Gamma_{t\text{CSM}}[i_m^*(I_m, \omega, J_c, n)], \end{aligned} \tag{1.189}$$

where h_m^* and i_m^* are those for the power law given by (1.187). Figure 1.19 shows the transport and magnetization loss for a thin strip as an example. The scaled results for the CSM fairly agree with the power-law results. Since the deduction is general, the scaled CSM results are a fair approximation of those of the power law for any sample shape.

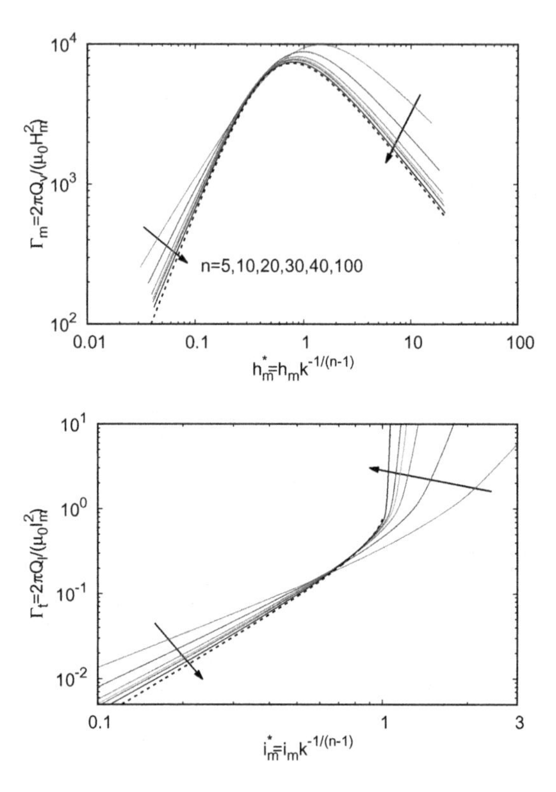

Figure 1.19. The master curves for the magnetization and transport loss factors as a function of $h_m^* = h_m k^{-1/(n-1)}$ and $i_m^* = i_m k^{-1/(n-1)}$ only depend on the power-law exponent n and the sample geometry, where k is the dimensionless parameter in Equation (1.171).

1.2.2. *Eddy currents*

Next, we provide several formulas and their deduction for the eddy current effects in normal metals submitted to varying magnetic fields. These equations are useful for both benchmarking numerical methods and providing quick estimates. By means of these analytical solutions, we also discuss the main behavior of the eddy current effects in superconductors.

1.2.2.1. *Low-frequency limit*

The eddy currents, or currents flowing entirely within normal metals, due to an applied magnetic field can be easily found analytically when the magnetic field that they create is much smaller than the applied magnetic field.

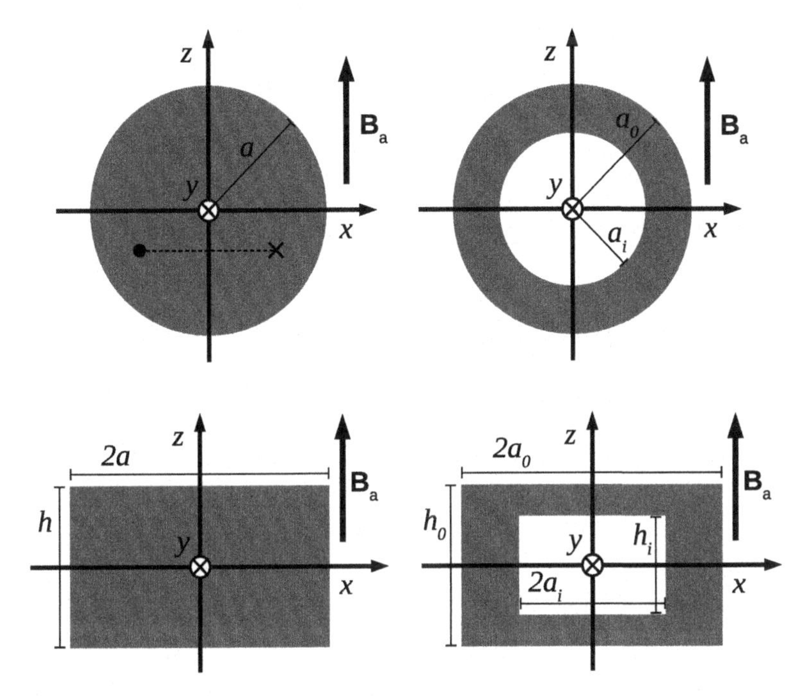

Figure 1.20. For the low-frequency limit, it is simple to analytically obtain the eddy current loss for the wire cross-sections above. The top-left plot also shows the cross-section of the closed circuit that we use to calculate the electric field.

As shown in the following, this occurs at low frequencies. In the following, we detail the case of a circular wire in transverse applied magnetic field, and later, we provide formulas for other geometries.

Let us take a circular wire of radius a that is infinitely long in the y direction and submitted to a uniform applied flux density $\mathbf{B}_a$ in the z direction (Figure 1.20). Then, $\mathbf{J}$ and $\mathbf{E}$ are in the y direction and do not depend on y. Using the integral Faraday's law,

$$\oint_{\partial S} \mathrm{d}\mathbf{r} \cdot \mathbf{E} = -\int_S \mathrm{d}\mathbf{s} \cdot \partial_t \mathbf{B},$$

and considering the circuit in Figure 1.20 (top-left), the electric field follows

$$E(x,z) = -\int_0^x \mathrm{d}x' \partial_t B(x',z), \tag{1.190}$$

where we use the fact that $E(-x,z) = -E(x,z)$ by symmetry. Now, we assume that the induced currents are low enough so that $\mathbf{B}$ created by the currents is much smaller than $\mathbf{B}_a$. Since $J = E/\rho$, this happens for high enough ρ or low enough $\partial_t B_a$, causing a low electric field. Since $\mathbf{B} \approx \mathbf{B}_a$, we obtain an electric field

$$E(x, z) = -x\partial_t B_a.$$

By integrating $JE = E^2/\rho$ over the whole volume, we obtain the instantaneous power loss:

$$P = \frac{4l}{\rho}(\partial_t B_a)^2 \int_0^a \mathrm{d}z \int_0^{x(z)} \mathrm{d}x' x'^2.$$

Here, $x(z)$ for a circular wire is $x = \sqrt{a^2 + z^2}$ for the first quadrant, where we find the integral. This integral becomes

$$\int_0^a \mathrm{d}z \int_0^{x(z)} \mathrm{d}x' x'^2 = \frac{1}{3}\int_0^a \mathrm{d}z[x(z)]^3 = \frac{1}{3}\int_0^a \mathrm{d}z\left(a^2 - z^2\right)^{\frac{3}{2}} = \frac{\pi a^4}{16},$$

and hence, the power loss is

$$P = \frac{\pi l a^4}{4\rho}(\partial_t B_a)^2. \tag{1.191}$$

Considering a sinusoidal applied field, $B_a = B_m \sin(2\pi f t)$, the average power loss in one cycle is

$$P_{\mathrm{av}} = \frac{\pi^3 l a^4 f^2 B_m^2}{2\rho},$$

resulting in a loss per cycle

$$Q = \frac{P_{\mathrm{av}}}{f} = \frac{\pi^3 l a^4 f B_m^2}{2\rho}. \tag{1.192}$$

Then, we see that for low $\partial_t B_a$, and hence low frequency f, the eddy current loss per cycle increases proportionally to l, f, B_m^2, and $1/\rho$. Since the loss per unit volume is $Q/\pi a^4$ and $Q \propto a^4$, the AC loss in a single large conductor is much higher than that of many small conductors with the same total cross-section.

The AC loss of a hollow circular wire of outer radius a_0 and inner radius a_i, similar to the shell of circular superconducting wires (Figure 1.20), can be easily found by taking Equations (1.192) and (1.191) and substracting the contribution from the hole, $Q = Q(a = a_0) - Q(a = a_i)$. Then,

$$P = \frac{\pi l}{4\rho}(\partial_t B_a)^2(a_0^4 - a_i^4),$$

$$Q = \frac{\pi^3 l}{2\rho} f B_m^2 (a_0^4 - a_i^4).$$

Using the same reasoning, we can find the AC loss of a single rectangular wire of side $2a$ and height h (Figure 1.20) as

$$P = \frac{2lh}{3\rho}(\partial_t B_a)^2 a^3,$$

$$Q = \frac{4\pi^2 lh}{3\rho} f B_m^2 a^3 \tag{1.193}$$

and the AC loss of a rectangular shell, resulting in

$$P = \frac{2l}{3\rho}(\partial_t B_a)^2(h_0 a_0^3 - h_i a_i^3),$$

$$Q = \frac{4\pi^2 l}{3\rho} f B_m^2 (h_0 a_0^3 - h_i a_i^3), \tag{1.194}$$

where $2a_0$ and $2a_i$ are the widths of the whole object and the inner hole, respectively, and h_0 and h_i are their respective heights. The case of two rectangular wires connected at the ends in the y direction corresponds to $h_i = h_0$.

1.2.2.2. *Whole frequency range*

For considerably large frequencies, the magnetic field generated by the currents partially shield the applied magnetic field, resulting in a more complex frequency dependence on the loss per cycle than the f scaling found above for low frequencies. Here, we focus on the simple example of an infinite slab with uniform applied field parallel to the surface. For this case, the solution is analytical and serves us to see the main features of the eddy currents and eddy current loss. Some analytical results for this situation are given by Kwasnitza [51] and Takács *et al.* [96], and examples of other situations are given by Íñiguez *et al.* [39] and Jackson [40, pp. 218–223]. In the following, we provide a comprehensive self-contained deduction.

For the eddy currents, a full analysis can be done by the magnetic diffusion equation. This equation can be found as follows. From Faraday's law, $\nabla \times \mathbf{E} = -\partial_t \mathbf{B}$. Using $\mathbf{E} = \rho \mathbf{J}$ and Ampere's law in the void, $\nabla \times \mathbf{B} = \mu_0 \mathbf{J}$, we obtain $\nabla \times (\mu_0^{-1}\rho\nabla \times \mathbf{B}) = -\partial_t \mathbf{B}$. For homogeneous linear materials (constant ρ), we obtain the differential equation

$$\partial_t \mathbf{B} + \frac{\rho}{\mu_0}\nabla^2 \mathbf{B} = 0.$$

For an infinitely long slab in the yz direction, width $2a$, and applied field along the z axis, this equation becomes

$$\partial_t B + \frac{\rho}{\mu_0}\partial_x^2 B = 0, \tag{1.195}$$

where $\mathbf{B} = B\mathbf{e}_y$. Next, we are interested in finding the solutions to uniform applied magnetic fields with any time dependence, $B_a(t)$. In order to find a general solution, we can decompose B into a Fourier transform in time:

$$B(x,t) = \int_{-\infty}^{+\infty} \mathrm{d}\omega\ b(x,\omega)e^{i\omega t},$$

where i is the imaginary unit. Now, Equation (1.195) becomes

$$i\omega b + \frac{\rho}{\mu_0}\partial_x^2 b = 0.$$

The general solution to this differential equation is

$$b(x,\omega) = c(\omega)e^{ikx} + d(\omega)e^{-ikx},$$

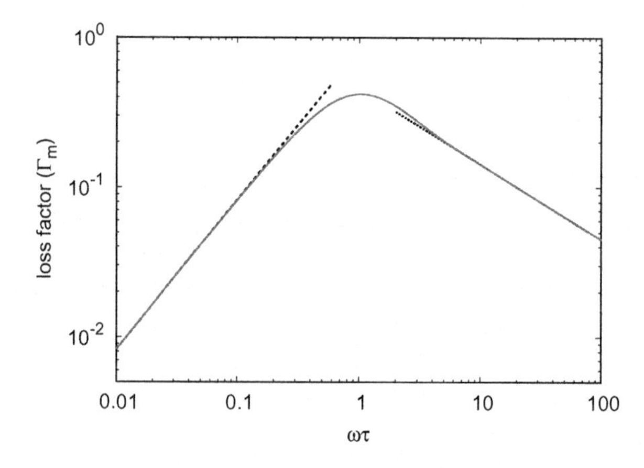

Figure 1.21. The frequency dependence of the loss factor, Γ_m, of a conducting slab under sinusoidal applied field (Equation (1.201)) shows a peak at $\omega \approx 1.03\tau$, with τ being the time constant given by (1.200). The coupling AC loss in a normal metal in between two superconducting slabs follows the same curve, but the meaning of the dimensions $2a$, h, and l is different (Section 1.2.3.2).

with

$$k = g(1 - i), \tag{1.196}$$

$$g = \sqrt{\frac{\omega\mu_0}{2\rho}},$$

where $c(\omega)$ and $d(\omega)$ are any functions of ω. Note the complex k and its frequency dependence through g. For uniform applied magnetic fields, the solution has to be symmetric with x, and hence, $c = d$ above. Then,

$$b(x,\omega) = 2c(\omega)\cos kx,$$

$$B(x,t) = \int_{-\infty}^{+\infty} \mathrm{d}\omega \; c(\omega)\cos(kx)e^{i\omega t},$$

where $\cos kx$ is complex due to the complex k of (1.196), with the cosine generally defined as $\cos Z = (e^{iZ} + e^{-iZ})/2 = \cos Z_r \cosh Z_i - i\sin Z_r \sinh Z_i$, where $Z = Z_r + iZ_i$. The value of $c(\omega)$ is found by imposing $B(x = a, t) = B_a(t)$, where a is the half-width of the slab. Taking the Fourier transform of $B_a(t)$,

$$B_a(t) = \int_{-\infty}^{+\infty} \mathrm{d}\omega \; b_a(\omega)e^{i\omega t},$$

we find that

$$B(x,t) = \int_{-\infty}^{+\infty} \mathrm{d}\omega \; b_a(\omega)\frac{\cos kx}{\cos ka}e^{i\omega t}.$$

For a sinusoidal applied field, $B_a(t) = B_m \sin\omega_a t$, the Fourier transform is

$$b_a(\omega) = \frac{B_m}{2i}\left[\delta(\omega - \omega_a) - \delta(\omega + \omega_a)\right],$$

resulting in the following solution to the magnetic field:

$$B(x,t) = \frac{B_m}{2i}\left[\frac{\cos kx}{\cos ka}e^{i\omega_a t} - \left(\frac{\cos kx}{\cos ka}\right)^* e^{-i\omega_a t}\right], \tag{1.197}$$

with

$$k = g(1 - i) \quad \text{and} \quad g = \sqrt{\omega_a\mu_0/(2\rho)}. \tag{1.198}$$

In spite of the complex formalism, the $B(x,t)$ above is real. The electric field in the slab can be found using (1.190). Thus,

$$E(x,t) = -\frac{B_m}{2}\omega_a\left[\frac{1}{k}\frac{\sin kx}{\cos ka}e^{i\omega_a t} + \left(\frac{1}{k}\frac{\sin kx}{\cos ka}\right)^* e^{-i\omega_a t}\right].$$

By integrating the power loss density $JE = E^2/\rho$ in one cycle and the whole volume, the loss per cycle becomes

$$Q = \frac{\pi}{\mu_0} lhB_m^2 \frac{1}{g} \frac{\sinh 2ga - \sin 2ga}{\cosh 2ga + \cos 2ga}, \tag{1.199}$$

where l and h are the dimensions in the y and z directions, respectively, which are assumed much larger than a. For the low frequency limit, $2ga \ll 1$, we recover the same expression as (1.193), with Q being proportional to the frequency. For large frequencies, or $2ga \gg 1$, the loss per cycle becomes

$$Q = \frac{\pi}{\mu_0} lhB_m^2 \frac{1}{g} = \frac{\pi}{\mu_0} lhB_m^2 \sqrt{\frac{2\rho}{\omega\mu_0}} \qquad \text{if } ga \gg 1,$$

which is proportional to $1/\sqrt{\omega}$. At intermediate frequencies, the loss per cycle reaches its maximum at frequency $\omega_m \approx 1.03/\tau$ (Figure 1.21), where the time constant, τ, is defined as

$$\tau \equiv \frac{\mu_0 4a^2}{\pi^2 \rho}. \tag{1.200}$$

This time constant also corresponds to the characteristic decay time of the magnetic field in the conductor after switching off the applied magnetic field [40, pp. 218–221]. In order to obtain universal results, it is convenient to use the dimensionless loss factor of (1.109), being also the imaginary susceptibility, which in our case is

$$\Gamma_m = \frac{1}{2ga} \frac{\sinh 2ga - \sin 2ga}{\cosh 2ga + \cos 2ga}. \tag{1.201}$$

It can be seen that for high frequencies ($2ga \gg 1$) and close to the surface ($x \to a_-$), the magnetic field of (1.197) becomes

$$B(x,t) \approx B_m e^{-(a-x)g} \sin\left[\omega_a t - g(a-x)\right] \qquad \text{for } x \to a_- \text{ and } ga \gg 1.$$

Then, for high frequencies, the magnetic field penetrates into the slab by a distance of the order of $1/g$, which is the penetration depth.

Summarizing, the main behavior of the eddy current loss can be found from the expressions above:

- At low frequencies, the AC loss per cycle is proportional to ω.
- At high frequencies, the AC loss per cycle decreases with ω as $1/\sqrt{\omega}$.
- The AC loss shows a peak at $\omega \approx 1/\tau$, with τ being the relaxation time constant.

- At high frequencies, the magnetic field penetrates into the sample by a distance of around $1/g$.

1.2.3. *Coupling currents*

Coupling currents, or currents joining two or more superconducting filaments through a normal metal conductor, have many features in common with eddy currents. However, their impact on the AC loss is usually higher since the total resistance of the closed current loops is much lower. The reason is that most of the loop is within the superconducting material, where the resistivity nearly vanishes if the filaments are not saturated with critical current density.

1.2.3.1. *On the decomposition of AC loss into eddy, coupling, and superconductor contributions*

We can always separate the AC loss into the contributions of where the local loss occurs as that from the normal metal or the superconductor as

$$Q = Q_{\text{normal}} + Q_{\text{SC}},$$

with Q_{SC} denoting the AC loss from the superconductor. In general, all currents – eddy, coupling, and superconductor – are both electrically and magnetically coupled, and hence, each type of current influences the other two.

For low enough frequencies, the currents in the normal metal are too low to influence the superconductor. Since the power-law exponent is usually high, the superconductor loss per cycle is virtually frequency independent. That is why the superconductor loss for this scenario is often called as hysteresis loss. However, for large enough frequencies, the coupling currents modify the superconductor loss per cycle, which becomes frequency dependent. That is why, here, we avoid using the terminology of "hysteresis loss" in favor of "superconductor loss." For low enough frequencies and low applied field amplitudes, we can always decompose the AC loss as follows:

$$Q = Q_{\text{eddy}} + Q_{\text{coupling}} + Q_{\text{SC}} \qquad \text{for } B_m \ll B_p,\ \omega \ll \tau_c, \tau_e, \tag{1.202}$$

where B_m is the alternating applied field amplitude, B_p is the penetration field of the superconducting filaments, and τ_c and τ_e are the time constants of the coupling and eddy currents, respectively. In the above equation, the following assumptions are made. For Q_{SC}, the current loops close within each superconducting filament separately, never crossing any normal metal.

Q_{coupling} assumes negligible magnetic flux and current density penetration in the superconductor as well as negligible effective resistance therein. It is also assumed that the magnetic flux density generated by the eddy and coupling currents is negligible compared to the applied flux density and B generated from the superconductor. Since Q_{SC} is virtually frequency independent and Q_{eddy} and Q_{coupling} are proportional to ω^2 at low frequencies, at extremely low frequencies, the superconductor contribution dominates.

When $B_m \geq B_p$, the resistivity in the superconductor is not negligible, thus increasing the resistance of the coupling current loops and hence suppressing the coupling currents.

When $\omega \sim \tau_c$ or $\omega > \tau_c$, the coupling currents are so large that they change the current density in the superconductor, changing Q_{SC}. For $\omega > \tau_c$, the coupling loss per cycle decreases with the frequency. Therefore, for large enough frequencies ($\omega \gg \tau_c$), the coupling loss is much lower than the superconductor loss. Then,

$$Q \approx Q_{\text{eddy}} + Q_{\text{SC}} \qquad \text{for } \omega \gg \tau_c,\ \omega \ll \tau_e, \text{any } B_m.$$

For this case, the coupling current loops tend to shield the whole sample (or the whole filamentary region), and the shape of the loops is dominated by the superconductor. Effectively, we can assume that the current distributes freely between all tapes. Note that Q_{SC} in the above equation is much larger than in Equation (1.202), where the current loops close within each superconducting filament separately.

If $\omega \sim \tau_e$ or $\omega > \tau_e$, the eddy currents could partially shield the AC loss in the superconductor, decreasing the other loss contributions. However, in the following, we always assume $\omega \ll \tau_e$ so that the eddy currents always add a contribution that does not interfere with the coupling and superconductor losses.

1.2.3.2. *Two slab filaments connected by normal conductor*

As an analytical example, we take the case of two superconducting slabs with normal conductor sandwiched between them (Figure 1.22). This could be the case, for example, of two coated conductors connected face-to-face. If the applied field amplitude is low enough, there is nearly no penetration of the magnetic field into the superconductor but only through the normal conductor. Thanks to this, the response of the metal part to alternating magnetic fields is the same as that of a slab. This can be seen as follows. Any slab of normal metal at a position x within a small length, $\mathrm{d}x$, has

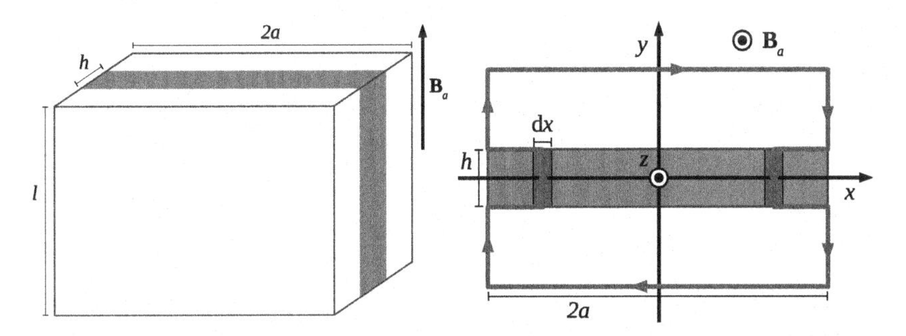

Figure 1.22. A changing applied magnetic field, $\mathbf{B}_a$, causes coupling currents in a normal conductor (blue shaded regions) sandwiched between two superconductors (white blocks): general 3D shape (left) and qualitative behavior of coupling current loops crossing a certain width $\mathrm{d}x$ of the normal region (right).

current $\mathrm{d}I = Jl\mathrm{d}x$, where l is the height of the slab in the applied field direction. This current of uniform sheet density $\mathrm{d}K = J\mathrm{d}x$ closes along the surface of the superconducting filaments (Figure 1.22). Thanks to the infinitely long geometry in the z direction, the magnetic flux density generated by the sheet current density of this circuit is uniform within the volume closed by the circuit with value $\mathrm{d}\mathbf{B} = J\mathrm{d}x\mathbf{e}_z$, regardless of the shape of the circuit. Then, the flux density $\mathrm{d}\mathbf{B}$ generated from the regions of width $\mathrm{d}x$ at $\pm x$ is the same as an infinite slab. As a consequence, the solution is the same as for the eddy currents in a slab, with all equations in Section 1.2.2.2 also applicable. The difference now is that $2a$ is the length of the sandwich, h is the normal metal thickness, and l is its width (Figure 1.22). Then, the qualitative behavior, such as the frequency dependence, is the same as for the eddy currents. From the reasoning above, we can also see that for a given length in the x direction, $2a$, the AC loss in the normal conductor is independent of the size of the superconducting filaments or their external shape.

Applying the results in Section 1.2.2.2 to our coupling configuration, we see the following key features:

- On increasing the frequency, f, the AC loss per cycle shows a peak at around the relaxation time constant, increasing proportionally to f at low frequencies and decreasing as $1/\sqrt{f}$ at high frequencies (Figure 1.21).
- At low frequencies, the AC loss is proportional to the cube of length, $(2a)^3$. Then, dividing the object into n parts along the x direction reduces the AC loss by a factor n^2.

Our coupling configuration is qualitatively similar to superconducting wires in a metal matrix that are transposed with a periodicity length $l_t = 4a$. Then, we see that the AC loss increases with the transposition length as l_t^2, and hence, low-loss wires should have low transposition length.

1.3. Numerical Methods

In this section, we detail the grounds of several numerical methods that have been successfully applied to modeling of HTS. These can be applied to model both HTS and normal metals, describing also eddy current and coupling effects. Some of these methods are also applicable to ferromagnetic materials interacting with HTS. Here, we present formulations that are valid for general 3D modeling, unless stated otherwise.

1.3.1. *Finite element methods*

FEMs solve a certain electromagnetic variable or variables in a region of the space, such as $\mathbf{H}$ or $\mathbf{A}-\phi$, usually including a relatively large portion of the air around the object of study. Within this region, the electromagnetic quantities obey a certain partial differential equation of second order. In addition, these methods require setting boundary conditions at the external surface of the modeled volume, usually of the Dirichlet type. Indeed, it can be shown that the solution to second-order differential equations with given differential equations is unique [42].

FEMs solve the governing partial differential equation by dividing the volume into elements and then solving the value of the electromagnetic quantities at each element by numerical methods, such as the Galerkin method.

Some efficient FEMs use different electromagnetic quantities in separated regions, such as the $\mathbf{H}$–$\mathbf{A}$ and $\mathbf{A}$–$\mathbf{T}$ formulations.

Here, we outline the formulations for general 3D shapes, detailing also the simplifications done for cross-sectional 2D models (either infinitely long wires or objects with cylindrical symmetry).

1.3.1.1. *H formulation*

In FEMs, we first need to write the partial differential equations of the problem. In the $\mathbf{H}$ formulation, we use $\mathbf{H}$ as the electromagnetic variable, from which all the other electromagnetic quantities are calculated, namely

$\mathbf{B}$, $\mathbf{J}$, and $\mathbf{E}$. In particular,

$$\mathbf{B} = \overline{\overline{\mu}}(\mathbf{H})\mathbf{H}, \tag{1.203}$$

$$\mathbf{J} = \nabla \times \mathbf{H}, \tag{1.204}$$

$$\mathbf{E} = \overline{\overline{\rho}}(\nabla \times \mathbf{H})\nabla \times \mathbf{H}, \tag{1.205}$$

where (1.203) is the constitutive relation between $\mathbf{B}$ and $\mathbf{H}$, (1.204) is Ampere's law, and (1.205) combines the constitutive relation between $\mathbf{E}$ and $\mathbf{J}$ through Equation (1.204). With these relations, Faraday's law, $\nabla \times \mathbf{E} = -\partial_t \mathbf{B}$, becomes

$$\nabla \times \left[\overline{\overline{\rho}}(\nabla \times \mathbf{H})\nabla \times \mathbf{H}\right] = -\partial_t \left[\overline{\overline{\mu}}(\mathbf{H})\mathbf{H}\right]. \tag{1.206}$$

In order to solve this differential equation, we need to set constraints and boundary conditions.

If we have a set of n simply connected bodies, which can be regarded as n conductors, each having a given current $I_i(t)$, we should impose the constraint

$$\int_{S_i} \mathrm{d}\mathbf{s} \cdot \mathbf{J} = I_i, \tag{1.207}$$

where S_i is the cross-section of conductor i. From (1.204) and using Stokes' theorem, we obtain

$$\oint_{\partial S_i} \mathrm{d}\mathbf{l} \cdot \mathbf{H} = I_i, \tag{1.208}$$

where $\mathrm{d}\mathbf{l}$ is the line differential. Then, the current at conductor i is determined by the circle integral of $\mathbf{H}$ around the edge of its cross-section. In 2D cross-sectional problems, the constraints of (1.208) are always well defined since a single cross-section describes the whole geometry. For general 3D objects, the current constraints should follow across any cross-section of the conductor (Figure 1.23(a)). The ends of the conductor are somewhat ill-defined because there should be current crossing these surfaces, while there is isolating air. Then, simply applying the current constraints at the ends clashes with the Biot–Savart law $\mathbf{J} = \nabla \times \mathbf{H}$, from which the continuity equation $\nabla \cdot \mathbf{J} = 0$ is obtained. Taking the current constraint directly at the open ends implicitly assumes that the electric current is created on one end and vanishes on the other, which could only be possible if there is an infinite charge reservoir at each end. However, this implies the general current conservation equation, $\nabla \cdot \mathbf{J} + \partial_t q = 0$, which requires the displacement current in the Biot–Savart law, $\nabla \times \mathbf{H} = \mathbf{J} + \partial_t \mathbf{D}$, to contradict the $\mathbf{H}$ formulation. Two simple solutions to the current constraints in 3D are to either consider closed loops (Figure 1.23(b)) or to assume periodic conditions (Figure 1.23(c)). For the latter, the modeled volume repeats indefinitely in at least one direction.

We may also define single constraints for more than one conductor. For example, the current constraint for two conductors perfectly sharing current is

$$\oint_{\partial S_i} \mathrm{d}\mathbf{l} \cdot \mathbf{H} + \oint_{\partial S_j} \mathrm{d}\mathbf{l} \cdot \mathbf{H} = I_{ij}.$$

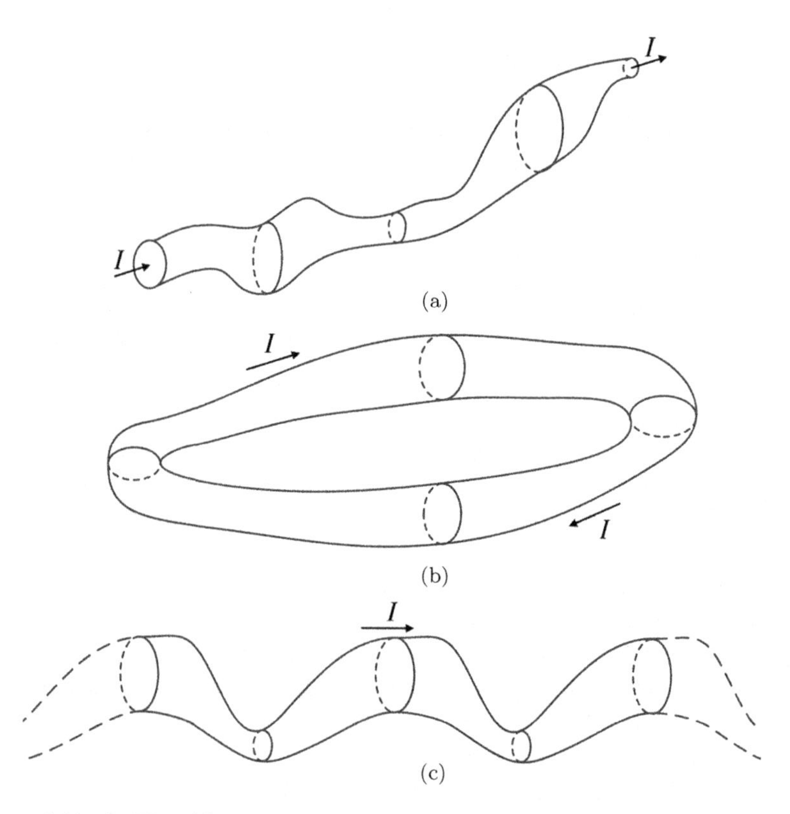

Figure 1.23. In 3D problems, the current constraints should be considered with care: (a) The current constraint of (1.208) should apply for any cross-section of the conductor. For nonzero transport currents, the 3D formulation is consistent with the quasimagnetostatic Maxwell equations for (b) closed loops or (c) periodic wire arrangements extending to infinite in one direction.

The boundary conditions of $\mathbf{H}$ are usually set at a distance much larger than the size of the modeled object based on the fact that the $\mathbf{H}$ generated by the object decays with distance. If there is no net transport current, the boundary condition at the whole modeled domain (or air box), s, can always be set as $\mathbf{H} = \mathbf{H}_a$ on the boundary, where $\mathbf{H}_a$ is the applied magnetic field, or magnetic field created by external sources of $\mathbf{H}$ independent of the solution of the modeled object. If there is some transport current, the same boundary condition can be taken for 3D problems where all the current loops close far away from the boundaries of the air box or for axisymmetrical 2D problems but not for infinitely long configurations, including 3D periodic problems also. The reason is that the net current, I, follows

$$I = \oint_{\partial S} \mathrm{d}\mathbf{s} \cdot \mathbf{H},$$

where S is any surface containing the cross-section of all the conductors. Taking S as a cross-section of the whole modeled region, $\mathbf{H} = \mathbf{H}_a$ at the boundary always results in zero net current since all current density creating $\mathbf{H}_a$ is outside the modeled volume, and hence,

$$\oint_{\partial S} \mathrm{d}\mathbf{l} \cdot \mathbf{H}_a = 0. \tag{1.209}$$

This could contradict the constraints on individual conductors of (1.207). The solution is to use the fact that the $\mathbf{H}$ generated by any infinitely long body approaches $I/(4\pi r_2)\mathbf{e}_\varphi$ far away from the conductor, where r_2 is the modulus of the position vector in 2D and $\mathbf{e}_\varphi$ is the unit vector in the angular direction. Then, we can set $\mathbf{H}$ in the boundary of the whole modeled domain as

$$\mathbf{H}_{\partial s} = \frac{I}{4\pi r_2}\mathbf{e}_\varphi + \mathbf{H}_a \tag{1.210}$$

for infinitely long problems (of periodic 3D problems) and as

$$\mathbf{H}_{\partial s} = \mathbf{H}_a \tag{1.211}$$

for axisymmetric problems or general 3D cases with closed current loops. Actually, it is enough to provide the tangential component of the magnetic field at the boundary, $\mathbf{H}_{\partial s} \times \mathbf{n}$, with $\mathbf{n}$ being the unit vector perpendicular to the surface. The solution to $\mathbf{H}$ is found by dividing into elements and later solving the discretized problem by a finite element technique, such as the Galerkin method [42].

Coming back to the initial equations on which the **H** formulation is based on, (1.203)–(1.206), the reader will note that we did not impose $\nabla \cdot \mathbf{B} = 0$ or $\nabla \cdot (\mu\mathbf{H}) = 0$. Actually, if one chooses the initial conditions such that $\nabla \cdot (\mu\mathbf{H}) = 0$, then the zero divergence of the magnetic field is automatically satisfied at successive instants, as explained by Kajikawa *et al.* [45] and Zermeno *et al.* [106]. In order to see how this comes about, one can start from Faraday's equations written in terms of the magnetic field:

$$\nabla \times (\rho \nabla \times \mathbf{H}) = -\frac{\partial(\mu\mathbf{H})}{\partial t}. \tag{1.212}$$

Taking the divergence of Equation (1.212) yields

$$\nabla \cdot [\nabla \times (\rho \nabla \times \mathbf{H})] = \nabla \cdot \left(-\frac{(\mu\mathbf{H})}{\partial t}\right). \tag{1.213}$$

The left-hand side of Equation (1.213) is identically zero and, after exchanging the order of time and spatial derivatives, it is easy to see that $\nabla \cdot \mathbf{B} = \nabla \cdot (\mu\mathbf{H})$ is constant in time. Consequently, if $\nabla \cdot \mathbf{B} = 0$ at a given time t_0, then $\nabla \cdot \mathbf{B} = 0$ will hold at any other instant. So, if the initial conditions are chosen such that

$$\nabla \cdot (\mu\mathbf{H})|_{t=t_0} = 0, \tag{1.214}$$

then $\nabla \cdot (\mu\mathbf{H}) = 0$ will hold at all times.

This shows that the use of edge elements, which are divergence-free by construction [42], is in principle not necessary to guarantee the zero divergence of the magnetic field. However, the edge elements seem to be more robust and efficient than Lagrange elements for the H formulation implemented in the commercial software COMSOL Multiphysics® [2].

1.3.1.2. A–ϕ *formulation*

We can also describe all electromagnetic fields by the vector and scalar potentials, **A** and ϕ, respectively. In particular, **B** and **E** follow Equations (1.22) and (1.24), respectively:

$$\mathbf{B} = \nabla \times \mathbf{A},$$

$$\mathbf{E} = -\partial_t \mathbf{A} - \nabla\phi.$$

J and **H** are found by inverting the constitutive relations (1.9) and (1.8), respectively, and hence,

$$\mathbf{J} = \overline{\overline{\rho}}^{-1}(\mathbf{E})\mathbf{E},$$
$$\mathbf{H} = \overline{\overline{\mu}}^{-1}(\mathbf{B})\mathbf{B},$$

where $\overline{\overline{\rho}}^{-1}$ is the conductivity tensor. Using all the equations above, the Biot–Savart law, $\nabla \times \mathbf{H} = \mathbf{J}$, becomes

$$\nabla \times [\overline{\overline{\mu}}^{-1}(\nabla \times \mathbf{A})\nabla \times \mathbf{A}] = \overline{\overline{\rho}}^{-1}(\partial_t \mathbf{A} - \nabla\phi)(\partial_t \mathbf{A} - \nabla\phi). \quad (1.215)$$

The current density also needs to follow the current conservation equation, $\nabla \cdot \mathbf{J} = 0$, and hence,

$$\nabla \cdot [\overline{\overline{\rho}}^{-1}(\partial_t \mathbf{A} - \nabla\phi)(\partial_t \mathbf{A} - \nabla\phi)] = 0. \quad (1.216)$$

The two equations above are the coupled partial differential equations to be solved for the general case, including 3D configurations.

Whether the problem contains current sources or not, the solution also needs to fulfill the following current constraint for each conductor (or simply connected region):

$$I_i = \int_{S_i} \mathrm{d}\mathbf{s} \cdot \mathbf{J} = \int_{S_i} \mathrm{d}\mathbf{s} \cdot \overline{\overline{\rho}}^{-1}(\partial_t \mathbf{A} - \nabla\phi)(\partial_t \mathbf{A} - \nabla\phi). \quad (1.217)$$

This constraint is unambiguously defined in 2D problems, either infinitely long or axisymmetrical. However, as is the case for the **H** formulation (see Section 1.3.1.1), this condition is only well defined in 3D problems where the conductors form closed loops or they show translational periodicity.

All the equations above are gauge independent. Therefore, in order to get a particular solution to **A**, we also need to impose the gauge. Next, we outline how to set the gauge and boundary conditions in 2D and 3D, respectively. When setting the boundary conditions, you are implicitly assuming a certain gauge for **A** and ϕ at the boundary. In the following, we assume Coulomb's gauge, and later, we also discuss Weyl's gauge, defined as $\phi = 0$.

Boundary conditions are usually set at the external boundary of a box that extends in air to a distance much larger than the size of the object of study. For infinitely long problems with no net transport current, axisymmetric problems, or 3D configurations with closed current loops, the **A** generated by the currents decays far away from the sample. Then, the boundary condition is

$$\mathbf{A} = \mathbf{A}_a, \tag{1.218}$$

where $\mathbf{A}_a$ is the vector potential due to currents external to the computed domain, which are assumed independent of the solution in this domain. For infinitely long problems and periodic 3D situations with net transport current I, the vector potential generated by the currents approaches $-I/(2\pi)\ln r_2 \mathbf{e}_\varphi$, where r_2 is $r_2 = \sqrt{x^2+y^2}$ for a Cartesian coordinate system with z as the long direction. Then, the boundary condition becomes

$$\mathbf{A} = \mathbf{A}_a - \frac{I}{2\pi}\ln r_2 \mathbf{e}_\varphi. \tag{1.219}$$

Alternatively, one could use Weyl's gauge, where

$$\phi = 0. \tag{1.220}$$

This defines a relation between the potentials in Coulomb's gauge, $\mathbf{A}_c$ and ϕ_c, as (see Equation (1.26))

$$\mathbf{A} = \mathbf{A}_c + \nabla \int_0^t \mathrm{d}t' \phi_c(t'). \tag{1.221}$$

As a consequence of $\phi = 0$, the electric field follows

$$\mathbf{E} = -\partial_t \mathbf{A}. \tag{1.222}$$

For infinitely long problems with no net current, axisymmetrical shapes, and 3D shapes with closed current loops, the electrostatic potential vanishes far away from the sample ($\phi_c \to 0$) so that the boundary condition becomes $\mathbf{A} = \mathbf{A}_a$, with $\mathbf{A}_a$ being the applied vector potential in Coulomb's gauge. For cases with net transport current, either infinitely long or periodic 3D, ϕ_c and $\nabla\phi_c$ do not vanish far away from the sample, and hence, the boundary condition is not straightforward.

For both Coulomb's and Weyl's gauges, the boundary condition for the scalar potential can always be set as zero. This is clear for axisymmetrical problems, 3D cases with closed loops, and infinitely long geometries with no net current. For infinitely long problems of periodic 3D shapes with transport current and insulating regions between conductors, we can always impose $\phi = 0$ at the mid-length in the transport direction. Therefore, $\phi = 0$ at this mid-length, although $\partial_z \phi$ does not generally vanish.

In the $\mathbf{A}$–ϕ formulation, we also need to ensure that the gauge is preserved in the whole domain. In 2D problems (infinitely long or axisymmetric), the gauge is fixed at the boundary condition since fixing the vector potential at a single point determines the gauge (Sections 1.1.3.1 and 1.1.3.2). In 3D, the gauge is usually ensured in the whole domain by the spanning tree technique [3, 56].

1.3.1.3. $\mathbf{T}$–Ω *formulation*

This formulation solves the current potentials $\mathbf{T}$ and Ω (Section 1.1.6), from which all electromagnetic quantities can be obtained. $\mathbf{J}$ and $\mathbf{H}$ are related to $\mathbf{T}$ and Ω as

$$\mathbf{J} = \nabla \times \mathbf{T},$$
$$\mathbf{H} = \mathbf{T} - \nabla\Omega,$$

and the constitutive relations yield $\mathbf{B}$ and $\mathbf{E}$:

$$\mathbf{B} = \overline{\overline{\mu}}(\mathbf{H})\mathbf{H},$$
$$\mathbf{E} = \overline{\overline{\rho}}(\mathbf{J})\mathbf{J}.$$

With these relations, Faraday's law, $\nabla \times \mathbf{E} = -\partial_t \mathbf{B}$, becomes

$$\nabla \times \left[\overline{\overline{\rho}}(\nabla \times \mathbf{T})\nabla \times \mathbf{T}\right] = -\partial_t \left[\overline{\overline{\mu}}(\mathbf{T} - \nabla\Omega)(\mathbf{T} - \nabla\Omega)\right], \tag{1.223}$$

and the magnetic Gauss' law, $\nabla \cdot \mathbf{B} = 0$, takes the form

$$\nabla \cdot \left[\overline{\overline{\mu}}(\mathbf{T} - \nabla\Omega)(\mathbf{T} - \nabla\Omega)\right] = 0. \tag{1.224}$$

These two coupled differential equations determine both $\mathbf{T}$ and Ω.

The current constraint at each conductor i is

$$I_i = \int_{S_i} \mathrm{d}\mathbf{s} \cdot \mathbf{J} = \oint_{\partial S_i} \mathrm{d}\mathbf{l} \cdot \mathbf{T} \tag{1.225}$$

for any cross-section of a conductor, S_i, with I_i being the current in conductor i. As for the other formulations, these boundary conditions are perfectly consistent with cross-sectional 2D problems, but special care needs to be taken in 3D (see Section 1.3.1.1 for the $\mathbf{H}$ formulation).

Next, we take the boundary conditions into account for a bounding box far away from the modeled object. If there is no net current expanding into infinity (such as 2D axisymmetric shapes, 3D closed current loops, and 2D infinite geometries with zero total current), the boundary condition is

$$\mathbf{T} - \nabla\Omega = \mathbf{H}_a, \tag{1.226}$$

with $\mathbf{H}_a$ being an applied magnetic field created by external sources. If there is net current I flowing toward infinity as in 2D infinitely long shapes or 3D periodic problems, the next current constraint needs to follow for any cross-section, s, of the whole modeled domain:

$$I = \int_s \mathrm{d}\mathbf{s} \cdot \mathbf{J} = \int_{\partial s} \mathrm{d}\mathbf{l} \cdot \mathbf{T} = \int_{\partial s} \mathrm{d}\mathbf{l} \cdot \mathbf{H}. \tag{1.227}$$

Since the sources of $\mathbf{H}_a$ are external to the modeled volume, $\int_{\partial s} \mathrm{d}\mathbf{l} \cdot \mathbf{H}_a = 0$ so that the above integral vanishes at the boundary. Then, we need to add an extra term to (1.225). Since $\mathbf{H}$ far away from an infinitely long conductor in the z direction follows $\mathbf{H} \approx I/(2\pi r)\mathbf{e}_\varphi$, at high enough distances from the conductor, the boundary condition becomes

$$\mathbf{T} - \nabla\Omega = \mathbf{H}_a + \frac{I}{2\pi r}\mathbf{e}_\varphi, \tag{1.228}$$

where $r = \sqrt{x^2 + y^2}$ and $\mathbf{e}_\varphi$ is the angular unit vector.

The boundary conditions (1.226) and (1.228) are valid for any gauge of $\mathbf{T}$ and Ω (gauge transformations given in Section 1.1.6). Next, we take particular gauges into account.

For the $\Omega = 0$ gauge, we get $\mathbf{T} = \mathbf{H}$, and hence, we reproduce the $\mathbf{H}$ formulation.

For the divergence-free gauge, $\nabla \cdot \mathbf{T} = 0$, Ω becomes the magnetic potential. Since there is always as many positive magnetic poles as negative ones, the leading term of Ω is dipolar, and hence, both Ω and $\nabla\Omega$ can be set as zero at the boundary. Then, the boundary conditions are

$$\begin{aligned} \mathbf{T} &= \mathbf{H}_a + \frac{I}{2\pi r}\mathbf{e}_\varphi, \\ \Omega &= 0. \end{aligned} \tag{1.229}$$

The gauge in the whole modeling region is ensured by the spanning tree technique on the meshing (or discretization), similar to $\mathbf{A}$–ϕ in 3D [3, 52]. For 2D problems, $\mathbf{T}$ still has two components, and hence, the spanning tree technique is also necessary.

1.3.1.4. *Combined formulations*

Any formulation or numerical approach has advantages and disadvantages. Combining different approaches, each applicable to a different region of space, can provide the best of both.

All the formulations presented in this book are dynamic since they take Faraday's law into account, and they are also valid for nonlinear conducting materials. However, static formulations or specific dynamic ones for linear conducting materials are more straightforward to solve [6, 42], being usually faster. In addition, they are often better suited for advanced meshing. This is the case of rotating mesh, which is crucial for rotating machines. These approaches are considered standard in electrical engineering, e.g. dynamic **A** formulation for normal conductors [6].

The **H**–**A** formulation uses the **A** formulation for linear conductors at normal conducting or magnetic materials, while using the **H** formulation in the region containing the superconductors and the surrounding air [19].

The **T**–**A** formulation uses the static **A** approach to conventional materials (normal conductors, magnetic materials, and air), while using **T** to describe the superconductor. This formulation is mostly used with the approximation of the superconductor as a thin shell [109], but it can also be applied to general geometries [35].

A different kind of mixed formulation is that based on **H**–ψ, which uses **H** in the superconducting or normal conducting regions and the magnetic scalar potential ψ in the air, where $\mathbf{H} = -\nabla\psi$. The problem is that the relation $\mathbf{H} = -\nabla\psi$ causes any closed-loop path integral of **H** to vanish, being incompatible to the current constraints of (1.208). Then, the $\mathbf{H} - \psi$ formulation is valid when any closed integral of **H** vanishes at the region described by ψ. This is the case of the air region for the magnetization problem (no current constraints) and no holes in the conductor as well as the air outside rotating machines. An improvement in this approach that enables current constraints is the **H**–ψ formulation with cohomological decomposition [53, 94], where **H** is decomposed as

$$\mathbf{H} = -\nabla\psi + \sum_i I_i \mathbf{h}_i,$$

with ψ being a scalar magnetic potential, the summation made over all possible current constraints, I_i the current relative to current constraint i, and $\mathbf{h}_i$ a base function related to the current constraint i. For this formulation, all base functions $\mathbf{h}_i$ are kept the same throughout the whole time evolution, and hence, ψ remains the sole quantity to be solved in the air.

Further developments in 3D modeling have been made by Arsenault *et al.* for applied magnetic fields [8, 9] and for transport current [7].

1.3.2. *Variational methods*

Variational methods find the electromagnetic quantities by minimizing a certain functional, which is equivalent to solving Maxwell's equations (or a set of differential or integro-differential equations based on them) [18, 69, 78]. Essentially, the solution is of the same kind as that obtained by FEMs, and hence, variational methods are an alternative. The variational methods based on solving electromagnetic properties that exist only in the materials (**J** for conductors and superconductors and **M** for magnetic materials) do not require us to solve the fields in the air, which greatly reduces the number of degrees of freedom. This property is also shared by integral methods (see Section 1.3.3). Another advantage of variational methods is their capability to solve any $\mathbf{E}(\mathbf{J})$ relation, including the multi-valued $\mathbf{E}(\mathbf{J})$ relation of the CSM (Section 1.1.2). Actually, only variational methods can solve the CSM in arbitrary shapes.

Similar to FEMs, variational methods also exist in several formulations, depending on what quantity is solved. First, we present the variational methods with no magnetic materials (Sections 1.3.2.1–1.3.2.3), and later, we present the extension to magnetic materials (Section 1.3.2.5). For no magnetic materials, the constitutive relation between **B** and **H** is simply $\mathbf{B} = \mu_0\mathbf{H}$, and hence, **B** and **H** are generated only by the currents.

Three formulations can be applied for variational methods in 3D: **T**, **H**, and $\mathbf{H}$–ψ. Although the formalism in Section 1.3.2.1 for the $\mathbf{J}$–ϕ formulation is also valid for 3D, the presence of the electrostatic potential in the functional highly complicates its applicability. However, the $\mathbf{J}$–ϕ formulation is very useful for cross-sectional problems (or 2D) because **J** can be reduced to a scalar, which greatly reduces the number of degrees of freedom and the computing time [66].

An interesting fact is that these variational methods are able to take any vectorial $\mathbf{E}(\mathbf{J})$ relation into account (including the double CSM or the anisotropic CSM), where **E** does not need to be parallel to **J** [10, 11, 46, 48, 69]. This is especially relevant for force-free configurations [25, 46].

Although the mathematical background of the first works on variational methods are based on variational inequalities [18, 77], the deduction in terms of the Euler equations of a functional is more understandable for engineering and physics backgrounds [11, 69, 73]. Therefore, we take the latter approach in this book.

1.3.2.1. J-ϕ *formulation*

Here, we present the functional with **J** and the scalar potential ϕ as the state variables. The starting point is that the equations that we wish to solve are

$$\mathbf{E}(\mathbf{J}) = -\partial_t \mathbf{A} - \nabla\phi,$$

$$\nabla \cdot \mathbf{J} = 0.$$

The first equation above joins the constitutive $\mathbf{E}(\mathbf{J})$ relation of the material with the relation of **E** with the electromagnetic potentials, while the second equation represents current conservation.

In time-discretized form at time t_k, these equations become

$$\mathbf{E}(\mathbf{J}_k) = -\frac{\Delta \mathbf{A}_k}{\Delta t_k} - \nabla\phi_k, \quad (1.230)$$

$$\nabla \cdot \mathbf{J}_k = 0, \quad (1.231)$$

where $\Delta \mathbf{A}_k = \mathbf{A}_k - \mathbf{A}_{k-1}$ and $\Delta t_k = t_k - t_{k-1}$.

Next, we need to take Coulomb's gauge into account in order to express **A** as an integral of **J** (see Section 1.1.3). We separate **A** from the contribution of the currents in the sample, $\mathbf{A}[\mathbf{J}]$, and those from external sources, $\mathbf{A}_a$, which are responsible for the external applied magnetic field. The square brackets in $\mathbf{A}[\mathbf{J}]$ denote functional dependence, with $\mathbf{A}[\mathbf{J}]$ being the integrals of **J** in the space of (1.32), (1.34), and (1.40) for 3D, infinitely long, and cylindrical geometries, respectively. Then, (1.230) becomes

$$\mathbf{E}(\mathbf{J}_{k-1} + \Delta\mathbf{J}_k) = -\frac{\mathbf{A}[\Delta\mathbf{J}_k]}{\Delta t_k} - \frac{\Delta\mathbf{A}_{ak}}{\Delta t_k} - \nabla\phi_k, \quad (1.232)$$

where $\Delta\mathbf{J}_k = \mathbf{J}_k - \mathbf{J}_{k-1}$ and $\Delta\mathbf{A}_{ak} = \mathbf{A}_{ak} - \mathbf{A}_{a,k-1}$. As a consequence of Coulomb's gauge, ϕ is the electrostatic potential (Section 1.1.3).

As a result, we need to solve the integral and differential equations of (1.232) and (1.231), respectively, in order to obtain the solution to the quasimagnetostatic problem.

As mathematically demonstrated by Pardo and Kapolka [69], Equations (1.232) and (1.231) are the Euler equations of the following functional with respect to $\Delta\mathbf{J}_k$ and ϕ_k, respectively:

$$L_k[\Delta\mathbf{J}_k, \phi_k] = \int_v \mathrm{d}^3\mathbf{r}\left[\frac{1}{2}\Delta\mathbf{J}_k \cdot \frac{\mathbf{A}[\Delta\mathbf{J}_k]}{\Delta t_k} + \Delta\mathbf{J}_k \cdot \frac{\Delta\mathbf{A}_{ak}}{\Delta t_k} + U(\Delta\mathbf{J}_k + \mathbf{J}_{k-1}) + \nabla\phi_k \cdot (\Delta\mathbf{J}_k + \mathbf{J}_{k-1})\right], \quad (1.233)$$

where $\Delta\mathbf{J}_k = \mathbf{J}_k - \mathbf{J}_{k-1}$ and U is the dissipation factor, defined as

$$U(\mathbf{J}) \equiv \int_0^{\mathbf{J}} \mathrm{d}\mathbf{J}' \cdot \mathbf{E}(\mathbf{J}').$$

This dissipation factor is uniquely defined for any physical $\mathbf{E}(\mathbf{J})$ relation because, due to irreversible thermodynamic principles, $\nabla_{\mathbf{J}} \times \mathbf{E}(\mathbf{J}) = 0$, and hence, the path integral in the U definition does not depend on the particular integration path [11, 73]. Then, the solution to the problem corresponds to an extreme of the functional above. As shown by Pardo and Kapolka [69], for any given ϕ_k, the extreme of this functional with respect to $\Delta\mathbf{J}_k$ is a minimum that always exists and is unique. In order to find the time evolution, we need to minimize the L_k above with respect to $\Delta\mathbf{J}_k$ for each time k, once the solution for time $k-1$ is known. The starting initial condition is usually $\mathbf{J} = 0$.

For general 3D configurations, obtaining $\mathbf{J}$ with this functional requires us to previously know ϕ, which is usually not the case. The general 3D case can be solved by the other formulations, $\mathbf{T}$, $\mathbf{H}$, or $\mathbf{H} - \psi$, in the following sections. An alternative is the $\mathbf{J} - q$ formulation, where $\mathbf{J}$ and the charge density q are iteratively solved [71], although we do not go into detail here.

For cross-sectional problems (or 2D), the $\mathbf{J} - \phi$ formulation does not require us to solve the scalar potential, thus becoming very powerful. In infinitely long problems, in the z coordinate, $\mathbf{J}$, $\mathbf{A}$, and $\nabla\phi$ have only one component so that $\mathbf{J}(\mathbf{r}) = J(x,y)\mathbf{e}_z$, $\mathbf{A}(\mathbf{r}) = A(x,y)\mathbf{e}_z$, and $\nabla\phi(\mathbf{r}) = \partial_z\phi\mathbf{e}_z$. Then, the functional takes the form

$$L_k[\Delta J_k] = l\int_s \mathrm{d}x\mathrm{d}y\left[\frac{1}{2}\Delta J_k \frac{A[\Delta J_k]}{\Delta t_k} + \Delta J_k \Delta A_{ak} \right.$$
$$\left. + U(\Delta J_k + J_{k-1}) + (\Delta J_k + J_{k-1}) \cdot \partial_z\phi_k\right],$$

where l is the conductor length and $A[J]$ is found using Equation (1.34). Thanks to the infinitely long geometry, it can be seen that $\partial_z\phi$ is uniform at each conductor i in the case where we are modeling multiple-tape conductor problems. Then,

$$L_k[\Delta J_k] = l\int_s \mathrm{d}x\mathrm{d}y\left[\frac{1}{2}\Delta J_k \frac{A[\Delta J_k]}{\Delta t_k} + \Delta J_k \Delta A_{ak} + U(\Delta J_k + J_{k-1})\right]$$
$$- \sum_i^{n_c} V_{ik} I_{ik},$$

where I_{ik} and V_{ik} are the current and voltage drop at conductor i at time k, respectively, with the latter following circuit sign convention (the current flows in the direction where there is a voltage drop). We also use $V_{ik} = -l(\partial_z \phi_k)_i$. Similarly, in axisymmetrical problems, we find that

$$L_k[\Delta J_k] = 2\pi \int_s \mathrm{d}r\mathrm{d}z \left[\frac{r}{2} \Delta J_k \frac{A[\Delta J_k]}{\Delta t_k} + r\Delta J_k \frac{\Delta A_{ak}}{\Delta t_k} + rU(\Delta J_k + J_{k-1}) \right] - \sum_i^{n_c} V_{ik} I_{ik},$$

where we use the fact that in axisymmetric problems, the angular derivative of the electrostatic potential, $\partial_\varphi \phi$, is constant in the cross-section of each conductor, and hence, the voltage drop in one closed conductor is $V_{ik} = -2\pi(\partial_\varphi \phi_k)_i$. For coils, the interpretation of V_i is the voltage applied by a source connected at a very thin cut in a loop. A coil can be seen as connecting these loops at the small cut (see Figure 1.24). This approximation of a round coil is realistic for closely wound pancakes.

If the given input variables are the voltages at each conductor (voltage sources), both the current density and the total currents are found via minimization. If the input is the total currents (current sources), we need to take the current constraints into account in the minimization process. This can be done by the minimization algorithm given by Pardo [63] and Pardo

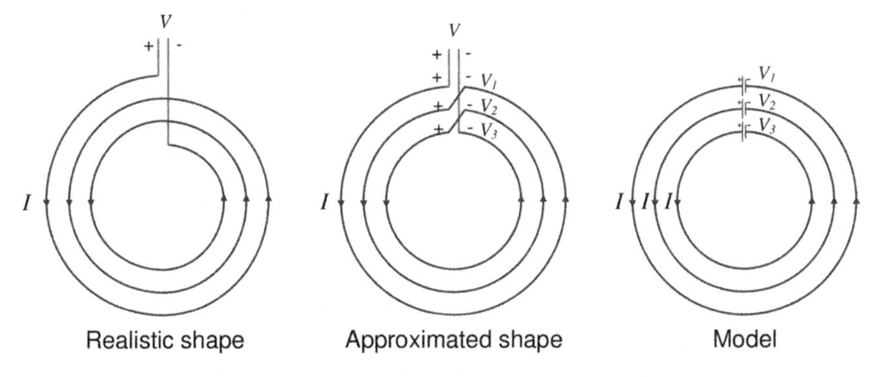

Figure 1.24. Closely packed pancake coils can be approximated by an axisymmetric geometry. The sketches above are for a pancake coil with its realistic spiral shape (left), assuming circular turns except a turn-to-turn crossover (center), and circular turns with independent voltage sources (right). For all configurations, the current at each turn, I, is the same. In the central sketch, V_1, V_2, and V_3 are the voltage drops between the beginning and end of turns 1, 2, and 3, respectively.

et al. [73] or by taking into account the current constraints via augmented functionals [80].

For a magnetic-field-dependent $\mathbf{E}(\mathbf{J})$ relation, where J_c or the power-law exponent n could depend on $\mathbf{B}$, we can still use the variational method. For this case, we need to find $\mathbf{J}$ and $\mathbf{B}$ iteratively. That is, we start with $\mathbf{B} = \mu_0 \mathbf{H}_a$, where $\mathbf{H}_a$ is the applied field, then calculate $\mathbf{J}$ by minimization at the given $\mathbf{B}$ and $\mathbf{B}$ with the $\mathbf{J}$ contribution, and later, we calculate $\mathbf{J}$ again for the new $\mathbf{B}$. We repeat this process until the difference in $\mathbf{J}$ between two iterations is below a certain tolerance.

1.3.2.2. T *formulation*

For the general 3D case, a way of dropping the scalar potential but keeping the modeled volume as the sample is to use the $\mathbf{T}$ potential with the interpretation of an effective magnetization so that $\mathbf{T}$ vanishes outside the conductors (see Section 1.1.6). If the studied object transports no current, $\mathbf{J} = \nabla \times \mathbf{T}$. A transport current can be taken into account if we add another term to $\mathbf{J}$ as

$$\mathbf{J} = \nabla \times \mathbf{T} + \mathbf{J}_t.$$

This transport current density is not an unknown of the problem, but it is given externally in order that the total current at each conductor, i, follows a given value at every time so that

$$\int_{S_i} \mathrm{d}\mathbf{s} \cdot \mathbf{J} = \int_{S_i} \mathrm{d}\mathbf{s} \cdot \mathbf{J}_t = I_i, \qquad (1.234)$$

where S_i is any cross-section of the 3D conductor. In the above equation, we used the fact that the term $\mathbf{J}$ with $\nabla \times \mathbf{T}$ does not contribute to the net current density since $\mathbf{T}$ outside the sample vanishes. An additional condition of the given $\mathbf{J}_t$ is that it should follow current conservation:

$$\nabla \cdot \mathbf{J}_t = 0. \qquad (1.235)$$

Any given $\mathbf{J}_t$ that follows conditions (1.234) and (1.235) could be selected. The particular choice of $\mathbf{J}_t$ only changes the value of $\mathbf{T}$ but not that of the final solution to $\mathbf{J}$, which is the observable quantity. With these definitions of $\mathbf{T}$ and $\mathbf{J}_t$, the functional in (1.233) at time step k becomes

$$L_k[\Delta\mathbf{T}_k] = \int_v \mathrm{d}^3\mathbf{r}\left[\frac{1}{2}\Delta\mathbf{T}_k \cdot \frac{\mathbf{B}[\Delta\mathbf{T}_k]}{\Delta t_k} + \Delta\mathbf{T}_k \cdot \frac{(\Delta\mathbf{B}_{ak} + \Delta\mathbf{B}_{tk})}{\Delta t_k} \right.$$
$$\left. + U\Big(\nabla \times (\Delta\mathbf{T}_k + \mathbf{T}_{k-1}) + \mathbf{J}_t\Big)\right], \qquad (1.236)$$

where $\mathbf{B}_a$ is the applied magnetic flux density, $\mathbf{B}_t$ is $\mathbf{B}$ generated by $\mathbf{J}_t$, and $\mathbf{B}[\mathbf{T}]$ is the magnetic flux density generated by $\mathbf{T}$, which is given by

$$\mathbf{B}[\mathbf{T}](\mathbf{r}) = \frac{\mu_0}{4\pi}\int_v \mathrm{d}^3\mathbf{r}'\frac{[\nabla' \times \mathbf{T}(\mathbf{r}')] \times (\mathbf{r} - \mathbf{r}')}{|\mathbf{r} - \mathbf{r}'|^3}$$
$$= \mu_0\mathbf{T}(\mathbf{r}) + \frac{\mu_0}{4\pi}\int_v \mathrm{d}^3\mathbf{r}'\overline{\overline{g}}(\mathbf{r} - \mathbf{r}')\mathbf{T}(\mathbf{r}'), \qquad (1.237)$$

with the symmetric tensor $\overline{\overline{g}}$ components as

$$g_{ij}(\mathbf{r}) = \frac{3r_ir_j}{r^5} - \frac{\delta_{ij}}{r^3}.$$

For infinitely long problems, $\mathbf{B}[\mathbf{T}]$ follows

$$\mathbf{B}[\mathbf{T}](\mathbf{r}_2) = \mu_0\mathbf{T}(\mathbf{r}_2) + \frac{\mu_0}{2\pi}\int_s \mathrm{d}^2\mathbf{r}_2'\,\overline{\overline{h}}(\mathbf{r}_2 - \mathbf{r}_2')\mathbf{T}(\mathbf{r}_2'), \qquad (1.238)$$

with

$$h_{ij}(\mathbf{r}_2) = \frac{2r_ir_j}{r_2^4} - \frac{\delta_{ij}}{r_2^2},$$

where $\mathbf{r}_2$ is the 2D position vector in the cross-section, $\mathbf{r}_2 = x\mathbf{e}_x + y\mathbf{e}_y$, and $r_2 = \sqrt{x^2 + y^2}$.

In (1.236), $\Delta\mathbf{T}_k = \mathbf{T}_k - \mathbf{T}_{k-1}$ and $\Delta\mathbf{B}_k = \mathbf{B}_k - \mathbf{B}_{k-1}$. With this formulation, it can be demonstrated that the minimum of the functional always exists, it is unique, and it corresponds to solving the time-discretized Faraday's law [69]:

$$\nabla \times \mathbf{E}(\nabla \times \mathbf{T}_k + \mathbf{J}_{tk}) = -\frac{\Delta\mathbf{B}_{ak}}{\Delta t_k} - \frac{\mathbf{B}[\Delta\mathbf{T}_k]}{\Delta t_k}.$$

Since $\mathbf{B}[\Delta\mathbf{T}_k]$ is an integral of $\Delta\mathbf{T}_k$, the above equation is an integro-differential equation of the state variable, $\Delta\mathbf{T}_k$.

1.3.2.3. **H** *formulation*

The electrostatic potential could also be avoided if we take the magnetic field **H** as the state variable. Actually, this was the first variational approach for superconductors [10, 18]. In this case, the equation that we wish to solve is Faraday's law in time-discretized form, which at time step k is

$$\nabla \times \mathbf{E}(\nabla \times \mathbf{H}_k) = -\mu_0 \frac{\Delta \mathbf{H}_k}{\Delta t_k}, \tag{1.239}$$

with $\Delta\mathbf{H}_k = \mathbf{H}_k - \mathbf{H}_{k-1}$. Solving this equation is the same as minimizing the functional [11, 73]

$$\begin{aligned} L_k[\Delta\mathbf{H}_{Jk}] = \int_{\mathbb{R}^3} \mathrm{d}^3\mathbf{r} \Bigg[&\frac{\mu_0}{2} \frac{(\Delta\mathbf{H}_{Jk})^2}{\Delta t_k} + \mu_0 \Delta\mathbf{H}_{Jk} \cdot \frac{\Delta\mathbf{H}_{ak}}{\Delta t_k} \\ &+ U(\nabla \times (\Delta\mathbf{H}_k + \mathbf{H}_{k-1})) \Bigg], \end{aligned} \tag{1.240}$$

with

$$U(\mathbf{J}) = \int_0^{\mathbf{J}} \mathrm{d}\mathbf{J}' \cdot \mathbf{E}(\mathbf{J}'),$$

where $\mathbf{H} = \mathbf{H}_J + \mathbf{H}_a$, with $\mathbf{H}_J$ being the magnetic field created by the current density $\mathbf{J}$ and $\mathbf{H}_a$ is the applied magnetic field. In the functional above, we need to set a certain $\mathbf{E}(\mathbf{J})$ relation in air, which can be $\mathbf{E} = \rho_{\mathrm{air}}\mathbf{J}$ with a large constant value for ρ_{air}. The functional above can also be directly deduced from (1.233) using electromagnetic relations and vector calculus. In the **H** formulation, the functional needs to be evaluated in the whole 3D space or a bounding box much larger than the object of study. This can be avoided if we study infinitely long geometries with $\mathbf{H}_a$ in an infinite direction and no transport current [10, 11], where $\mathbf{H}_J$ outside the sample vanishes.

1.3.2.4. **H**–ψ *formulation*

For the general case, a way to minimize the number of degrees of freedom outside the sample is to use a mixed formulation of **H** and the magnetic scalar potential ψ, the latter being the state variable in the air [48]. Taking into account that $\mathbf{H} = -\nabla\psi$ in the air, the functional in (1.240) becomes

$$L_k[\Delta\mathbf{H}_{Jk}, \psi_{sk}] = \int_v \mathrm{d}^3\mathbf{r}\left[\frac{\mu_0}{2}(\Delta\mathbf{H}_{Jk})^2 + \mu_0\Delta\mathbf{H}_{Jk}\cdot\Delta\mathbf{H}_{ak} + \Delta t_k U(\nabla\times(\Delta\mathbf{H}_k + \mathbf{H}_{k-1}))\right] + \int_{v'} \mathrm{d}^3\mathbf{r}\left[\frac{\mu_0}{2}|\nabla\psi_{sk}|^2 + \mu_0\nabla\psi_{sk}\cdot\nabla\psi_{ak}\right], \quad (1.241)$$

where v is the region of the sample, v' is the whole 3D space volume excluding the sample volume, ψ_s is the contribution to ψ from the sample, and ψ_a is the scalar magnetic potential from the applied magnetic field, which is given by $\psi_a = -\mathbf{r}\cdot\mathbf{H}_a$ for uniform applied fields. Minimizing this functional corresponds to solving Faraday's law in the sample and $\nabla^2\psi = 0$ in the air. An additional condition is that at the surface of the sample, $\mathbf{H}\times\mathbf{n} = -\nabla\psi\times\mathbf{n}$, where $\mathbf{n}$ is the normal unit vector to the surface. As mentioned in this paragraph and by Kashima [48], this formulation is applicable to the magnetization case only (no transport current).

For nonzero transport currents, the situation is more complicated because using $\mathbf{H} = -\nabla\psi$ results always in zero closed-loop integrals of $\mathbf{H}$, which are not compatible with the current constraints of (1.208). For this case, we could use similar concepts as the cohomological decomposition [53]. Here, we provide an alternative comprehensive approach as follows. In the air, we decompose $\mathbf{H}$ as

$$\mathbf{H} = -\nabla\psi + \sum_i I_i\mathbf{h}_i, \quad (1.242)$$

where ψ is a magnetic scalar potential to be solved, I_i are the given current constraints for each conductor i, and $\mathbf{h}_i$ is the magnetic field per unit current generated by a wire with a particularly given path C_i within the conductor that follows the main direction of the transport current (Figure 1.25). If a conductor has holes, we should take an additional $\mathbf{h}_i$ for a path flowing within the conductor but closing around the hole (path C_3 in Figure 1.25). If no current constraint is imposed for some closed loops, such as C_3 in Figure 1.25, the current corresponding to this loop becomes an additional variable to be minimized. For the decomposition of (1.242), the current constraints of (1.208) are automatically satisfied, and hence, they do not require any additional numerical treatment. The particular choice of the paths may change the solution to the scalar potential but not the total

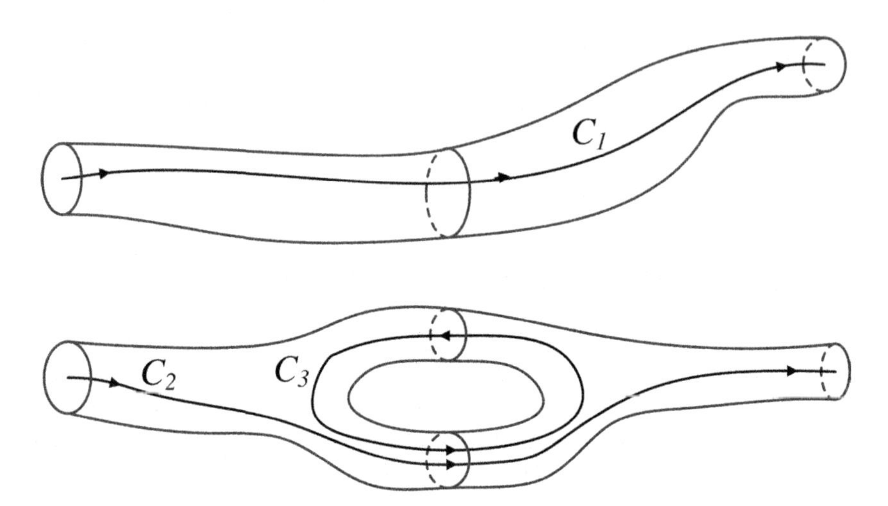

Figure 1.25. The current constraints of the variational method in the $\mathbf{H}$–ψ formulation require us to take certain paths into account along each conductor and closing around the holes (Section 1.3.2.4). The chosen paths (C_1, C_2, and C_3) are fixed at the beginning of the calculations and kept the same afterward.

solution to $\mathbf{H}$. With the decomposition of (1.242), the functional to be solved at any time step k becomes

$$L_k[\Delta\mathbf{H}_{Jk}, \psi_{sk}] = \int_v \mathrm{d}^3\mathbf{r}\left[\frac{\mu_0}{2}(\Delta\mathbf{H}_{Jk})^2 + \mu_0\Delta\mathbf{H}_{Jk}\cdot\Delta\mathbf{H}_{ak} + \Delta t_k U(\nabla\times(\Delta\mathbf{H}_k + \mathbf{H}_{k-1}))\right] + \mu_0\int_{v'}\mathrm{d}^3\mathbf{r} \times\left[\frac{1}{2}\left|\nabla\psi_{sk} - \sum_i I_{ik}\mathbf{h}_i\right|^2 + \left(\nabla\psi_{sk} - \sum_i I_{ik}\mathbf{h}_i\right)\cdot\nabla\psi_{ak}\right], \tag{1.243}$$

where ψ_s and ψ_a are the magnetic scalar potentials from the sample and applied magnetic field, respectively.

1.3.2.5. *Interaction with nonlinear magnetic materials*

A magnetic material can also be taken into account by variational principles. Here, we propose a formulation that uses the magnetization, $\mathbf{M}$, as the state variable, limiting the modeled region to the magnetic materials.

First, we assume that the free current density, $\mathbf{J}$, is given and later, we outline how to obtain $\mathbf{J}$ in superconductors (or normal conductors) interacting with the magnetic materials.

The starting point is the constitutive relation between $\mathbf{B}$ and $\mathbf{H}$; since $\mathbf{B} = \mu_0(\mathbf{H}+\mathbf{M})$, this also defines a constitutive relation between $\mathbf{B}$ and $\mathbf{M}$ as

$$\mathbf{M} = (1 - \overline{\overline{\mu}}^{-1})\mathbf{B} \tag{1.244}$$

or

$$\mathbf{B} = \mathbf{G}(\mathbf{M}),$$

where $\mathbf{G}(\mathbf{M})$ is the inverse relation of (1.244). In this book, we restrict ourselves to single-valued $\mathbf{B}$–$\mathbf{M}$ relations, either linear or not [68]. An interesting approach taking hysteresis into account can be found in the work by Prigozhin *et al.* [83].

Independent of the constitutive relation, $\mathbf{M}$ directly generates magnetic flux density because it is the density of magnetic dipoles. The magnetic flux density generated by $\mathbf{M}$ can be found by taking into account the fact that $\mathbf{M}$ generates the same $\mathbf{B}$ as an effective current density $\nabla \times \mathbf{M}$. Using the Biot–Savart integral law and vector calculus relations, we obtain

$$\mathbf{B}[\mathbf{M}](\mathbf{r}) = \mu_0\mathbf{M}(\mathbf{r}) + \frac{\mu_0}{4\pi}\int_v \mathrm{d}^3\mathbf{r}'\overline{\overline{g}}(\mathbf{r}-\mathbf{r}')\mathbf{M}(\mathbf{r}'), \tag{1.245}$$

with the components of the $\overline{\overline{g}}$ tensor being $g_{ij}(\mathbf{r}) = (3r_ir_j)/r^5 - \delta_{ij}/r^3$. Above, we use square brackets to denote the functional relation between $\mathbf{B}$ and $\mathbf{M}$. For infinitely long problems, the $\mathbf{B}[\mathbf{M}]$ relation becomes

$$\mathbf{B}[\mathbf{M}](\mathbf{r}_2) = \mu_0\mathbf{M}(\mathbf{r}_2) + \frac{\mu_0}{2\pi}\int_s \mathrm{d}^2\mathbf{r}_2'\,\overline{\overline{h}}(\mathbf{r}_2-\mathbf{r}_2')\mathbf{M}(\mathbf{r}_2'), \tag{1.246}$$

with the components of the $\overline{\overline{h}}$ tensor being $h_{ij}(\mathbf{r}_2) = 2r_ir_j/r_2^4 - \delta_{ij}/r_2^2$. In addition to $\mathbf{M}$, there are other sources of $\mathbf{B}$. These are the free current density, generating $\mathbf{B}_J$, and any source external to the domain of study (superconductors, conductors, and magnetic materials), which create the applied flux density $\mathbf{B}_a$.

In the magnetic material, the constitutive relation needs to be consistent with the total flux density generated by all sources, and hence,

$$\mathbf{G}(\mathbf{M}) = \mathbf{B}[\mathbf{M}] + \mathbf{B}_a + \mathbf{B}_J. \tag{1.247}$$

Therefore, we can obtain $\mathbf{M}$ if we solve this integral equation.

Following the same methodology adopted by Pardo and Kapolka [69], it can be seen that Equation (1.247) is the Euler equation of the following functional, and hence, minimizing this functional is the same as solving Equation (1.247):

$$L_M[\mathbf{M}] = \int_v \mathrm{d}^3\mathbf{r} \left[U_M(\mathbf{M}) - \frac{1}{2}\mathbf{B}[\mathbf{M}] \cdot \mathbf{M} - \mathbf{B}_a \cdot \mathbf{M} - \mathbf{B}_J \cdot \mathbf{M} \right], \quad (1.248)$$

with

$$U_M(\mathbf{M}) = \int_0^{\mathbf{M}} \mathrm{d}\mathbf{M}' \cdot \mathbf{G}(\mathbf{M}').$$

Again, it can be shown with the same argumentation as that of Pardo and Kapolka [69] that the minimum of this functional always exists, and it is unique for any material with permeability greater than or equal to one.

A superconductor or normal conductor interacting with a magnetic material can be solved by minimizing both the functional L_M in (1.248) for $\mathbf{M}$ and the functional L_k for the free current density of (1.233) or (1.236) for the $\mathbf{J}$ and $\mathbf{T}$ formulations, respectively. In this way, we can also model materials that are *both* superconducting and magnetic. One way to simultaneously minimize both functionals is to minimize first L_M with $\mathbf{J}$ (or $\mathbf{T}$) fixed, later minimize L_k with $\mathbf{M}$ fixed, and finally iterate until the difference between iterations is below a certain tolerance.

Examples of calculations using this variational method are provided by Pardo and Kapolka [68] for REBCO superconducting coils with magnetic substrate and by Pardo *et al.* [72] for superconducting motors.

1.3.3. *Integro-differential methods*

The aim of the integral methods is to restrict the modeling volume to the superconducting or conducting regions, thus avoiding meshing the air. In this way, it is possible to save many unnecessary degrees of freedom, potentially reducing the computing time. Another advantage of avoiding meshing the air is that modeling moving objects (such as in levitation configurations or rotating machines) is simpler. All these advantages are shared by the variational methods in the $\mathbf{J} - \phi$ and $\mathbf{T}$ formulations, which could be also considered as integral methods.

The names of these methods come from the fact that they solve integral or integro-differential equations, with the latter being the case of FEM and variational principles in the $\mathbf{H}$ and $\mathbf{H} - \psi$ formulations.

A drawback compared to methods using differential equations, such as FEMs, is that integral methods contain dense interaction matrices, which causes a large consumption of computer RAM and slows down the calculations. However, advanced integral methods use algorithms to reduce the size of the interaction matrices and severely speed up the calculations, such as multipole expansion of $\mathbf{A}$ (or $\mathbf{B}$) [98] or hierarchical matrices [97].

1.3.3.1. J *integral formulation*

Most integro-differential approaches take directly the current density, $\mathbf{J}$ as the state variable [21, 22, 58, 59, 85, 88, 99]. For any nonlinear $\mathbf{E}(\mathbf{J})$ characteristics, the basic equation is the relation between the vector and scalar potentials of (1.24). When using Coulomb's gauge, the vector potential from the sample is given by the integral of (1.32). Joining both equations and taking a known applied vector potential into account, $\mathbf{A}_a$, the integro-differential equation of the general 3D problem is

$$\mathbf{E}(\mathbf{J}) = -\frac{4\pi}{\mu_0}\int_v \mathrm{d}^3\mathbf{r}'\frac{\partial_t \mathbf{J}(\mathbf{r}')}{|\mathbf{r}-\mathbf{r}'|} - \partial_t \mathbf{A}_a - \nabla\phi, \tag{1.249}$$

where ϕ is the electrostatic potential. Here, we also need to take the continuity equation into account:

$$\nabla \cdot \mathbf{J} = 0 \tag{1.250}$$

since we have four variables for each point of the space (three components of $\mathbf{J}$ and one component of ϕ). In order to avoid calculating the scalar potential also (and inherently satisfy (1.250)), it is possible to use closed-loop currents as local variables since the integral of $\nabla\phi$ in a closed loop vanishes [88].

Certain types of $\mathbf{J}$ integral equations are called circuit methods since they solve the integro-differential Equations (1.249) and (1.250) using solvers for circuits [99].

For (infinitely) long shapes in the z direction, the integral equation becomes

$$E(J) = \frac{\mu_0}{2\pi}\int_s \mathrm{d}^2\mathbf{r}_2'\partial_t J(\mathbf{r}_2') \ln|\mathbf{r}_2 - \mathbf{r}_2'| - \partial_t A_a - \partial_z\phi, \tag{1.251}$$

where $\mathbf{E} = E\mathbf{e}_z$, $\mathbf{J} = J\mathbf{e}_z$, $\mathbf{A} = A\mathbf{e}_z$, $\nabla\phi = \partial_z\phi\mathbf{e}_z$, and $\mathbf{r}_2'$ is the 2D position vector at the cross-section. Thanks to the infinitely long shape, (1.250) is automatically satisfied, and hence, it does not need to be explicitly taken into account by the solver.

A pioneering integral method is the one known as Brandt's method for 2D infinitely long or axisymmetric shapes [21, 22], later extended by Rhyner [85]. That method solves Equation (1.249) or (1.251) by matrix inversion and Euler time integration, while Equation (1.250) does not need to be solved thanks to symmetry. In addition, for the 2D geometries above, $\nabla\phi$ is uniform in the cross-section, which simplifies the problem. Indeed, $\nabla\phi$ can be ignored for magnetization configurations (no transport current). For infinitely long shapes with nonzero transport currents, $\nabla\phi$ corresponds to the input voltage per unit length.

1.3.3.2. **T** *integral formulation*

This approach uses the current vector potential $\mathbf{T}$ as the state variable (Section 1.1.6). The starting point is Faraday's law:

$$\nabla \times \mathbf{E} = -\partial_t \mathbf{B}.$$

By inserting the constitutive E–J relation, $\mathbf{E} = \bar{\bar{\rho}}(\mathbf{J})\mathbf{J}$, into Faraday's law and using $\mathbf{J} = \nabla \times \mathbf{T}$, we obtain

$$\nabla \times [\bar{\bar{\rho}}(\nabla \times \mathbf{T})\nabla \times \mathbf{T}] = -\partial_t \mathbf{B}[\mathbf{T}] - \partial_t \mathbf{B}_a, \tag{1.252}$$

where $\mathbf{B}_a$ is the external applied field and $\mathbf{B}[\mathbf{T}]$ is the magnetic field generated by $\mathbf{T}$, given by the integral relation of (1.237). This results in the following integro-differential equation:

$$\nabla \times [\bar{\bar{\rho}}(\nabla \times \mathbf{T})\nabla \times \mathbf{T}] = -\frac{\mu_0}{4\pi}\int_v \mathrm{d}^3\mathbf{r}' \frac{[\nabla' \times \partial_t \mathbf{T}(\mathbf{r}')] \times (\mathbf{r} - \mathbf{r}')}{|\mathbf{r} - \mathbf{r}'|^3} - \partial_t \mathbf{B}_a. \tag{1.253}$$

The current potential also needs to follow the current constraints for each conductor, given by (1.225).

To solve the integro-differential equation of (1.253), we need to set Dirichlet boundary conditions for $\mathbf{T}$. For general 3D bodies, we may need to set $\mathbf{T}$ far away from the sample, as done by Amemiya *et al.* [4], which requires us to mesh the air. However, for thin films, where $\mathbf{T}$ is chosen to be perpendicular to the film surface, it is enough to set the boundary conditions at the narrow edges of the film. We can arbitrarily choose $\mathbf{T} = 0$ on one edge and $T_n = I/d$ on the other, with T_n being the (single) normal component of $\mathbf{T}$ and d being the film thickness, in order to satisfy the current constraint. This enables us to model any surface with 3D bending [61, 89].

This integral formulation became very popular because for thin films, only the perpendicular component of **T** to the surface needs to be taken into account, reducing the number of degrees of freedom.

1.3.4. *Spectral methods*

Spectral methods are another set of interesting techniques. These are based on separating the solution into the base functions that are not localized on a single element but expand on the whole sample and often also a relatively large portion of the surrounding air. This contrasts with all methods presented up to now, where the state variable is decomposed into base functions localized in one or very few elements. The earliest methods were based on Fourier expansions [20, 81, 82, 100] and, later, on Chevichev polynomial decomposition [91].

Although the Fourier expansion method requires us to mesh the air surrounding the sample also, it is able to solve bulk 3D shapes in very short computing times and with a relatively simple implementation [81]. Another particularity of Fourier expansion is that it requires regular rectangular meshing, which limits the applicability for shapes containing thin superconducting layers, especially for cross-field demagnetization or surfaces with 3D bending. Another advantage of the Fourier expansion method is that it can be very efficiently linked to electro-thermal problems [100].

Chevichev polynomial decomposition is very promising for REBCO superconducting windings, where the thin-film assumption can usually be made [91]. Indeed, this method requires a one-dimensional (1D) approximation of the conductor cross-section. Therefore, this method is not suitable for 3D modeling.

1.3.5. *Particular issues for three dimensions*

The modeling of 3D bodies is usually considered a special topic because the calculations require the number of degrees of freedom orders of magnitude to be higher than for 2D axisymmetric or surface modeling, even when the studied surfaces are bent into complex shapes. Besides, there are numerical methods that are inherently only applicable to 2D or even 1D mathematical regions. In addition, there is a particular phenomenology that appears only in 3D, which we focus on in this section.

First, certain 3D shapes, such as bulk rectangular prisms, contain a current density, **J**, with a magnitude below the critical current density, $|\mathbf{J}| < J_c$, also in the CSM [69, 70]. This contrasts with cross-sectional 2D modeling, both from axisymmetric and infinitely long problems, where $|\mathbf{J}| = J_c$ or 0 always. Indeed, that was even considered to be a requirement of the CSM, while both its original and general forumulations enable intermediate values of $|\mathbf{J}|$, $|\mathbf{J}| \leq J_c$ (see Equation (1.20) and Figure 1.1). In addition, only the presence of intermediate values of $|\mathbf{J}|$ in the superconductor volume can explain the observed intermediate values of $|\mathbf{J}|$ in thin films and thin bulks and the experimentally found rounded current paths [70].

Another interesting issue is force-free effects. These appear when there is a component of the magnetic field, $\mathbf{B}$, that is parallel to the current density, causing no macroscopic contribution to the driving force density on superconducting vortices, $\mathbf{f} = \mathbf{B} \times \mathbf{J}$ [25]. As a consequence, there is an increase in the critical current density compared to configurations where the magnetic field is perpendicular to the current density, as assumed in usual axisymmetric and infinitely long configurations [25]. This translates into an anisotropic relation between J_c and the magnetic field that contains also the direction of $\mathbf{J}$, $J_c(\mathbf{B}, \mathbf{J})$, also when the material presents isotropic pinning for the $\mathbf{B} \perp \mathbf{J}$ configuration. As a consequence of this feature, several effects appear. For example, rectangular thin films exposed to tilted magnetic fields shows regions of different current densities [46, 102]. Another effect is that the critical current in spiral cables increases when there is a $\mathbf{B}$ component in the direction of the cable axis [49].

1.4. Modeling of Power Applications

This section describes some examples of electromagnetic modeling of superconducting applications by numerical methods. The purpose is not to provide a complete review of the state of the art but to give the reader a glimpse of what is currently possible to simulate with well-established numerical methods.

Most examples concern rare-earth-based HTS wires because they are the most promising wires for future superconducting applications. In addition, the high width-to-thickness ratio of their superconducting layer makes the simulation more challenging.

1.4.1. *Numerical modeling of individual wires*

1.4.1.1. *Dependence of J_c on magnetic field*

The critical current of superconductors depends on the magnetic field. Sometimes, as in HTS, J_c depends not only on the magnetic field amplitude but also on its direction. Figure 1.26 shows the angular dependence of a

4 mm wide HTS-coated conductor manufactured by SuperOx for different temperatures.

Typical expressions of the $J_c(B)$ dependence for HTS and investigations of their influence on AC losses are given by Robert *et al.* [86]. In the case of HTS-coated conductors, the $J_c(B)$ dependence can take quite complex forms as a result of the introduction of artificial pinning centers aimed at reducing the anisotropy of the superconductor. Some examples are given by Long [55], Pardo *et al.* [76], and Hilton *et al.* [37]. A database of the field dependence of commercial HTS wires in a wide range of fields and temperatures has been made publicly available by the Robinson Research Institute [87, 104].

In general, numerical models need the dependence of the critical current density J_c on the local magnetic flux density, which we can indicate as $J_c(B_{local})$. What is typically measured in experiments is the dependence of the critical current I_c on the applied magnetic field, which we can indicate as $I_c(B_{applied})$. This means that the experimental data also contain the effect of the field generated by the current flowing in the sample during the characterization (the so-called self-field effect). If the applied field is large, the self-field can be neglected, and the $J_c(B_{local})$ dependence can be obtained by simply dividing $I_c(B_{applied})$ by the cross-section of the superconductor. If, on the other hand, one is interested in the field dependence for low values of B (for example, to simulate low-field applications such as HTS power cables or fault current limiters), one needs to take the self-field into account in order to have a sensible $J_c(B_{local})$ dependence as input for the simulations.

An example of how the self-field can be accounted for is given by Pardo *et al.* [76]. Zermeno *et al.* [108] developed a method to take the self-field

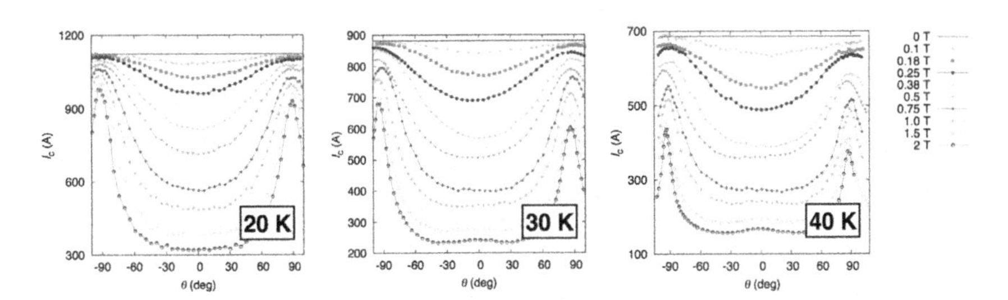

Figure 1.26. Angular dependence of I_c at 20, 30, and and several magnetic fields of a SuperOx-coated conductor sample. 0° refers to the field direction perpendicular to the tape's wide face.

Source: Reproduced from Grilli *et al.* [33].

into account that produces a set of discrete data points for the $J_c(B_{local})$ to be used as input for successive simulations, without resorting to complex mathematical expressions.

1.4.1.2. *Dependence of J_c on position*

HTS tapes experience a variation in J_c along their width as a consequence of the manufacturing process [93]. Typically, J_c is lower near the edges of the tape than in its center. This has some influence on the AC loss characteristics for low AC transport currents or a low AC applied field. In those cases, the field (self-generated by the current or externally applied) penetrates over a short distance from the edges. In those regions, the current density easily reaches the (low) local value of J_c and causes a high power dissipation. As the AC current or AC field increases, the current starts flowing in the central region of the tape where J_c is higher, and the loss behavior becomes similar to that of a tape with uniform J_c across the width. A position-dependent J_c can be easily inserted in simulations and allows us to obtain a better match with experimental data. However, the effect is limited to low current and field amplitudes, so including such dependence does not usually have a very important practical impact.

The critical current J_c also varies along the length of the tape [31]. The statistical distribution of J_c in good-quality, long pieces of commercial HTS tape is typically of a few percentage points. This dependence is often ignored in simulations, although the presence of points with very low J_c can give rise to dangerous "hot spots."

1.4.1.3. *Simulation of magnetic materials*

In certain cases, for reasons related to the manufacturing process, superconducting wires, such as MgB_2 wires or HTS-coated conductors, contain magnetic materials. This represents a challenge for simulations because those materials have highly nonlinear magnetic characteristics, which may cause difficulties in the implementation of the models or in the numerical solution.

An example of implementation of a model with ferromagnetic materials and a study of its influence on tape arrangements for fault current limiter applications is given by Nguyen *et al.* [60].

Figure 1.27 shows an example of AC loss calculation in an HTS wire with magnetic substrate carrying transport current. The substrate makes

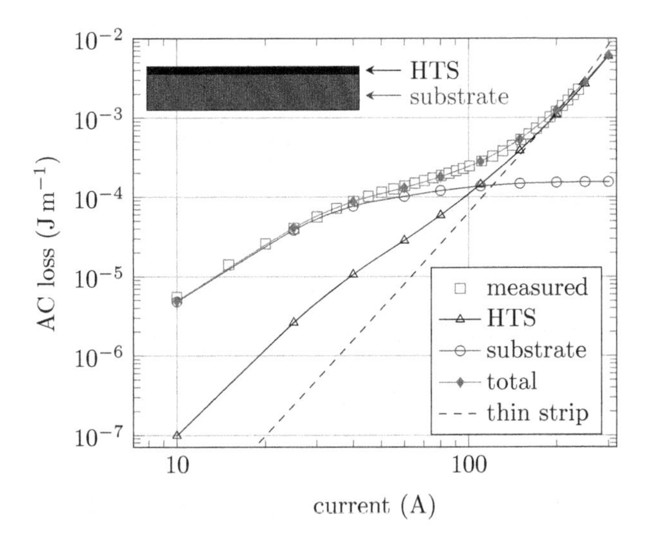

Figure 1.27. Transport losses of an HTS-coated conductor with ferromagnetic substrate: comparison of experimental and simulation data. The numerical model gives access to the different loss contributions.

a significant contribution to the losses, especially at low and medium currents. This example demonstrates that numerical models allow us to access quantities (such as the individual loss contributions in the different materials) that are not directly available from measurements. It also shows that the magnetic material modifies the losses of the superconductor part, which are different from those that would occur if the substrate was not magnetic ("thin strip" line in the figure).

Magnetic materials can be added to superconducting tapes or windings in order to guide the magnetic flux away from the superconductor, thus decreasing its AC loss [32, 50]. It should be remarked that, while AC losses in the superconductor are decreased, a new loss contribution — the AC losses in the magnetic materials — appears. So, the effectiveness of these arrangements has to be carefully evaluated. An example of the use of magnetic flux diverters in a HTS-coated conductor coil is given by Pardo *et al.* [75].

1.4.1.4. *Dynamic resistance*

The term dynamic loss refers to the energy dissipation caused by the DC electrical resistance appearing when a superconductor carries a DC transport current in the presence of an AC background magnetic field. HTS flux pumps — which allow charging of DC magnets without the need for large current sources — are actually based on dynamic resistance [23, 29, 41]. The effect is also important in other HTS applications, such as superconducting electrical machines, where the superconducting coils carry DC current while subjected to varying magnetic fields [44].

1.4.2. *Interacting tapes*

In general, superconducting applications are composed of wires that are closely packed and thus electromagnetically interacting. Numerical models are now able to cope with a large number of superconductors. A typical case is represented in Figure 1.28, which shows the current distribution and the AC losses of four pancake coils made of HTS tapes. The model is able to calculate the current distribution of the individual turns. Simulations up to 10,000 individual turns have been carried out [64].

In order to reduce the computation time and to handle even larger systems, alternative approaches to the detailed and simultaneous simulation of all turns have been developed. These approaches can be divided into three categories:

- Homogenization,
- Multi-scale,
- Densification.

A comprehensive review of these three methods applied to the H and T–A formulations was presented by Berrospe-Juarez *et al.* [15] for the case of HTS wires.

In short, with the homogenization, the details of the tapes are "washed out" and stacks of tapes (like the turns of a pancake coil in 2D) are simulated as a rectangular bulk. Specific boundary conditions need to be applied in order to account for the fact that all the tapes (or turns) carry the same current. The computational advantage stems from a much reduced number of degrees of freedom (due to the simplified geometry and mesh) and from the use of a smaller number of current constraints.

The multi-scale method is based on separating the calculation of the current density in the tape and of the magnetic field of the "environment" around each tape. The main idea is to simulate individual tapes with boundary conditions for the field generated by all the other tapes. This approach has several advantages: First, the details of current and

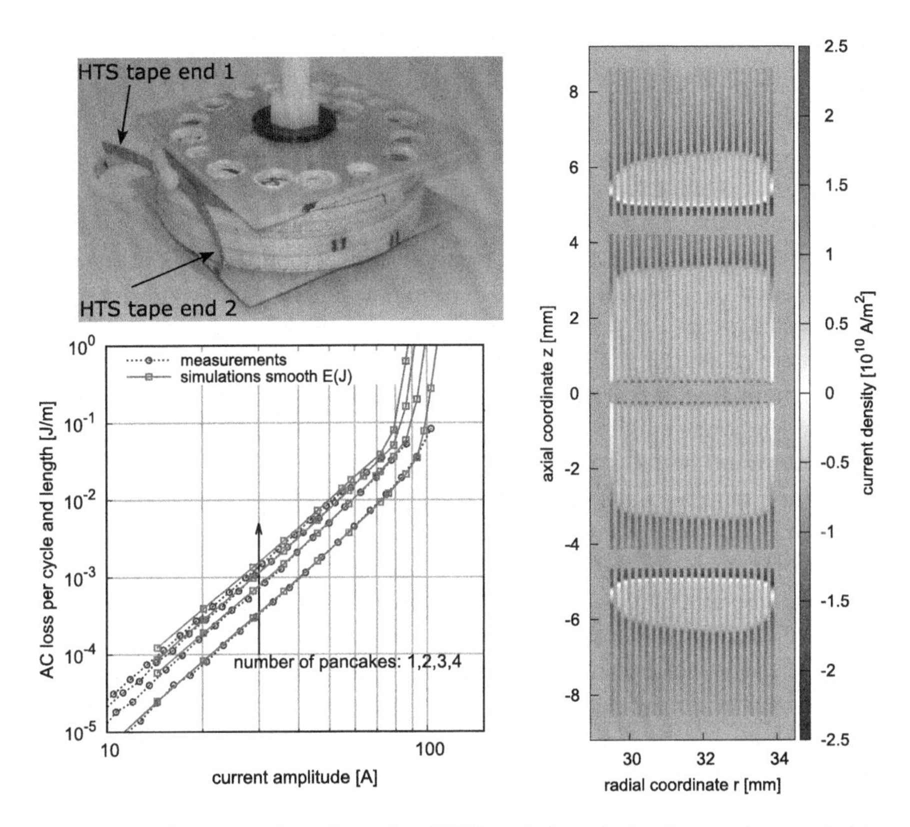

Figure 1.28. Four pancake coils made of HTS-coated conductor tape and connected in series (top-left). AC losses as a function of the transport current for different assemblies, ranging from a single coil to four coils: comparison of experimental results and numerical calculations (bottom-left). Details of the current density distribution in the individual turns of the four coils (right). See Refs. [73, 74] for further details.

field distributions inside the superconductor are calculated for very small problems, one tape at the time. Then, the simulation of the various tapes can be truly parallelized. Finally, there are virtually no limits on the size of the device to be simulated. The drawbacks are that the results depend, to a certain extent, on the choice of the initial current density distribution in the tape and that the model implementation is rather complex due to the exchange of information between the model for the whole device and that for the individual tapes.

With the densification method, a certain number of tapes are merged into a single tape. The densified tapes concentrate the transport current of their surrounding tapes, while the surrounding tapes are erased. The densification allows us to build models with a smaller number of elements. However, similar to multi-scale, this method has quite an elaborate implementation.

As mentioned by Berrospe-Juarez *et al.* [15], the homogenization is probably the best compromise in terms of gained speed (with respect to a full model simulating all the tapes) and ease of implementation. However, it is limited to the case of simulation of infinitely thin superconductors because, by construction, it cannot take into account the currents induced across the thickness of a tape.

In most cases, the simulation of superconducting coils assumes that the same current flows in each turn. In other words, the turns are assumed to be electrically uncoupled. This is schematically represented in Figure 1.29 for a coil wound from a single tape: The (known) current is injected at one end of the tape and comes out at the other end of the tape, and neighboring turns carry the same current.

However, in reality, some degree of coupling may exist, and models can take this into account by appropriately setting constraints on where the current flows. Let us consider a similar coil with the same number of tape turns (12) but obtained by winding three turns of a "cable"; each turn is composed of four tapes in parallel (Figure 1.30, top).

The tapes can be electrically connected along the entire length of the coil (Figure 1.30, middle). The (known) current is injected at the level of the cable turns: In each turn, it is left free to distribute between the tapes. Such distribution, in general, changes along the length of the coil. We can call this situation "fully coupled tapes."

In another scenario, the tapes can be electrically insulated along the coil but joint together at the ends (Figure 1.30, bottom). In this situation too, the (known) current is injected at the level of the cable turns, and one does not know in advance how it distributes between the tapes. However, current

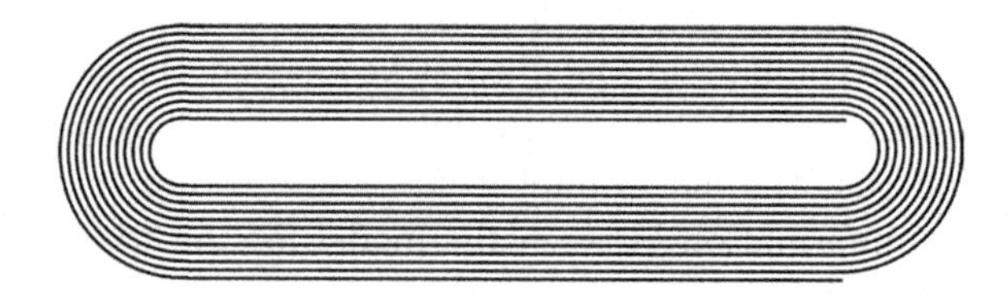

Figure 1.29. Racetrack coil wound with one tape. The turns are electrically isolated, with the current entering at one end of the coil and coming out at the other end.

Source: Figure courtesy of Tara Benkel.

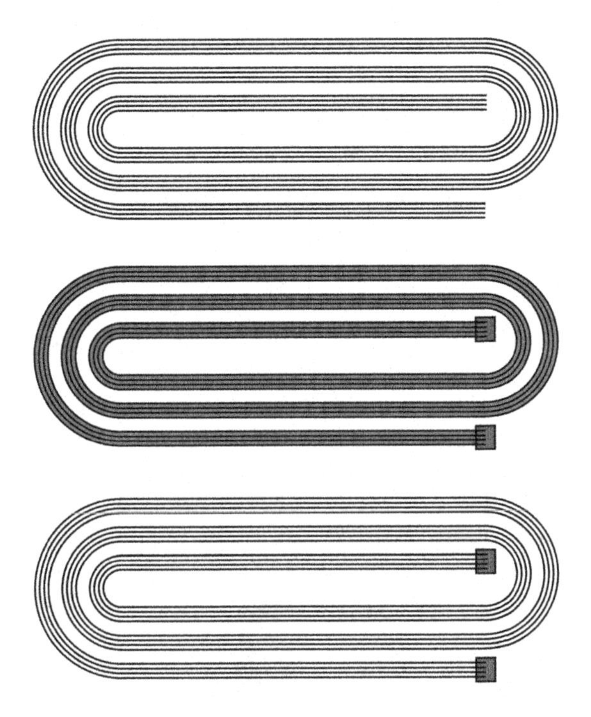

Figure 1.30. Racetrack coil wound with a cable composed of tapes in parallel (top). The tapes can be electrically connected along the entire length of the coil (middle) or just at the end of the coil (bottom).

Source: Figure courtesy of Tara Benkel.

transfer from one tape to another along the length of the coil is not allowed: The current flowing in each tape, which is unknown *a priori*, must be preserved along the length. We can call this situation "coupled-at-ends tapes."

Each situation described above requires the application of particular current constraints. An application of these coupling scenarios for the coils of a superconducting motor was carried out by Pardo *et al.* [67]. It was found that coupled-at-ends turns have similar losses to the uncoupled ones. This was encouraging because the former is easier to be realized in practice than the latter.

1.4.3. *3D modeling*

Superconducting bulks are the ideal geometry to test 3D models due to their relatively simple geometry, which allows limiting the total number of degrees of freedom, even for 3D simulations. In order to test the accuracy and speed of 3D numerical models, the research community has setup a benchmark for the magnetization of a cube under the action of an AC magnetic field. The results of three models were compared, and a report of the results is available on the website of the HTS Modeling Workgroup.[1]

More complex arrangements than simple bulks have been studied. The A–ϕ formulation was used to simulate the magnetization of a superconducting ring subjected to the field generated by a racetrack coil in 3D [92]. Another 3D example is shown in Figure 1.31, which shows the currents induced in a cup-shaped superconductor by a slanted magnetic field. Such a case could not be simulated by 2D models. Vestgården *et al.* [101] developed a model for studying nonlocal electrodynamics in superconducting films, which was later applied to the case of a right-angled corner.

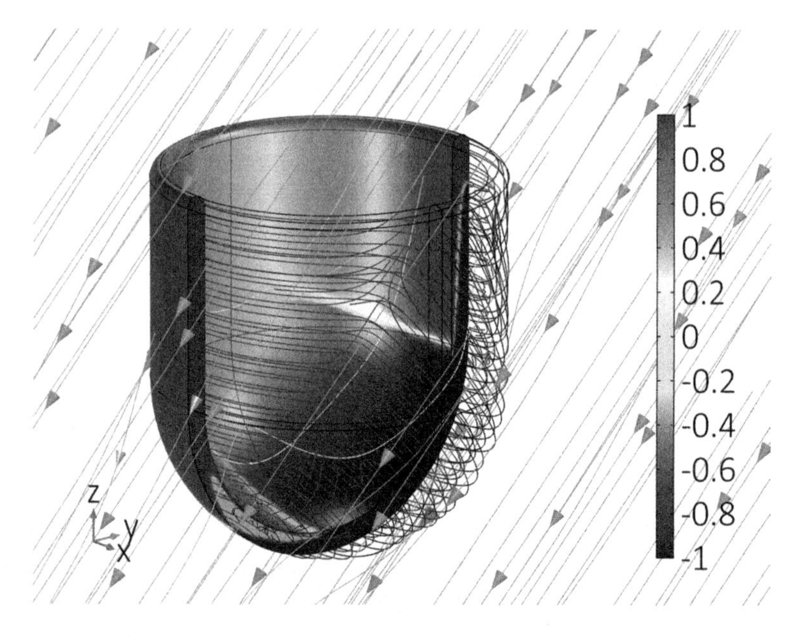

Figure 1.31. Current induced in a cup-shaped superconductor by a slanted magnetic field (arrows). Such a case can only be handled with full 3D models. The plot shows the x component of J, normalized with respect to J_c.

Source: Figure courtesy of Mykola Solovyov, Institute of Electrical Engineering, Slovak Academy of Sciences.

[1]http://www.htsmodelling.com.

Another example of complex 3D simulations can be found in the work by Fagnard *et al.* [28], in which the authors studied the effectiveness of magnetic shielding of various geometries with the A–ϕ formulation implemented in the open-source software GetDP. In particular, they investigated semi-closed superconducting bulk cylinders (obtained by bringing together a superconducting tube and a disk) subjected to axial and transverse fields, as schematically represented in Figure 1.32

Full 3D models also allow exploring the so called force-free configurations, i.e. situations where the magnetic field and the current density are parallel to each other [46] Another effect for which 3D simulations are useful (or even necessary) is the study of demagnetization of bulks or tape stacks. Magnetization can be trapped inside bulks (or tape stacks) in order to use them as permanent magnets in several applications; however, in real applications, these "magnets" are often also subjected to small transverse fields, for example created by the rotation in electrical machines. These transverse fields have the practical effect of decreasing the trapped magnetic flux, and hence, they can significantly reduce the benefits of using superconductors instead of conventional permanent magnets. Modeling these effects in 3D is very challenging because it is necessary to simulate a large number of

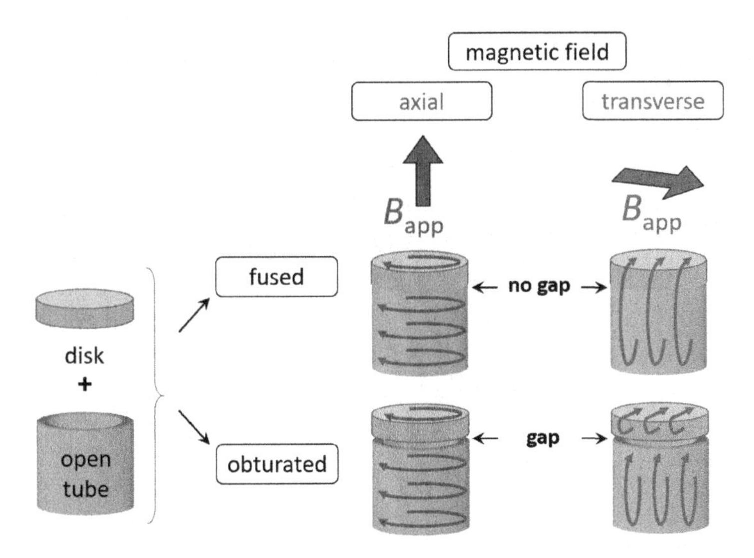

Figure 1.32. Schematic illustration of the bulk semi-closed superconducting cylinders subjected to axial and transverse fields investigated by 3D simulations in [Fagnard *et al.*].

Source: Figure courtesy of J.-F. Fagnard, University of Liege, Belgium.

cycles. In the case of HTS tape stacks, there is the additional challenge of handling superconducting layers with length and width in the range of centimeters, as opposed to a thickness of a few micrometers (which also needs to be discretized) [47]. For these reasons, an alternative approach based on the effective $E(J)$ relation from dynamic magneto-resistance has been developed. This allowed the simulation of cross-field demagnetization of superconducting stacks and bulks for up to 100 tapes and two million cycles [26].

Three-dimensional simulations can be used to investigate levitation forces, for example in superconducting magnetic bearings. In the model presented by Quéval *et al.* [84], the model consists of an unidirectional coupling between the permanent assembly model and the HTS assembly model. An example of a typical geometry is given in Figure 1.33.

Three-dimensional models can be used to study the coupling between superconducting filaments through a normal metal between them. For example, the influence of the aspect ratio of the filaments on the onset of coupling was studied by Grilli *et al.* [34]. The coupling AC loss in soldered tapes and striated coated conductors was investigated by Pardo *et al.* [71]. Losses in MgB_2 wires with twisted filaments were calculated by Escamez *et al.* [27].

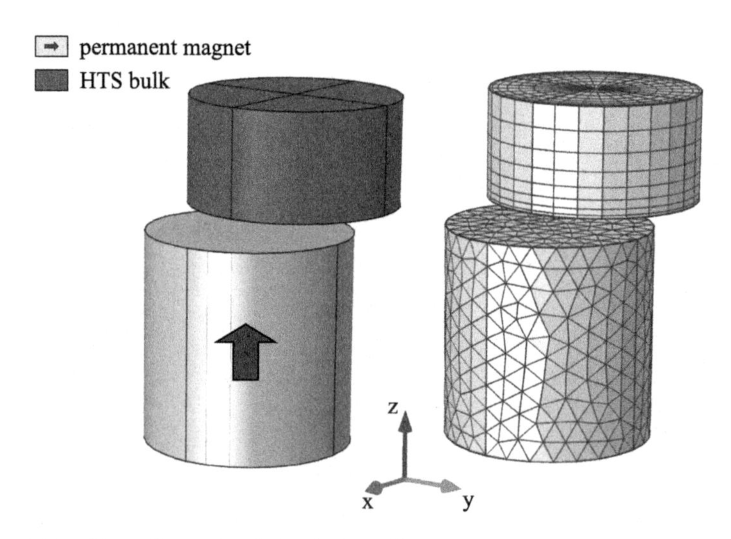

Figure 1.33. Three-dimensional simulation of levitation of an HTS bulk above a permanent magnet: example of the geometry and mesh considered by Quéval *et al.* [84].

Source: Loïc Quéval, University of Paris-Saclay, France.

The modeling of complex 3D structures remains quite challenging to this day. One example is represented by the calculation of transport and magnetization losses in a Roebel cable presented by Zermeno *et al.* [106]. In that work, the superconducting strands were also discretized along the thickness, although an artificial expansion of the thickness was necessary in order to have a reasonable mesh and number of degrees of freedom. For the modeling of complex cable geometries involving HTS-coated conductors, the infinitely thin tape approximation is often used. With this approximation, the superconducting strand is modeled as a sheet: In other words, the variation in the electromagnetic properties across the superconductor's thickness is neglected. The 3D features of the cable (such as the meander-shaped intertwined strands in Roebel cables or twisted strands in CORC cables) are preserved. Two notable examples are the simulation of Roebel-cable coils [90] and multi-layer CORC cables [103].

1.4.4. *Rotating machines*

Recently, researchers from different groups have benchmarked their models for the simulation of an HTS dynamo, which has the potential for several applications [1].

When it comes to simulating rotating machines, such as motors and generators, and calculating the dissipation of superconductors inside them, one possible approach is to split the problem into two parts: First, calculate the magnetic field maps with well-established FEM approaches without simulating the superconductor material; second, use the calculated magnetic field as a boundary condition for a detailed calculation of the AC losses in superconductors in a separate model dedicated to that purpose. This approach was used by Zermeno *et al.* [107] for evaluating the losses in a generator for wind turbine applications and by Pardo *et al.* [67] for a superconducting motor. The recently developed T–A formulation enabled the simulation of the whole machine and the calculation of the AC losses in the same simulation environment [14].

Acknowledgments

E. Pardo wrote Sections 1.1, 1.2 and 1.3 of this chapter; F. Grilli wrote Section 1.4. Both authors revised the text of the whole chapter. Enric Pardo acknowledges the financial support of the Slovak Research and Development Agency (APVV) under contract No. APVV-19-0536 and the Grant Agency of the Ministry of Education of the Slovak Republic and the Slovak Academy of Sciences (VEGA) under contract No. 2/0097/18.

References

[1] Ainslie M, Grilli F, Quéval L, Pardo E, Perez-Mendez F, Mataira R, Morandi A, Ghabeli A, Bumby C and Brambilla R. 2020. A new benchmark problem for electromagnetic modelling of superconductors: The high-T_c superconducting dynamo. *Supercond. Sci. Technol.* 33(10), 105009.

[2] Ainslie MD, Flack TJ, Hong Z and Coombs TA. 2011. Comparison of first- and second-order 2D finite element models for calculating AC loss in high temperature superconductor coated conductors. *COMPEL Int. J. Comput. Math. Electr. Electr. Eng.* 30(2), 762–774.

[3] Albanese R and Rubinacci G. 1990. Magnetostatic field computations in terms of two-component vector potentials. *Int. J. Numer. Methods Eng.* 29(3), 515–532.

[4] Amemiya N, Murasawa S, Banno N and Miyamoto K. 1998. Numerical modelings of superconducting wires for AC loss calculations. *Phys. C* 310(1–4), 16–29.

[5] Anderson P and Kim Y. 1964. Hard superconductivity: Theory of the motion of abrikosov flux lines. *Rev. Mod. Phys.* 36(1), 39–43.

[6] Arkkio A. *et al.* 1986. *Analysis of Induction Motors based on the Numerical Solution of the Magnetic Field and Circuit Equations* (Helsinki University of Technology), phD thesis. http://lib.tkk.fi/Diss/198X/isbn95122607 6X/).

[7] Arsenault A, de Sousa Alves B and Sirois F. 2021a. COMSOL implementation of the H$-\phi$-formulation with thin cuts for modeling superconductors with transport currents. *IEEE Trans. Appl. Supercond.* 31(6), 6900109.

[8] Arsenault A, Sirois F and Grilli F. 2021b. Efficient modeling of high-temperature superconductors surrounded by magnetic components using a reduced H$-\phi$ formulation. *IEEE Trans. Appl. Supercond.* 31(4), 6800609.

[9] Arsenault A, Sirois F and Grilli F. 2021c. Implementation of the H$-\phi$ formulation in COMSOL Multiphysics for simulating the magnetization of bulk superconductors and comparison with the H-formulation. *IEEE Trans. Appl. Supercond.* 31(2), 6800111.

[10] Badía A and López C. 2001. Critical state theory for nonparallel flux line lattices in type-II superconductors. *Phys. Rev. Lett.* 87(12), 127004.

[11] Badía-Majós A and López C. 2012. Electromagnetics close beyond the critical state: Thermodynamic prospect. *Supercond. Sci. Technol.* 25(10), 104004.

[12] Bean C. 1964. Magnetization of high-field superconductors. *Rev. Mod. Phys.* 36(1), 31–38.

[13] Bean CP. 1962. Magnetization of hard superconductors. *Phys. Rev. Lett.* 8(6), 250–253.

[14] Benkel T, Lao M, Liu Y, Pardo E, Wolfstädter S, Reis T and Grilli F. 2020. T–A-formulation to model electrical machines with HTS coated conductor coils. *IEEE Trans. Appl. Supercond.* 30(6), 5205807.

[15] Berrospe-Juarez E, Trillaud F, Zermeño VMR and Grilli F. 2021. Advanced electromagnetic modeling of large-scale high-temperature superconductor systems based on H and T-A formulations. *Supercond. Sci. Technol.* 34(4), 044002.

[16] Beth R. 1967. An integral formula for 2-dimensional fields. *J. Appl. Phys.* 38(12), 4689.

[17] Blatter G, Feigelman MV, Geshkenbein VB, Larkin AI and Vinokur VM. 1994. Vortices in high-temperature superconductors. *Rev. Mod. Phys.* 66, 1125.

[18] Bossavit A. 1994. Numerical modelling of superconductors in three dimensions: A model and a finite element method. *IEEE Trans. Magn.* 30(5), 3363–3366.

[19] Brambilla R, Grilli F, Martini L, Bocchi M and Angeli G. 2018. A finite-element method framework for modeling rotating machines with superconducting windings. *IEEE Trans. Appl. Supercond.* 28(5), 5207511.

[20] Brandt E. 1995. Square and rectangular thin superconductors in a transverse magnetic field. *Phys. Rev. Lett.* 74(15), 3025–3028.

[21] Brandt EH. 1996. Superconductors of finite thickness in a perpendicular magnetic field: Strips and slabs. *Phys. Rev. B* 54(6), 4246.

[22] Brandt EH. 1998. Superconductor disks and cylinders in an axial magnetic field. I. Flux penetration and magnetization curves. *Phys. Rev. B* 58(10), 6506.

[23] Campbell A. 2019. Dynamic resistance and flux pumping. *Preprints*, 050253.

[24] Clem J and Sanchez A. 1994. Hysteretic ac losses and susceptibility of thin superconducting disks. *Phys. Rev. B* 50(13), 9355.

[25] Clem J, Weigand M, Durrell JH and Campbell AM. 2011. Theory and experiment testing flux-line cutting physics. *Supercond. Sci. Technol.* 24, 062002.

[26] Dadhich A and Pardo E. 2020. Modeling cross-field demagnetization of superconducting stacks and bulks for up to 100 tapes and 2 million cycles. *Sci. Rep.* 10(1), 19265.

[27] Escamez G, Sirois F, Lahtinen V, Stenvall A, Badel A, Tixador P, Ramdane B, Meunier G, Perrin-Bit R and Bruzek C-E. 2016. 3-D numerical modeling of AC losses in multifilamentary mgb_2 wires. *IEEE Trans. Appl. Supercond.* 26(3), 1–7.

[28] Fagnard JF, Vanderheyden B, Pardo E and Vanderbemden P. 2019. Magnetic shielding of various geometries of bulk semi-closed superconducting cylinders subjected to axial and transverse fields. *Supercond. Sci. Technol.* 32(7), 074007.

[29] Geng J and Coombs T. 2015. Mechanism of a high-Tc superconducting flux pump: Using alternating magnetic field to trigger flux flow. *Appl. Phys. Lett.* 107(14), 142601.

[30] Goldfarb RB, Lelental M and Thompson, edited by R. A. Hein, p. 49, C. 1991. *Magnetic Susceptibility of Superconductors and Other Spin Systems*, Plenum, New York.

[31] Gömöry F, Šouc J, Adámek M, Ghabeli A, Solovyov M and Vojenčiak M. 2019. Impact of critical current fluctuations on the performance of a coated conductor tape. *Supercond. Sci. Technol.* 32(12), 124001.

[32] Gömöry F, Vojenčiak M, Pardo E, Solovyov M and Šouc J. 2010. AC losses in coated conductors. *Supercond. Sci. Technol.* 23, 034012.

[33] Grilli F, Benkel T, Hänisch J, Lao M, Reis T, Berberich E, Wolfstädter S, Schneider C, Miller P, Palmer C, Glowacki B, Climente-Alarcon V, Smara A, Tomkow L, Teigelkötter J, Stock A, Büdel J, Jeunesse L, Staempflin M, Delautre G, Zimmermann B, van der Woude R, Perez A, Samoilenkov S, Molodyk A, Pardo E, Kapolka M, Li S and Dadhich A. 2020. Superconducting motors for aircraft propulsion: The advanced superconducting motor experimental demonstrator project. *J. Phys. Conf. Ser.* 1590, 012051.

[34] Grilli F, Costa Bouzo M, Yang Y, Beduz C and Dutoit B. 2003. Finite element method analysis of the coupling effect between superconducting filaments of different aspect ratio. *Supercond. Sci. Technol.* 16(10), 1228.

[35] Grilli F, Pardo E, Morandi A, Gömöry F, Solovyov M, Zermeño VM, Brambilla R, Benkel T and Riva N. 2021. Electromagnetic modeling of superconductors with commercial software: Possibilities with two vector potential-based formulations. *IEEE Trans. Appl. Supercond.* 31(1), 5900109.

[36] Grilli F, Pardo E, Stenvall A, Nguyen DN, Yuan W and Gömöry F. 2014. Computation of losses in HTS under the action of varying magnetic fields and currents. *IEEE Trans. Appl. Supercond.* 24(1), 8200433.

[37] Hilton D, Gavrilin A and Trociewitz U. 2015. Practical fit functions for transport critical current versus field magnitude and angle data from (RE)BCO coated conductors at fixed low temperatures and in high magnetic fields. *Supercond. Sci. Technol.* 28(7), 074002.

[38] Huebener R, Kampwirth R and Clem J. 1972. Meissner shielding currents and magnetic flux penetration in thin-film superconductors. *J. Low Temp. Phys.* 6(3–4), 275–285.

[39] Íñiguez J, Raposo V, Zazo M, García-Flores A and Hernández-Gómez P. 2009. The electromagnetic field in conductive slabs and cylinders submitted to a harmonic longitudinal magnetic field. *Am. J. Phys.* 77(11), 1074–1081.

[40] Jackson JD. 1999. *Classical Electrodynamics*, Hoboken: John Wiley & Sons Inc., 3rd edition.

[41] Jiang Z, Hamilton K, Amemiya N, Badcock R and Bumby C. 2014. Dynamic resistance of a high-Tc superconducting flux pump. *Appl. Phys. Lett.* 105(11), 112601.

[42] Jin J-M. 2002. *The Finite Element Method in Electromagnetics*, 2nd edn., Wiley, New York.

[43] Johnson RC. (ed.) 1993. *Antenna Engineering Handbook,* McGraw-Hill, New York.

[44] Kails K, Zhang H, Mueller M and Li Q. 2020. Loss characteristics of HTS coated conductors in field windings of electric aircraft propulsion motors. *Supercond. Sci. Technol.* 33(6), 064006.

[45] Kajikawa K, Hayashi T, Yoshida R, Iwakuma M and Funaki K. 2003. Numerical evaluation of AC losses in HTS wires with 2D FEM formulated by self magnetic field. *IEEE Trans. Appl. Supercond.* 13(2), 3630–3633.
[46] Kapolka M and Pardo E. 2019. 3D modelling of macroscopic force-free effects in superconducting thin films and rectangular prisms. *Supercond. Sci. Technol.* 32(5), 054001.
[47] Kapolka M, Pardo E, Grilli F, Baskys A, Climente-Alarcon V and Glowacki BA. 2020. Cross-field demagnetization of stacks of tapes: 3D modelling and measurements. *Supercond. Sci. Technol.* 33(4), 044019.
[48] Kashima Y. 2008. On the double critical-state model for type-II superconductivity in 3D. *ESAIM Math. Model. Numer. Anal.* 42(3), 333–374.
[49] Kinoshita Y, Yonenaka T, Ichiki Y, Akasaka T, Otabe E, Kiuchi M, Matsushita T, Hu N, Ni B and Ma T. 2021. Design and evaluation of 10-kA class superconducting DC power cable based on longitudinal magnetic field effect. 1975(1), 012037.
[50] Krueger P, Grilli F, Vojenčiak M, Zermeño VMR, Demenčík E and Farinon S. 2013. Superconductor/ferromagnet heterostructures exhibit potential for significant reduction of hysteretic losses. *Appl. Phys. Lett.* 102, 202601.
[51] Kwasnitza K. 1977. Scaling law for the AC losses of multifilament superconductors. *Cryogenics* 17(11), 616–620.
[52] Lahtinen V, Lyly M, Stenvall A and Tarhasaari T. 2012. Comparison of three Eddy current formulations for superconductor hysteresis loss modelling. *Supercond. Sci. Technol.* 25(11), 115001.
[53] Lahtinen V, Stenvall A, Sirois F and Pellikka M. 2015. A finite element simulation tool for predicting hysteresis losses in superconductors using an H-oriented formulation with cohomology basis functions. *J. Supercond. Novel Mag.* 28(8), 2345–2354.
[54] Landau LD, Lifshitz EM and Pitaevskii LP. 2008. *Electrodynamics of Continuous Media,* Elsevier Butterworth Heinemann, Amsterdam.
[55] Long NJ. 2008. Model for the angular dependence of critical currents in technical superconductors. *Supercond. Sci. Technol.* 21(2), 025007.
[56] Lousberg G, Ausloos M, Geuzaine C, Dular P, Vanderbemden P and Vanderheyden B. 2009. Numerical simulation of the magnetization of high-temperature superconductors: A 3D finite element method using a single time-step iteration. *Supercond. Sci. Technol.* 22, 055005.
[57] Maxwell JC. 1881. *A Treatise on Electricity and Magnetism,* Oxford: Clarendon Press.
[58] Morandi A. 2004. *Circuit methods for three dimensional field analysis in large scale superconducting systems*, Ph.D. thesis, University of Bologna (online at: www.die.ing.unibo.it/dottorato_it/index_en.htm).
[59] Morandi A and Fabbri M. 2015. A unified approach to the power law and the critical state modeling of superconductors in 2D. *Supercond. Sci. Technol.* 28(2), 024004.
[60] Nguyen DN, Ashworth SP and Willis JO. 2009. Experimental and finite-element method studies of the effects of ferromagnetic substrate

on the total ac loss in a rolling-assisted biaxially textured substrate $YBa_2Cu_3O_7$ tape exposed to a parallel ac magnetic field. *J. Appl. Phys.* 106, 093913.

[61] Nii M, Amemiya N and Nakamura T. 2012. Three-dimensional model for numerical electromagnetic field analyses of coated superconductors and its application to Roebel cables. *Supercond. Sci. Technol.* 25(9), 095011.

[62] Norris W. 1970. Calculation of hysteresis losses in hard superconductors carrying AC: Isolated conductors and edges of thin sheets. *J. Phys. D: Appl. Phys.* 3, 489–507.

[63] Pardo, E. 2008. Modeling of coated conductor pancake coils with a large number of turns. *Supercond. Sci. Technol.* 21, 065014.

[64] Pardo E. 2016. Modeling of screening currents in coated conductor magnets containing up to 40000 turns. *Supercond. Sci. Technol.* 29(8), 085004.

[65] Pardo E, Gömöry F, Šouc J and Ceballos J. 2007. Current distribution and AC loss for a superconducting rectangular strip with in-phase alternating current and applied field. *Supercond. Sci. Technol.* 20(4), 351–364.

[66] Pardo E and Grilli F. 2012. Numerical simulations of the angular dependence of magnetization AC losses: Coated conductors, roebel cables and double pancake coils. *Supercond. Sci. Technol.* 25, 014008.

[67] Pardo E, Grilli F, Liu Y, Wolftaedler S and Reis T. 2019a. AC loss modelling in superconducting coils and motors with parallel tapes as conductor. *IEEE Trans. Appl. Supercond.* 29(5), 5202505.

[68] Pardo E and Kapolka M. 2016. Modeling of superconductors interacting with non-linear magnetic materials: 3D variational principles, force-free effects and applications. *5th Internatinal Workshop on Numerical Modelling of High Temperature Superconductors.* doi: 10.5281/zenodo.56322.

[69] Pardo E and Kapolka M. 2017a. 3D computation of non-linear eddy currents: Variational method and superconducting cubic bulk. *J. Comput. Phys.* 344, 339–363.

[70] Pardo E and Kapolka M. 2017b. 3D magnetization currents, magnetization loop, and saturation field in superconducting rectangular prisms. *Supercond. Sci. Technol.* 30, 064007.

[71] Pardo E, Kapolka M, Kováč J, Šouc J, Grilli F and Piqué RNA. 2016. Three-dimensional modeling and measurement of coupling AC loss in soldered tapes and striated coated conductors. *IEEE Trans. Appl. Supercond.* 26(3), 4700607.

[72] Pardo E, Li S, Grilli F, Liu Y, Benkel T, Wolfstaedler S, Berberich E and Reis T. 2019b. AC loss in the distributed stator winding of a 1 MW motor for aviation. *14th European Conference on Applied Superconductivity.* doi:10.5281/zenodo.3517048, available at https://doi.org/10.5281/zenodo.3517048.

[73] Pardo E, Šouc J and Frolek L. 2015. Electromagnetic modelling of superconductors with a smooth current-voltage relation: Variational principle and coils from a few turns to large magnets. *Supercond. Sci. Technol.* 28, 044003.

[74] Pardo E, Šouc J and Kováč J. 2012. AC loss in rebco pancake coils and stacks of them: Modelling and measurement. *Supercond. Sci. Technol.* 25(3), 035003.
[75] Pardo E, Šouc J and Vojenčiak M. 2009. AC loss measurement and simulation of a coated conductor pancake coil with ferromagnetic parts. *Supercond. Sci. Technol.* 22, 075007.
[76] Pardo E, Vojenčiak M, Gömöry F and Šouc J. 2011. Low-magnetic-field dependence and anisotropy of the critical current density in coated conductors. *Supercond. Sci. Technol.* 24, 065007.
[77] Prigozhin L. 1996. The Bean model in superconductivity: Variational formulation and numerical solution. *J. Comput. Phys.* 129(1), 190–200.
[78] Prigozhin L. 1997. Analysis of critical-state problems in type-II superconductivity. *IEEE Trans. Appl. Supercond.* 7(4), 3866–3873.
[79] Prigozhin L. 1998. Solution of thin film magnetization problems in type-II superconductivity. *J. Comput. Phys.* 144(1), 180–193.
[80] Prigozhin L and Sokolovsky V. 2011. Computing AC losses in stacks of high-temperature superconducting tapes. *Supercond. Sci. Technol.* 24, 075012.
[81] Prigozhin L and Sokolovsky V. 2018a. Fast Fourier transform-based solution of 2D and 3D magnetization problems in type-II superconductivity. *Supercond. Sci. Technol.* 31(5), 055018.
[82] Prigozhin L and Sokolovsky V. 2018b. Solution of 3D magnetization problems for superconducting film stacks. *Supercond. Sci. Technol.* 31(12), 125001.
[83] Prigozhin L, Sokolovsky V, Barrett JW and Zirka SE. 2016. On the energy-based variational model for vector magnetic hysteresis. 52(12), 1–11.
[84] Quéval L, Liu K, Yang W, Zermeño VMR and Ma G. 2018. Superconducting magnetic bearings simulation using an H-formulation finite element model. *Supercond. Sci. Technol.* 31(8), 084001.
[85] Rhyner J. 1998. Calculation of AC losses in HTSC wires with arbitrary current voltage characteristics. *Phys. C* 310(1–4), 42–47.
[86] Robert BC, Fareed MU and Ruiz HS. 2019. How to choose the superconducting material law for the modelling of 2G-HTS coils. *Materials* 12(17), 2679.
[87] Robinson Research Institute — Victoria University of Wellington. 2022. HTS wire database. https://htsdb.wimbush.eu
[88] Rozier B, Badel A, Ramdane B and Meunier G. (2019). Calculation of the local current density in high-temperature superconducting insulated rare earth–barium–copper oxide coils using a volume integral formulation and its contribution to coil protection. *Supercond. Sci. Technol.* 32(4), 044008.
[89] Sogabe Y and Amemiya N. 2018. AC loss calculation of a cosine-theta dipole magnet wound with coated conductors by 3D modeling. *IEEE Trans. Appl. Supercond.* 28(4), 1–5.
[90] Sogabe Y, Nii M, Tsukamoto T, Nakamura T and Amemiya N. 2014. Electromagnetic field analyses of rebco roebel cables wound into coil configurations. *IEEE Trans. Appl. Supercond.* 24(3), 4803005.

[91] Sokolovsky V, Prigozhin L and Kozyrev AB. 2020. Chebyshev spectral method for superconductivity problems. *Supercond. Sci. Technol.* 33(8), 085008.

[92] Solovyov M and Gömöry F. 2019. A-V formulation for numerical modelling of superconductor magnetization in true 3D geometry. *Supercond. Sci. Technol.* 32(11), 115001.

[93] Solovyov M, Pardo E, Souc J, Gömöry F, Skarba M, Konopka P, Pekarčíková M and Janovec J. 2013. Non-uniformity of coated conductor tapes. *Supercond. Sci. Technol.* 26, 115013.

[94] Stenvall A, Lahtinen V and Lyly M. 2014. An H-formulation-based three-dimensional hysteresis loss modelling tool in a simulation including time varying applied field and transport current: The fundamental problem and its solution. *Supercond. Sci. Technol.* 27(10), 104004.

[95] Swan G. 1968. Current distribution in a thin superconducting strip. *J. Math. Phys.* 9(8), 1308–1312.

[96] Takács S, Kaneko H and Yamamoto J. 1994. Time constants of normal metals and superconductors at different ramp rates during a cycle. *Cryogenics* 34(8), 679–684.

[97] Tominaga N, Mifune T, Ida A, Sogabe Y, Iwashita T and Amemiya N. 2018. Application of hierarchical matrices to large-scale electromagnetic field analyses of coils wound with coated conductors. *IEEE Trans. Appl. Supercond.* 28(3), 4900305.

[98] Ueda H, Imaichi Y, Wang T, Ishiyama A, Noguchi S, Iwai S, Miyazaki H, Tosaka T, Nomura S, Kurusu T. *et al.* 2016. Numerical simulation on magnetic field generated by screening current in 10-T-class REBCO coil. *IEEE Trans. Appl. Supercond.* 26(4), 4701205.

[99] van Nugteren J. 2016. *High Temperature Superconductor Accelerator Magnets*, Ph.D. thesis, University of Twente, Netherlands. https://research.utwente.nl/en/publications/high-temperature-superconductor-accelerator-magnets.

[100] Vestgården J, Mikheenko P, Galperin Y and Johansen T. 2013. Nonlocal electrodynamics of normal and superconducting films. *New J. Phys.* 15(9), 093001.

[101] Vestgården JI, Mikheenko P, Galperin YM and Johansen TH. 2013. Nonlocal electrodynamics of normal and superconducting films. *New J. Phys.* 15(9), 093001.

[102] Vlasko-Vlasov V, Koshelev A, Glatz A, Phillips C, Welp U and Kwok W. 2015. Flux cutting in high-t_c superconductors. *Phys. Rev. B* 91(1), 014516.

[103] Wang Y, Zhang M, Grilli F, Zhu Z and Yuan W. 2019. Study of the magnetization loss of CORC cables using 3D T-A formulation. *Supercond. Sci. Technol.* 32(2), 025003.

[104] Wimbush SC and Strickland NM. 2017. A public database of high-temperature superconductor critical current data. *IEEE Trans. Appl. Supercond.* 27(4), 8000105.

[105] Zeldov E, Clem JR, McElfresh M and Darwin M. 1994. Magnetization and transport currents in thin superconducting films. *Phys. Rev. B* 49(14), 9802–9822.

[106] Zermeno V, Grilli F and Sirois F. 2013. A full 3-D time-dependent electromagnetic model for Roebel cables. *Supercond. Sci. Technol.* 26(5), 052001.

[107] Zermeno VMR, Abrahamsen AB, Mijatovic N, Sorensen MP, Jensen BB and Pedersen NF. 2012. Simulation of an HTS synchronous superconducting generator. *Phys. Proc.* 36, 786–790. doi:10.1016/j.phpro.2012.06.043.

[108] Zermeño VMR, Habelok K, Stepien M and Grilli F. 2017. A parameter-free method to extract the superconductor's $J_c(B, \theta)$ field-dependence from in-field current-voltage characteristics of HTS tapes. *Supercond. Sci. Technol.* 30(3), 034001.

[109] Zhang H, Zhang M and Yuan W. 2016. An efficient 3D finite element method model based on the T–A formulation for superconducting coated conductors. *Supercond. Sci. Technol.* 30(2), 024005.

Chapter 2

Introduction to Stability and Quench Protection

Antti Stenvall

Basware Corporation, Linnoitustie 2, 02600 Espoo, Finland

Tiina Salmi

Tampere University, Kalevantie 4, 33100 Tampere, Finland

Erkki Härö

Sweco Finland, Ilmalanportti 2, 00240 Helsinki, Finland

Stability typically refers to maintaining an existing state or the ability to attain it with control. In superconductors, stability is related to maintaining the superconducting state and therefore to the ability to transport lossless direct current. The *stability analysis* of superconductors includes considerations of the following questions: When is stability lost? How easily is it lost and what happens after it is lost? The answers to these questions contribute to selecting the operation conditions for the device and the design of its protection as well as the system for detecting the loss of stability. Definitely, the system needs to be able to go through the loss-of-stability situation so that after the operation conditions are recovered, it can be utilized again. In the latest, the role of the protection is crucial.

In large-scale superconducting applications, *quench* is an event of loss of stability in which the heat generation is such that temperature rise within the system cannot be stopped by the available cooling. Consequently, the system needs to be de-energized, or taken out from its normal use, to prevent damage. Often, it is necessary to design an adequate quench detection and protection system for the superconducting device in order to perform the de-energization fast enough.

This chapter begins by introducing margins of quench in Section 2.1. The purpose is to introduce elementary concepts related to quench. First, we consider numerical simulations of the minimum quench energy in a case that can be utilized as a benchmark to develop or compare the tools at hand. Second, we discuss how prone magnets are to quench by considering various margins in the load line of a magnet.

Section 2.2 introduces two well-established classifications of quenches. These are important for understanding how quenches originate and why, especially, superconducting magnets behave like they do from a quench point of view. In Section 2.3, we consider the methodology of quench simulations in general. We consider the design of quench protection systems, quench modeling, and quench experiments.

Finally, Sections 2.4 and 2.5 present two case studies encountered in superconducting magnet R&D projects. The first scrutinizes the quench modeling of an R&D REBCO coil. The second presents design and modeling of quench protection heaters for a Nb_3Sn accelerator magnet prototype.

Cooling is a necessity for making superconducting devices functional. Cooling is very important, for example, in solving the heat balance of a system or, naturally, when estimating the duration of a cool down phase. Quench is often a very fast event: of the order of 1 s. Consequently, the quench analyses of impregnated windings typically neglect the effect of cooling and consider an adiabatic situation in which the magnet is completely isolated from the coolant. However, when the cooling fluid is in direct contact with the strands, the adiabatic approach is not necessarily adequate. The effect of cooling in quench has been reviewed by Bottura [10]. This chapter neglects the analysis of cooling during quench.

2.1. Margins to Quench

In the system design and comparison of different conductors or cables, it is important to know how prone they are to quench. This can reflect on the selection of a device's operation point or on the choice between available conductor options. Here, we present two different approaches for the available safety margin. In Section 2.1.1, we consider when an energy release in a conductor is sufficient to cause a quench.

In magnets, the margin is a sum of several contributions because an increase in current also increases the field the conductors are exposed to. Correspondingly, it is not reasonable to only consider how far away the operation current is from the critical one. The magnet's load line must be taken into account too. This is the topic of Section 2.1.2.

2.1.1. *Minimum quench energy*

Ideally, when distributed power is focused on a superconductor and cooling is neglected, the margin to quench represents the energy required to increase the temperature to such a value that the operation current gets above the critical one somewhere in the device. However, this is not the only way quench originates. A possible cause of quench is a local energy release somewhere in the device. Then, if the dissipated energy is high enough, it will locally cause the operation point to shift above the critical surface. Correspondingly, Joule heat generation occurs in the superconductor in the originated normal zone. If this normal zone propagates, quench occurs. However, if the normal zone is small enough and heat conducts away from it powerfully enough, the normal zone shrinks, superconductivity is recovered, and stability is maintained.

The smallest volume that is required to generate a propagating normal zone and quench is defined as *minimum propagation zone* (MPZ). The energy required to generate an MPZ is called *minimum quench energy* (MQE). The stability of different conductors can be compared at defined operation conditions, *inter alia*, by comparing their minimum quench energies. An analytical approach to the MPZ and MQE in low-temperature superconductors was detailed by Wilson [75]. Its applicability to HTS coils was considered by Härö *et al.* [33]. The analytical approach is often a good first throw into the ballpark. However, when one considers, for example, the influence of n values on the MQE, a more detailed, and typically numerical, approach is required. We consider this next.

2.1.1.1. *Numerical modeling of MQE*

Numerical modeling of the MQE allows one to consider flexibly, for example, which kinds of heat pulses cause quenches. The time duration and spatial distribution of the heat pulses can be selected in these simulations, unlike in analytical computations. Also, one is not restricted in the choice of the critical-current–temperature–magnetic-flux-density relation in the superconducting domain. Furthermore, through numerical modeling, one has the option to scrutinize the temperature distribution within the whole modeling domain at every necessary time instant. This can be then utilized to compute, for example, thermal stresses during the event.

Here, we consider one-dimensional (1D) numerical simulations of the MQE, where the current diffusion within the conductor is neglected. Thus, we describe the dynamics of the relevant physics with the *heat diffusion equation* along the conductor. In doing this, we assume cross-sectional isotherms and material-component-wise homogeneous current densities, and due to the temperature increase and change in local resistivities, the current redistributes instantaneously according to the minimum power principle. Before going to the simulations, we formulate the thermodynamical problem we are solving.

To go into more detail about the physics, i.e. to, first, abandon the assumption of homogeneous currents in each component, one needs to solve simultaneously the heat diffusion equation and the *magnetoquasistatic problem*, in which the current diffusion from the superconductor to the stabilizer is modeled too. Because instantaneous current diffusion is the minimum power solution to a net-current-constrained problem, the diffusion problem will lead to higher total heat generation at the quench propagation frontier. However, after this diffusion process has leveled off, the stabilizer will carry a homogeneous current distribution. An important consideration is whether or not the difference in the heat generation is meaningful with respect to the more laborious approach that a simultaneous consideration of current diffusion requires.

The described *multiphysical* approach cannot be based solely on a 1D modeling domain because in that case, the current diffusion does not occur. One can either use the same modeling domain for both physics or an approximation of isothermal cross-sections and a different modeling domain for the magnetoquasistatic problem. The heat diffusion equation couples with the magnetoquasistatics via the temperature-dependent resistivity, and the solution to the magnetoquasistatic problem gives the heat generation as an input to the heat diffusion equation.

In the MQE simulation where only the thermal diffusion is considered, we solve only the heat diffusion equation:

$$\nabla \cdot \lambda(T, \mathbf{B}, \mathrm{RRR}) \nabla T + Q = \gamma C(T) \frac{\partial T}{\partial t}, \tag{2.1}$$

where λ is the heat conductivity, T the temperature, $\mathbf{B}$ the magnetic flux density, RRR the residual resistivity ratio, Q the heat generation, γ the density, and C the specific heat. One may also consider cooling in Q. Heating in Q consists of the heat in the conductor and the external energy input — the disturbance Q_{dist}; therefore, Q may depend on temperature, magnetic flux density, operation current, and time.

Thermal time constant and magnetic diffusion time can be utilized to estimate whether the 1D thermal model is adequate to find the MQE or if magnetoquasistatic physics should also be considered. For a single conductor, the time constant of the thermal problem τ_T is

$$\tau_T = \frac{\gamma C r^2}{\lambda}, \tag{2.2}$$

where r is the radius of the conductor. The magnetic diffusion time τ_M is

$$\tau_M = \frac{\mu r^2}{\rho}, \tag{2.3}$$

where ρ is the resistivity and μ the permeability. In a simplified case, we neglect the contact resistance between the superconductor and the stabilizer and take a conductor with a radius of 0.5 mm and consider the material properties of copper with a residual resistivity ratio (RRR) of 100 at 4 K and 5 T: $\gamma = 9.0\,\mathrm{g/cm^3}$, $C = 0.09\,\mathrm{J/kg/K}$, $\rho = 0.35\,\mathrm{n\Omega m}$, and $\lambda = 260\,\mathrm{W/mK}$. The thermal time constant is $3\,\mu$s. Correspondingly, only the disturbances at this time scale are required to be considered in approaches other than the 1D approach from the temperature perspective. The corresponding magnetic diffusion time is 0.9 ms. Again, if the disturbances last for several ms, we can assume immediate homogenization of the current distribution in the stabilizer arising from the excess current in the superconducting region due to the increase in the local temperature. However, in superconducting cables, perhaps with a stabilizing aluminum or copper jacket, the influence that current diffusion inertia has on heat generation cannot be neglected when high accuracy is required. One should note that the 1D approach also neglects the turn-to-turn heat conduction in the coils. To compare individual conductors, this is adequate, but when the quench propagation in a magnet is considered, one must include turn-to-turn heat transfer in the model.

The next important modeling decision involves the heat generation model. There are various ways to model the heat generation in the superconductor and stabilizer. The three main methods rely on *current-sharing model*, *power-law model for the whole conductor*, and *power-law model for the superconducting domain*.

In the current-sharing model, one assumes that the superconducting domain carries at most its critical current I_c in every cross-section of the conductor. This assumption relies on isothermal conductor cross-sections. The excess current I_s flows in the stabilizer and determines the electric field based on the stabilizer's resistivity. The operation current I_{op} is thus the sum of I_s and I_c. The same electric field that exists in the stabilizer is also present in the superconducting region in the current-sharing model. The average heat generation Q_{cs} on the conductor cross-section is computed as

$$Q_{cs}(T, \mathbf{B}) = \begin{cases} 0 & I_{op} < I_c(T, \mathbf{B}), \\ \rho_s(T)\frac{(I_{op}-I_c)I_{op}}{\alpha A_{tot}^2} & I_{op} \geq I_c(T, \mathbf{B}), \end{cases} \tag{2.4}$$

where ρ_s is the stabilizer's resistivity, A_{tot} the cross-sectional area of the conductor, and α the fraction of the stabilizer in the conductor's cross-section. Note that immediately above the current-sharing temperature T_{cs}, i.e. the temperature at which the operation current corresponds to the critical one, losses occur in the superconductor too.

In the power-law model for the whole conductor, one assumes that the effective resistivity of the whole conductor behaves in a power-law-like fashion until the situation where all the current flows in the stabilizer is reached. Consequently, the heat generation Q_{pl} is computed from

$$Q_{pl}(T, \mathbf{B}) = \min\left\{\rho_s(T)\frac{I_{op}^2}{\alpha A_{tot}^2}, \frac{E_c I_{op}^{n+1}}{A_{tot} I_c(T, \mathbf{B})^n}\right\}, \tag{2.5}$$

where E_c is the critical electric field criterion utilized in the I_c characterization of the wire and n is the conductor's index number characterizing the steepness of the resistive transition. n is a function of T and $\mathbf{B}$, though in simulations, one often approximates it with a constant. With min, one takes into account that the resistivity does not result in higher losses than the situation in which all the current flowing in the stabilizer would.

In the power-law model for the superconducting domain, one assumes that the electric field develops according to power law in the superconducting domain. Then, I_{op} is shared between the stabilizer and the superconducting domain in a way that their electric fields are equal. The same is assumed in the current-sharing model too. But now, to solve the current sharing between the matrix and the superconductor, one needs to solve the current of the stabilizer I_s from

$$\rho_s(T)\frac{I_s}{\alpha A_{tot}} = E_c\left(\frac{I_{op} - I_s}{I_c(T, \mathbf{B})}\right)^n, \tag{2.6}$$

when the critical current is not zero. After this, the computation of the electric field is straightforward, and the average heat generation can be computed by weighting the heat generations of the matrix and superconducting region by their volumetric fractions.

When considering numerical MQE simulations, the important parameters of the heat input pulse are its duration, spatial size, and time variation. However, for a comparison of conductors, it is typically enough to compare square-wave heat pulses. Thus, for a given duration and spatial size, one needs to find the minimum constant magnitude of the applied external heat generation that causes quench. However, if one anticipates specific energy releases that may occur, for example in the interaction region (IR) of particle accelerators, a more detailed analysis is necessary. Also, one needs to consider the length of the modeling domain and its discretization. To solve the heat diffusion equation with a given energy input, we discretize the modeling domain in space and solve the ordinary differential equation in time. One can easily get involved in adjusting several tolerances and parameters that control the linear solvers. Therefore, one needs to familiarize oneself with the details of the available tools to solve problems reliably.

2.1.1.2. *MQE simulations*

This example presents a reference case, or a benchmark, for performing 1D MQE simulations. Therefore, we selected a simulation case that is reasonable but easy to replicate. For the material properties — heat conductivity, thermal conductivity, and resistivity — we utilized the data of copper. We considered an RRR of 100 for thermal conductivity and resistivity and $||\mathbf{B}||$ of 1 T to include the magnetoresistance. Various operation temperatures were considered. The thermal conductivity and heat capacity were taken from Ref. [43] and the resistivity from the work of Kim [35].

The volumetric heat capacity C_v (J/m^3) is given by

$$C_v(T) = \gamma \cdot 10^{a+b\ln(T)+c\ln(T)^2+d\ln(T)^3+e\ln(T)^4+f\ln(T)^5+g\ln(T)^6+h\ln(T)^7}, \tag{2.7}$$

where γ is the density (which is 8960 kg/m^3 for copper) and following are the values of the constants: $a = -1.91844$, $b = -0.15973$, $c = 8.61013$,

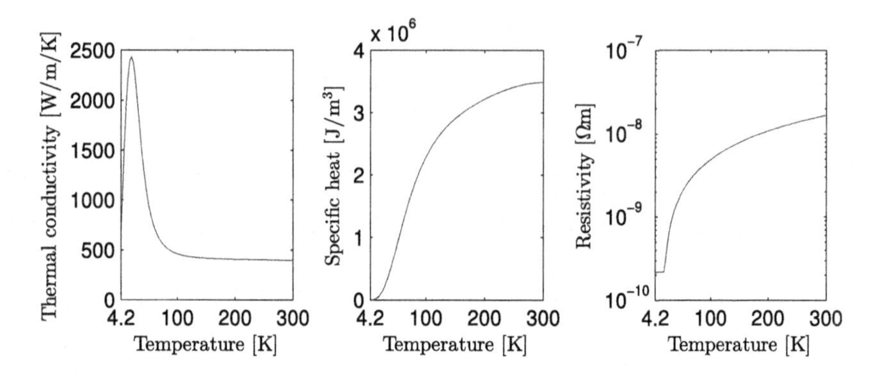

Figure 2.1. Material properties of copper with $RRR = 100$ and $||\mathbf{B}|| = 1\,\mathrm{T}$.

$d = -18.996$, $e = 21.9661$, $f = -12.7328$, $g = 3.54322$, and $h = -0.3797$. The resistivity (Ωm) is given by

$$\rho(T, RRR, \mathbf{B}) = \max\left\{\frac{1.67 \cdot 10^{-8}}{RRR} + 5 \cdot 10^{-11}||\mathbf{B}||,\right. \\ \left. 5.9 \cdot 10^{-11}T - 1 \cdot 10^{-9}\right\}. \tag{2.8}$$

The thermal conductivity (W/m/K) is given by

$$\lambda(T) = 10^{\frac{a+b\sqrt{T}+cT+dT^{\frac{3}{2}}+eT^2}{1+f\sqrt{T}+gT+hT^{\frac{3}{2}}+iT^2}}, \tag{2.9}$$

where $a = 2.215$, $b = -0.88068$, $c = 0.29505$, $d = -0.04831$, $e = 0.003207$, $f = -0.47461$, $g = 0.13871$, $h = -0.02043$, and $i = 0.001281$. One should note that for RRR other than 100, different constants must be utilized. The material properties are displayed in Figure 2.1.

We considered a conductor with a radius of 0.5 mm. It was assumed that 70% of the conductor's cross-section was made up of the stabilizer. Insulation was neglected. We utilized the current-sharing model for computing the heat generation. The critical current of the wire was determined with the equation

$$I_c(T) = 500\left(1 - \frac{T - 4.2}{20 - 4.2}\right). \tag{2.10}$$

This may approximate a MgB_2 conductor. The length of the considered piece of conductor was 8 m and quench was initiated with a 20 mm long heater. We utilized the discrete symmetry of this particular case, and correspondingly, it was enough to utilize a modeling domain having a length of 4 m (see Figure 2.2). The heater's pulse duration was 10 ms, and its power

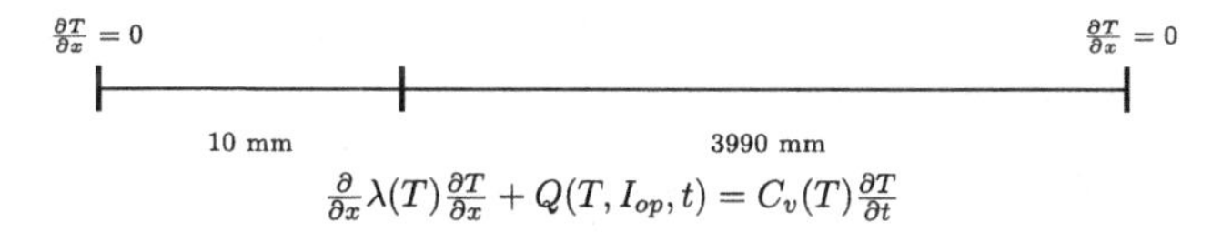

Figure 2.2. Modeling domain for the MQE problem (not to scale): The 10 mm on the left represents the area of disturbance, where the heat generation is the sum of disturbances given by the current-sharing model. Also, adiabatic boundary conditions are shown.

was uniformly applied to its volume. The simulation was carried out for 2 s, and if at the end of the simulation, the maximum temperature was at least 5 K above the T_{cs}, it was considered that quench originated. However, in the case of quench, a temperature of 300 K is reached faster when excluding very low currents. Therefore, we also terminated the simulation if the maximum temperature exceeded the T_{cs} by 30 K and deducted that a quench has occurred. We show that, for this particular case also, this quench criterion was adequate. The minimum energy that caused a quench was determined to be the MQE.

Stability-related simulations are time-consuming because they include highly nonlinear material parameters, must be performed over a certain time interval, and often include parametric studies that are of interest. Therefore, in the numerical analysis, one wants to have as few degrees of freedom (DoFs) as possible to get adequate results. However, the risk of having too few is that one does not attain accurate enough results. To minimize this risk, it is often useful to perform, first, a convergence analysis for a particular study. If one utilizes a method in which the modeling domain is meshed, one begins with a sparse mesh, proceeds toward denser ones, and determines the adequate mesh density for the rest of the simulations. In the case of finite element methods (FEMs), the influence of the polynomial degree p of the basis function on computation can be studied too. Often, in FEMs one attains better results with the same number of DoFs in the case of higher-order basis functions. Furthermore, above a certain threshold related to the number of DoFs, the linear solvers tend to solve these problem faster. However, in 1D, that is not necessarily the behavior.

In addition to the MQE, the *normal zone propagation velocity* v_{nzp} is often of great interest. Essentially, v_{nzp} tells how fast the volume where heat is generated extends. Whereas finding out the MQE under specific operation conditions requires several simulations, v_{nzp} can be determined from a single simulation. To compute the v_{nzp}, one finds the curve $T(t, x) = T_{cs}$ from the solution, including the spatial points x, time t, and temperature T. Then, one can restrict x to a certain space where the end of the modeling domain and the quench onset do not have influence. In that region, $T(t, x)$ is a straight line on the (t, x)-plane and correspondingly can be expressed as $x = v_{nzp}t + a$, where a is a constant. a depends on the duration of the heat pulse and its magnitude, whereas v_{nzp} depends only on the operation conditions. Therefore, by fitting the first-order polynomial to the curve on the (t, x) plane, v_{nzp} can be calculated.

For the reference case, which we utilized to find adequate discretization, we considered an operation temperature of 4.2 K and a current of 350 A (i.e. $0.7I_c(4.2\,\mathrm{K})$). Our solution was based on a homemade FEM software in MATLAB. We used 100 times denser elements in the hot spot than in the other end and let the element size gradually increase in between. For the convergence analyses (one for MQE and one for v_{nzp}), we used a scaling parameter to control the absolute element size, and correspondingly, we got a different number of DoFs for different meshes. We also investigated the effect of changing the interpolation polynomial degree from 1 to 4.

The results of the MQE convergence analysis are shown in Figure 2.3 and those of v_{nzp} in Figure 2.4. As can be noted, the MQE computation converged only to $\pm1\%$ limits. This is due to the numerous parameters in the MQE computation and the discontinuous nature of the square-wave power that we applied. The simulations are sensitive to time-stepping, the particular mesh, and the algorithm, which searches for the MQE by making educated guesses of the energy to be utilized in each simulation. Therefore, a small fluctuation can be tolerated. However, the fluctuation is kept within 2% of the average value of the previous four computations for each value of p in every case when one has more than 100 DoFs. In the case of v_{nzp}, the same limit required about 250 DoFs, but no fluctuation occurred. Whereas the MQE is a single-point property of the simulation that cannot be expected to even converge in FEM-based analysis, v_{nzp} is more a global quantity. The utilized polynomial degree did not have any notable effect on the convergence. This could have been a property of this 1D simulation only and cannot be generalized to simulations in higher spatial dimensions. For example, in magnetostatic problems where the total energy is of interest, an increase in p from 1 to 2 is typically considerably more beneficial than a corresponding densification of the mesh.

Based on the convergence analysis, we utilized in all other MQE simulations a mesh that resulted in 116 DoFs. All the simulations involving v_{nzp} were executed with a mesh that resulted in 775 DoFs. In both cases, the polynomial degree of the basis functions was 1. Based on these simulations, the MQE for the reference study was 1.38 mJ and for the v_{nzp}

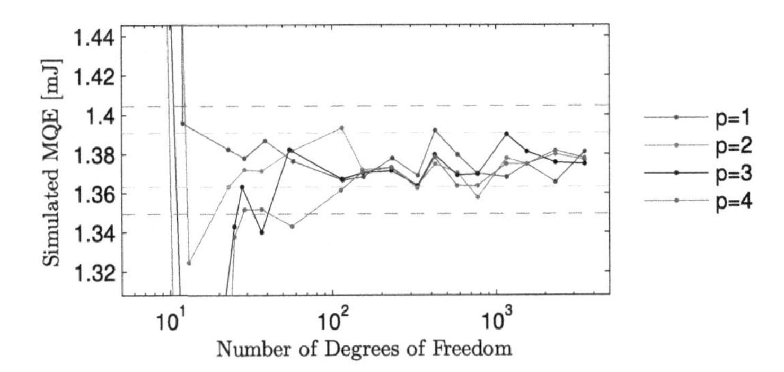

Figure 2.3. Results from the convergence analysis of the MQE: Light and dark dashed lines represent the 1% and 2% differences, respectively, from the average value of the previous four computations for each p.

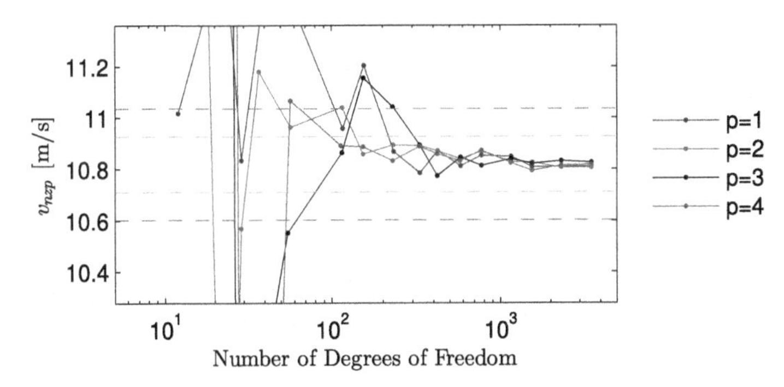

Figure 2.4. Results from the convergence analysis of normal zone propagation velocity. Light and dark dashed lines represent the 1% and 2% differences, respectively, from the average value of the previous four computations for each p.

study, 10.8 m/s. The simulated energy input was multiplied by two to get the MQE for the conductor because the symmetry of the modeling domain was utilized.

In general, the solution to this kind of thermal problem is very sensitive to the disturbance energy E_{dist}. Figure 2.5 presents the maximum temperatures in the modeling domain as a function of E_{dist} at different time instants in the reference case. The clear limit between the quenching and non-quenching simulations is visible. At 10 ms, i.e. when the heater was switched off, the temperature distributions were very similar for the investigated energy range. However, at 20 ms, there was already a clear difference. In the reference case, the maximum temperature started to increase

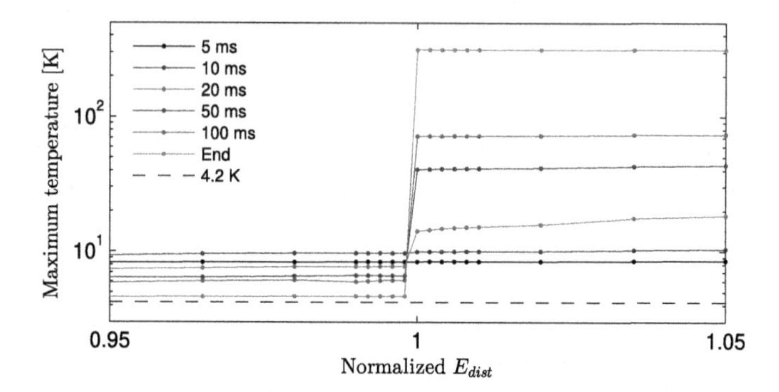

Figure 2.5. Maximum temperature in the modeling domain at different time instants as a function of normalized disturbance energy. Normalization was done to the MQE.

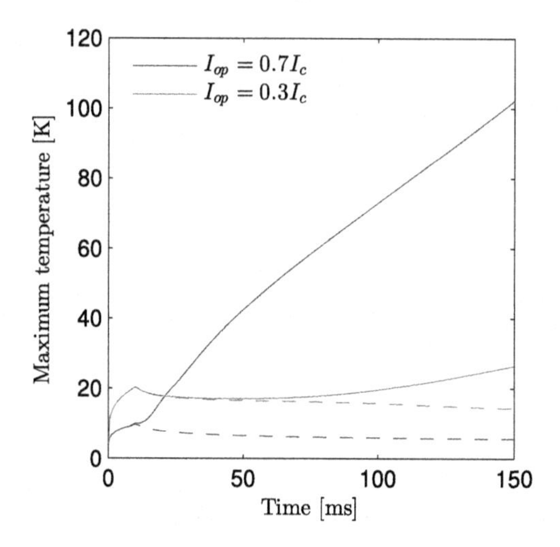

Figure 2.6. Maximum temperature as a function of time when energy disturbance was the MQE (solid lines) and 1% below the MQE (dashed lines) for two different operation currents.

immediately after the heater was switched off when the disturbance energy was the MQE or higher.

The derivative of the maximum temperature, however, cannot be utilized as a criterion to determine whether the given disturbance causes a quench or not. Figure 2.6 compares two different simulations at 4.2 K with the operation currents of $0.3I_c$ and $0.7I_c$. As can be seen, in the case of $I_{op} = 0.3I_c$, it took more than 100 ms before the quench really started to propagate. In both the non-quenching and quenching cases, the maximum temperature started to decrease immediately after the heater was powered off. Therefore, the simulation must include a significant time after

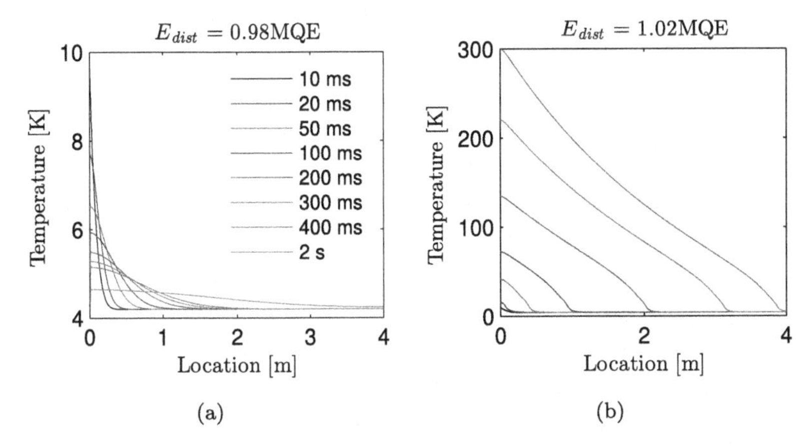

Figure 2.7. Temperature evolution in the reference case when the disturbance energy was slightly (a) below and (b) above the MQE.

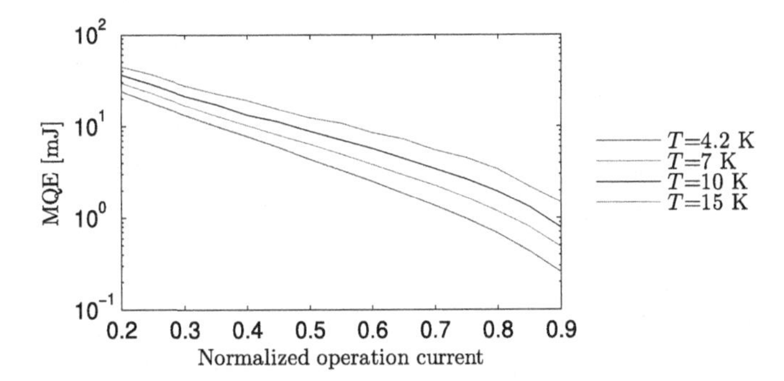

Figure 2.8. Minimum quench energies at various operation temperatures as a function of normalized operation current. Normalization was done to the I_c of given temperature.

the heater is powered off in order to determine whether the quench originated or not, especially if the current is considerably below the critical one.

In Figure 2.7, the temperature evolution in the case of disturbances 2% above and below the MQE is shown at the reference operation conditions. The very rapid increase in the temperature as a function of time is noticeable. It took only 400 ms to reach 300 K in the case of a quench. This also emphasizes the need for and importance of quench detection and protection.

Parametric studies considering the MQE and v_{nzp} at different temperatures with different normalized operation currents are shown in Figures 2.8 and 2.9, respectively. The most important observation is that the higher the MQE, the lower the v_{nzp}. This is also a general conclusion because

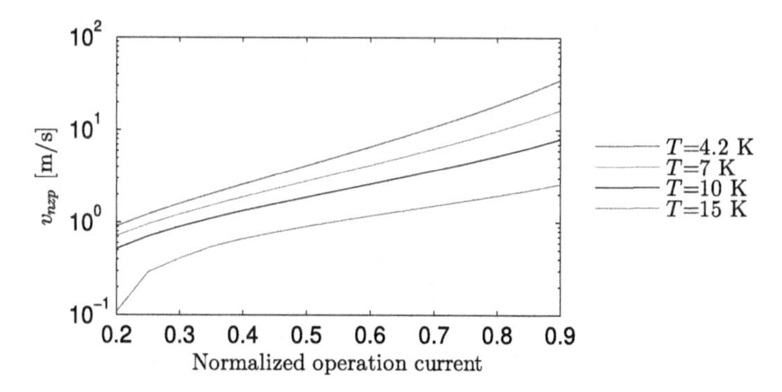

Figure 2.9. Normal zone propagation velocities at various operation temperatures as a function of normalized operation current. Normalization was done to the I_c of given temperature.

a higher MQE means more energy, or more time, to make the quench propagate. At higher operation temperatures, the MQE tends to increase because the heat generation is lower due to the diminished current densities and the substantially increased specific heat. At the same time, the normal zone propagation velocity decreases. All this means that if the operation temperature is increased, it becomes more and more difficult to detect the quench from a resistive voltage signal. So, a high MQE and v_{nzp} are both desirable from the stability point of view.

Finally, we considered the maximum temperatures that could be reached in the investigated hot spot, i.e. the highest temperature in the modeling domain, without causing a quench. The results are depicted in Figure 2.10. As can be noted, a temperature increase to 30 K above T_{cs} always caused a quench. The maximum reachable temperature is of interest when one compares the volumetric MQE in the hot spot to the cable's enthalpy margin. In a volumetrically large enough disturbance, these coincide. Therefore, if homogeneous power was dissipated (for example, due to AC losses), the one matching the enthalpy margin would cause a quench. However, in the case of localized disturbance, this is not the case, as demonstrated. When the disturbance is concentrated in a small volume for a short time, it is possible to overstep the current-sharing temperature, notably — naturally, greatly depending on the operation current — without causing a quench. When the I_c is approached, this possibility decreases to insignificant values.

As demonstrated, there is a lot to study even in simple MQE simulations, including:

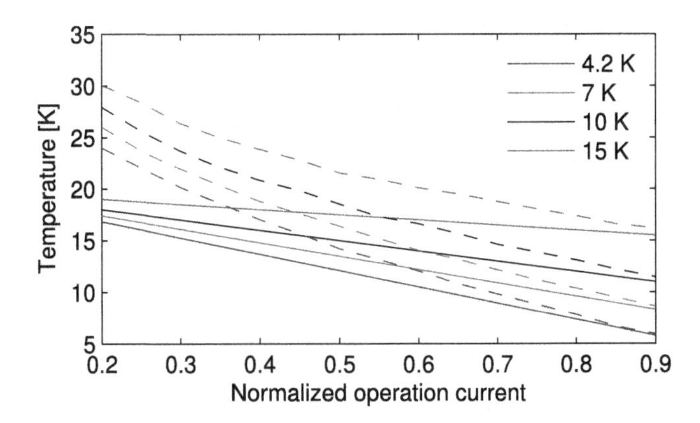

Figure 2.10. The current-sharing temperatures (solid lines) and the maximum temperatures T_{max} that could be reached in the considered hot spot without causing a quench (dashed lines).

- a study of the minimum number of DoFs needed to acquire simulation results that do not suffer from numerical noise;
- a study to find an adequate criterion to determine whether a given disturbance caused a quench or not during simulation;
- parametric analyses as a function of temperature, current, and magnetic field to understand the sensitivity to disturbances at different operation conditions;
- a study of the parameters of the disturbances (at least duration and spatial size) that are relevant to a given system.

Here, we considered, first, the DoFs and the quench criterion. Then, we presented parametric studies. We considered for the disturbance only a square-wave heat pulse at fixed time and space. However, it is not often possible to know beforehand what the disturbances that cause quenches are like. Therefore, a reference disturbance can be utilized to compare different conductors or cables.

When this analysis is extended to three-dimensional finely structured modeling domains, the simulations easily get computationally intensive. Therefore, it is always important to find out how the situation under study can be simplified without losing accuracy. Also, it might be important to study only the effect of some conductor manufacturing-related detail. Then, one should focus on the issues under study and remove the uncertainty and influence of the simulation parameters that are not of interest in that case.

2.1.2. *Margins in magnet load line*

In a superconducting coil, the magnetic field applied to the turns of the coil is directly proportional to the operation current when the magnetization possible in the iron yoke and filaments is neglected. The proportionality gives a load line for the magnet, which can be utilized, together with the critical current characteristic, to determine how much the current or temperature can be increased before a quench. Here, we detail these margins.

The linear approximation between the current and the magnetic field is typically very good for the purpose of stability considerations but not necessarily adequate when the field in the magnet bore is of concern with 10 ppm accuracy. Also, in many applications, it is not reasonable to estimate the current-carrying entity, a cable or a conductor, that is exposed to a homogeneous magnetic flux density. A more detailed analysis is needed, especially in particle accelerator magnets, from the perspective of the bore field. Also, in magnets wound from cables whose cross-sectional area is proportional to the cross-section of the whole magnet, the homogeneous magnetic flux on the cable's cross-section is not a reasonable estimation for determining how much current the cable can carry.

In a series-coupled magnet system,[1] the applied magnetic field is completely determined in a linear situation[2] by the magnetic flux density distribution at unit current $\mathbf{B}_{unit}$ and by the operation current as $\mathbf{B}(I_{op}, x) = \mathbf{B}_{unit}(x)I_{op}$, where x refers to a point in the magnet. In the case of an isotropic critical current characteristic, all the margins of the magnet can be computed with a single real number, the internal magnet constant K, which gives $\mathbf{B}(I_{op}, x') = KI_{op}$, where x' is the location where the critical surface is pierced first when the operation current is increased in the linear magnet. A magnet's *load line*, whose slope is K, relates the magnetic flux density at x' in an ideal linear magnet to its operation current.

We now consider an example of the margins of a magnet by using the load line. *Current margin* means how much the current can be increased before the critical current surface is pierced when the magnet load line

[1] In this chapter, we refer to coil as a single winding having one lead in and one lead out. A magnet can be either a single coil or an assembly of coils, such as a dipole magnet made from two coils. In the case of a resistive fault current limiter, a non-inductive coil is not a magnet.

[2] Linear situation means that there are no materials present that magnetize and that there are no magnetization currents in the conductors. Consequently, the current density is always homogeneous in conductors.

is taken into account. When the operation current is gradually increased, there is a location in the magnet where the critical surface is reached first. This location is called the *critical point* of the magnet — the previously mentioned x'. *Temperature margin* means how much the operation temperature can be increased before the critical surface is pierced. *Short-sample margin* means how much below the short-sample I_c the magnet operates at given operation conditions. The load line is not taken into account in the short-sample margin. However, in some literature, the short-sample margin and the current margin both mean the current margin. The short-sample margin presented in this way is beneficial when one considers how much the coil's critical current characteristic has degraded due to the manufacturing of the magnet if the theoretically maximum possible current I_{max} cannot be reached in a constructed magnet. I_{max} can also be called the *short-sample limit* of the magnet.

In the example case that we present here, we have utilized the following expression for the critical current [35]:

$$I_c(\mathbf{B}, T) = I_{c0}\left(1 - \frac{||\mathbf{B}||}{B_{c0}\left(1 - \left(\frac{T}{T_{c0}}\right)^{1.7}\right)}\right)\left(1 - \frac{T}{T_{c0}}\right), \tag{2.11}$$

where $I_{c0} = 500\,\text{A}$, $B_{c0} = 15\,\text{T}$, and $T_{c0} = 9.5\,\text{K}$. Also, we consider a K of $20\,\text{mT/A}$.

Figure 2.11 presents the load line of the magnet and the critical current characteristics at few selected temperatures. We have considered that the magnet operates at an I_{op} of 127 A and at a T_{op} of 4.2 K. Then, according to the K at its critical point, the magnet produces a field of 2.53 T. This is denoted as B_{op}.

When we follow the magnet's load line to the critical current characteristic at an operation temperature, we reach the maximum current that the magnet can operate with, i.e. I_{max}. In this case, that is 187 A. Thus, our current margin is 187 A−127 A=60 A.

If we start increasing the temperature, I_{max} gradually slides down on the load line. When we have increased the temperature to such a value that I_{max} equals I_{op}, i.e. to the current-sharing temperature, we have lost all the temperature margin. Here, that would mean increasing the temperature to 6 K, which means the temperature margin is 6 K−4.2 K=1.8 K.

The *enthalpy margin* is closely related to the temperature margin. It is the integral of the heat capacity from the operation temperature to the

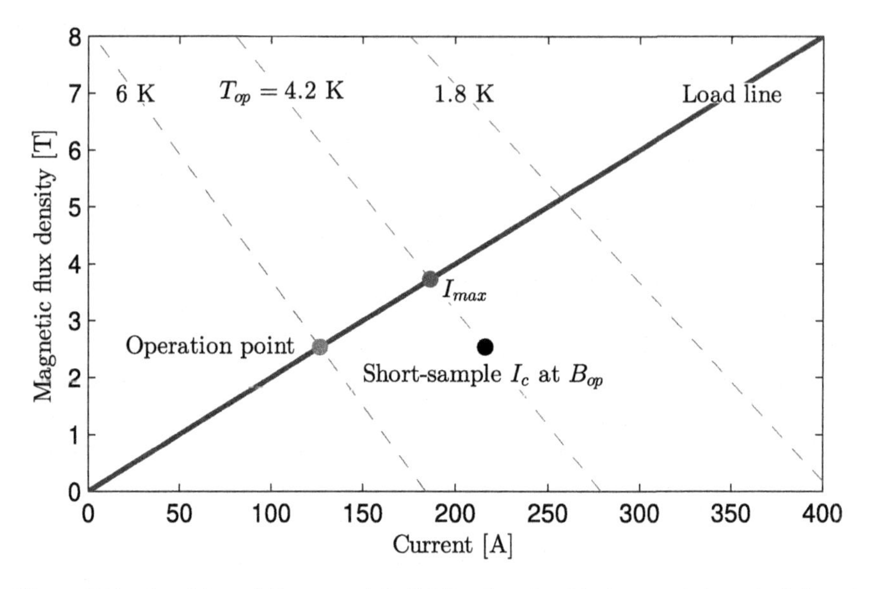

Figure 2.11. Load line of the magnet (solid lined) and critical current characteristics at selected temperatures (dashed lines).

current-sharing temperature. Naturally, when the operation temperature increases and the operation point remains the same, the enthalpy margin decreases. Correspondingly, smaller and smaller disturbances can cause a quench.

As we can see, the short-sample I_c at B_{op} is larger than I_{max}. This is obvious because an increase in $||\mathbf{B}||$ decreases I_c. In this case, I_c at B_{op} and T_{op} is 216 A. Therefore, our magnet operates at 59% of I_c, but one should note that in a magnet, I_c at B_{op} of I_{op} cannot be reached because $||\mathbf{B}||$ increases as a function of current. The operation point of the magnet is 68% of its maximum current I_{max}.

Before proceeding from the analysis of the single conductor (or cable) to the stability analysis of a magnet system, one must consider the magnet's load line. Whereas in the parametric MQE or v_{nzp} analyses, a constant field can be applied to the whole modeling system, for the entire magnet, this is hardly the case. Therefore, it is important to understand the margins related to the magnet design and how they influence stability.

In this section, we first analyze the stability of a single conductor parametrically and then look at the margins that can be deduced from the magnet's load line and the conductor's critical current characteristic. Often, however, the case is more complicated — but the particular methods can be easily extended from the ones presented. In the case of REBCO-based magnets, the critical current characteristic can be anisotropic, meaning that it depends on the direction of **B** at a given point. Furthermore, if large cables are utilized to wind the coils, the current is not homogeneously distributed across the cable's cross-section. Then, one needs to solve the current distribution to find I_{max} for the magnet. An approach to solve the critical current for YBCO tapes in such situations was presented by Rostila *et al.* [51].

2.2. Classifying Quenches

Altogether, quench is an undesirable event, though some devices, such as resistive fault current limiters, utilize it to provide dynamic resistance to the power grid. Often, quenches occur abruptly when failure occurs in some location in the system. In order to improve future performance, it is important to know why the quench originated. Consequently, quenches can be classified to study the different possibilities that may have caused the undesired event. Here, we present two different classifications. The first one was presented by Devred [16] and the other one by Wilson [75, p. 70].

2.2.1. *Devred's classification of quenches*

Devred developed a qualitative classification based on the relation between the current at which the quench is detected and the magnet's short-sample limit. Even though this classification was initially developed for magnets wound from isotropic low-temperature superconductors, it can be used as such to qualify any superconducting system.

The classification divides quenches into two groups as follows. $\hat{I}_{max}$ is the temperature-dependent maximum operation current that the magnet has reached, sooner or later, in the experiments or during operation. In general, a magnet quench at current I_{quench} satisfies either $I_{quench} = \hat{I}_{max}$ or $I_{quench} < \hat{I}_{max}$ — with adequate margin for defining the equality. At an operation temperature T_{op}, the conductor that has been utilized to wind the magnet can reach at most I_{max}, the current at the junction of the critical current characteristic at the operation temperature and the magnet's load line (see Figure 2.11).

If a quench occurs at $\hat{I}_{max}$, a *conductor-limited quench* has occurred. If $\hat{I}_{max}$ is the same as I_{max}, a *short-sample quench* has occurred. However, if $\hat{I}_{max} < I_{max}$ holds, then one can say that the coil had *degraded.* Some degradation is very common because during the winding process, the

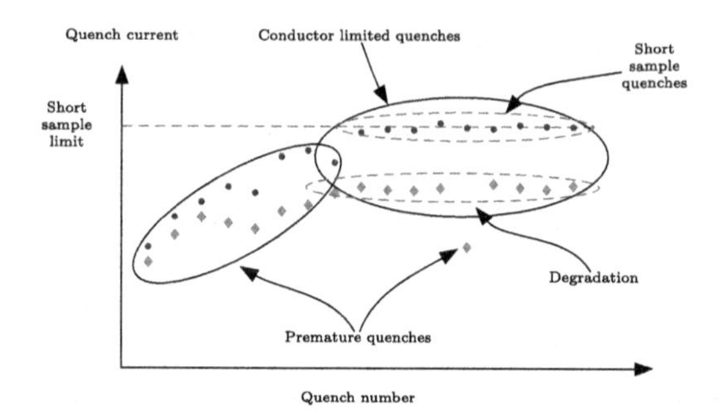

Figure 2.12. Quench history of two fictitious magnets. Quenches are classified according to Devred's classification.

conductor is exposed to a mechanical load. Degradation means that the short-sample limit cannot be reached.

If a quench occurs at a current below $\hat{I}_{max}$, it is typically due to an energy release inside the coil that locally increases the temperature enough for a quench to originate. This kind of quench is called as *energy-deposited quench* or *premature quench.* Figure 2.12 presents the quench currents of consecutive quenches in two fictitious magnets and the relation of the quenches to Devred's classification.

In Figure 2.12, the inclined ellipse represents a region where the magnet is *training.* In this region, quenches occur before any quench has happened at $\hat{I}_{max}$. Stresses are produced in the magnet during manufacturing. Consequently, mechanical energy is loaded into the magnet. If these stresses are released during magnet energization, for example in the epoxy, causing mechanical cracking, premature quenches can occur. Once these energies are released, they no longer cause quenches, Hence, the training.

It is noteworthy that, sometimes, in magnets, it is possible to reach currents above the short-sample limit. The reason for this is that stress has an influence on the critical current [17]. With some conductors, the stress distribution induced during the heat treatment changes during the cooling and powering of the magnet. The short-sample limit may be defined without the presence of an external stress, and correspondingly, it may be slightly lower.

Sometimes, the maximum quench current of a magnet stays considerably below the short-sample limit. Then, we say that the magnet has *degraded.* There can be overall degradation, especially if the conductor is brittle (such as those made of Nb_3Sn) or if a single spot in the winding is damaged so that it limits the performance of the whole magnet. Very often, even if some degradation is noticed, good magnets can achieve more than 95% of the short-sample limit.

Devred's classification mainly distinguishes whether a quench occurred for normal, predictable reasons, such as too high an operation current, or if it occurred without any clear explanation. Based on the experience in magnet operation, Devred's classification helps us to determine whether the given magnet is a successful piece of work or is degraded. The important criteria in assessing the quality of a magnet are the number of quenches needed to reach the conductor-limited quenches, i.e. how long training takes, the amount of degradation that has occurred, and the frequency of premature quenches after the training has finished.

2.2.2. *Wilson's classification of quenches*

Wilson presented another classification of quenches in superconducting magnets earlier than Devred. These two classifications are not exclusive but complementary. The classification by Wilson answers questions such as how the quench originated, but it does not reveal the particular reason that caused the quench.

Wilson considered what kinds of energy releases, *disturbances*, cause quenches. Correspondingly, he added an attribute to the quenches he considered: *energy-deposited quenches* — the quenches occurring below $\hat{I}_{max}$ in Devred's classification. According to Wilson's point of view, even close to $\hat{I}_{max}$, a quench originates due to a disturbance. When $\hat{I}_{max}$ is approached, the enthalpy margin becomes so negligible that, in principle, any disturbance can cause a quench.

In Wilson's classification, the time and space scales of the disturbances are of interest. Wilson's classification divides quenches into four classes, combining two classes in time and two in space. The overview of this classification can be presented with a *disturbance spectrum*. Table 2.1 summarizes this spectrum and shows what kinds of energies or powers are related to the given disturbance category.

The disturbances that cause quenches are divided into two, in time and space. In time, the *transients* are something that occur only once, typically for an unknown reason. If these can be considered to happen at a single *point*, as that occurring on the cross-section of a single conductor or in the insulation between two conductors, then one is interested in the energy that is released. When the transient disturbance is spread out to a larger volume, for example due to a false beam in a particle accelerator, then

Table 2.1. Disturbance spectrum in Wilson's classification of quenches.

		Space	
		Point	*Distributed*
Time	*Transient*	Joule	J/m^3
	Continuous	Watt	W/m^3

the quantity of interest is the energy per volume and the disturbance is *distributed* in space.

Continuous disturbances are something that accumulates and slowly causes a thermal runaway, an increase in temperature that at some time instant is detectable as a quench. In these situations, the cooling cannot match the heat generation inside the cryostat for a long period of time. If these occur throughout the coil, for example due to AC losses or low conductor n value, which causes sub-critical losses in the superconductor, the continuous disturbance is *distributed.* The continuous disturbance is related to a *point* if the location can be clearly positioned and the heat dissipation does not stop during system operation. A resistive joint is an example of such a disturbance.

Often, transient quenches are something that one cannot prepare for. Therefore, a quench detection system is required to protect the magnet from a quench. The quenches originating from continuous disturbances may take a considerable amount of time to develop. Often, delicate modeling helps to understand if continuous disturbances are a threat to normal operation. It is important to know on what time scales the different disturbances occur. Figure 2.13 presents the typical disturbances that can occur in accelerator magnets and their related time scales and energy densities.

Disturbances in the ms range and below are transient ones, and longer ones are of continuous character. For example, wire motion, flux jump, and insulation crack are point disturbances. AC losses and heat leaks are typically continuous distributed disturbances.

2.3. Engineering Methodology in Quench Protection

Analyzing quench requires a multiphysics approach. After simple, and often pessimistic, analytical paper-and-pen calculations have taken the quench

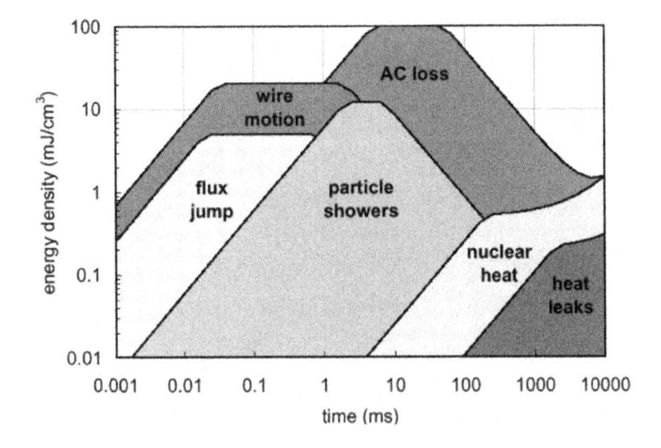

Figure 2.13. Overview of time and energy scales of different disturbances. *Source*: Reproduced from Bottura [10].

protection engineer to the ballpark, she has to start considering how severe the quench protection problem really is via numerical modeling.

The first thing in quench analysis is to consider which physics is relevant for reliably describing the quench event. We call this phase by the name *model*. After the appropriate model is composed, one considers *design*. Typically, a device's layout comes from another engineer with some constraints related to assembling the device and its intended operation, and the task of the quench protection engineer is to design the quench protection circuit with quench detection. For simplicity, we refer to this as quench protection, but the detection is included too.

The design typically requires an interplay with the whole electric circuit of the device: The desired operation needs to be achievable with adequate quench protection. To achieve that, the quench engineer comes up with a quench protection system and makes a simulation model representing it. Then, it is important to simulate particular quench scenarios, such as quench at nominal current and quench during loading the magnet. For these, one represents the modeling domain in a fashion required by the simulation tool and its underlying method. Then, one solves the equations governing the physics in the specific scenarios that are determined by the operation conditions and quench onset. The simulation results give feedback to the design phase and also to the model. Some results may reveal that the model was not adequate. After the final design, the device is to be constructed and tested. In the *experiments*, one replicates the operation conditions that were used in the simulations and records the output

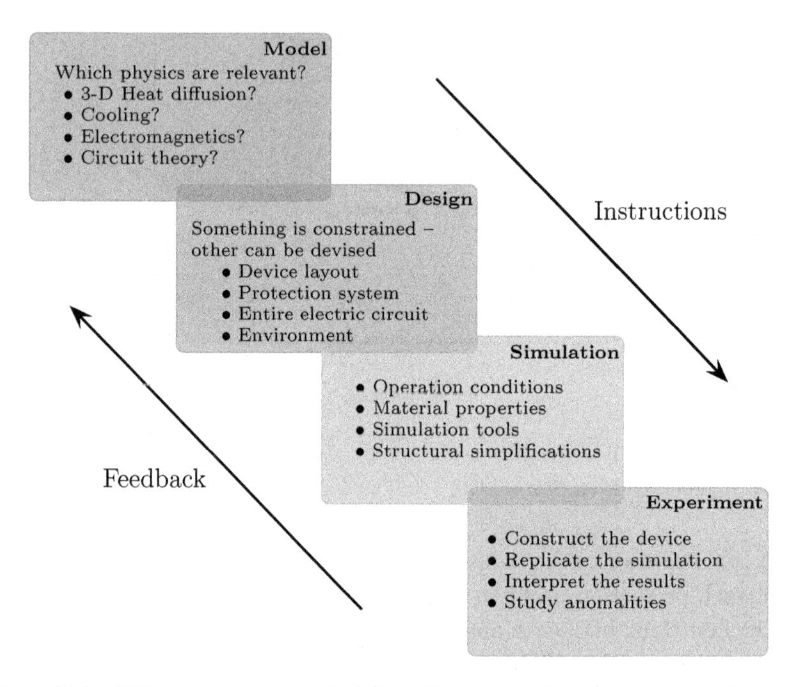

Figure 2.14. Different components that form the methodology for quench analysis and the two main workflows: instructions and feedback.

signals that can be compared against the simulated ones. Correspondingly, the feedback goes all the way from the experiments back to the model via the intermediate steps. Vice versa, the model gives input to the design, which influences the simulation, and finally to the experiment. The simulation governs the experiments because feedback is required for the other stages of the quench analysis methodology. However, often the designed device is not precisely implemented, and the simulations must be rerun to benchmark them against the experiments.

Figure 2.14 schematically presents the interrelated components: model, design, simulation, and experiment of quench analysis. Next, we look at the contents of each box in detail. We note that the feedback and instructions are between each box, not just between the model and the experiment.

2.3.1. *Model*

The model is about selecting the physics that the engineer deals with. It enforces, or allows, specific design decisions and guides later the simulation. This phase is of special interest if designs based on new principles are developed. However, in the conventional workflow, the model may be self-evident to the engineer and does not need much attention. But before we get to the physics, let us start from the primitive needs.

The primary target of quench analysis is to ensure that the device can safely go through a quench event. Therefore, the chosen model needs to be able to answer a few carefully selected questions. Depending on the complexity of the device, these questions somewhat differ. For example, in small devices, the stored energy may not play a role. In large magnets that are connected in series, the stored energy poses challenges to, for example, bypassing the magnet that is quenching. Here, we focus on a single coil in such a way that, without loss of generality, the methodology can be extended to a more complicated situation or to other devices.

To get started, one needs to be able to answer, at least, the following questions:

(a) How rapidly does the current need to be brought down in order not to overstep the maximum tolerable temperature? A short answer is that the current integral

$$\int_0^\infty I(t)\mathrm{d}t \tag{2.12}$$

plays a crucial role. Here, the time instant 0 s refers to the quench onset, not to the time of detection or activation of the protection system.

(b) If the coil is not self-protective or if the energy extraction via a dump resistor is not adequate to protect the coil, how large a fraction of the coil needs to be quenched and how fast?

Clearly, the first answers to these questions do not require considerations where the heat diffusion is analyzed in the coil volume.

To consider the rapidity required for the current decay, one can first neglect the heat diffusion completely and compute the *MIITs*[3] of the cable, or conductor, $\Gamma(T)$ [71]:

$$\Gamma(T_{max}) = A^2 \int_{T_{op}}^{T_{max}} \frac{\gamma C(T)}{\rho(T)} \mathrm{d}T, \tag{2.13}$$

[3]The MIIT abbreviation comes from mega (M), current (I), current (I), and time (T). If the integral equation (2.12) results in $5 \cdot 10^6$ A^{2}s, the MIITs equal 5.

with T_{max} being the maximum tolerated temperature. Here, all the material properties are those effective for the coil unit cell, and A is its area. This integration gives the maximum value that Equation (2.12) can take. The instruction for the design phase is then to consider if the energy extraction can be made in such a way that the current decay integral remains below the MIITs at T_{max}.

If the energy extraction is not enough, a non-negligible part of the energy must be dissipated inside the coil. For this, one needs to consider what degree of resistance must be achieved in the coil. This does not address the heat diffusion yet, but macroscopic considerations are enough. However, the input from these considerations to the design is crucial. For example, if quench protection heaters are used, the surface area that they must cover can already be estimated.

After one knows how severe the problem is and what kind of design is needed for the protection, one starts to approach the modeling of particular quench events at particular operation conditions. All of the previous analyses can be carried out under nominal operation conditions.

To reveal the challenge of modeling, we look at the spectrum of physics involved. Perhaps the most crucial role is played by the heat diffusion in the superconducting device. Furthermore, superconductors are very finely structured components. One necessarily needs to do some homogenization approximation, the structural simplification shown in Figure 2.14, to make the simulation feasible with larger entities emulating the finely structured device. Whether this is a task to be done during modeling or simulation is up to the engineer's standpoint. This is true in many issues that we discuss here.

When the heat flow reaches the interface between the coolant and the superconductor, thermal hydraulics become relevant. Very often, one considers the coil's surface to be adiabatic and neglects the cooling during the quench. However, this is not appropriate in all kinds of devices. In magnets made from cable-in-conduit superconductors, this needs to be considered.

The coolant–magnet interface does not represent the only important connection of the coil with the surroundings. The support structure also offers thermal stabilization. Furthermore, during the current decay, the magnetic field in the support may change rapidly. This causes eddy currents in the conducting parts, which also contribute to the energy extraction. The heating in the support structure may cause a quench-back, i.e. ignite additional quenches at various locations on the coil's surface or in its interior. This can be especially meaningful in conduction-cooled coils, where high-purity copper is in close contact with the superconducting device. In order to take this into account, one needs to consider magnetoquasistatics. Of course, the same effect is also present in superconductors when AC losses originate due to the decaying magnetic field. Correspondingly, the simple approximation of heating losses arising only locally from the transport current may not be enough. Furthermore, these methods can be utilized to protect the magnet [45].

When the quench originates, there is some diffusion time for the current to homogenize in the stabilizer. During this time, the current density in the stabilizer cannot necessarily be expected to be uniform. For coils made of single strands, this does not play a role, as noted in Section 2.1.1, but for large magnets, especially those made from jacketed cables [5], the influence can be substantial not only on local heat generation but also because this bursts the normal zone propagation. In particular, with models based on field theory, the fine cross-sectional structures of cables made from LTS conductors make a faithful analysis of these phenomena time-consuming and complicated. One possibility is to support the heat generation models with experimental data.

The technical superconductors and the devices made of them consist of very different materials, characterized by different thermal expansion. Because of this, stresses originate in the system during cool down, possibly including high-temperature heat treatment [1]. In the case of a magnet, this stress state considerably changes during the magnet powering due to the Lorentz forces. The stress has an influence on the local critical current density. In quench simulations, this is very often neglected.

Especially in HTS magnets, the normal zone propagation is slow if the temperature margin, or equivalently the enthalpy margin, is high. Therefore, the hot-spot temperature may be very localized. This can cause high localized thermal stresses, which can lead to magnet damage. In coated conductors, the risk is in the delamination of the layers [83]. The maximum tolerable hot-spot temperature may not be the only limiting factor for the quench, but voltages (internal and voltage to ground) as well as the quench stresses need to be considered too [61, 78, 79].

Because many superconducting devices operate with currents, the modeling of the electric circuit is, of course, very important too. Though essential, often, but not always [80], the electric circuit model is merely a small increment to the otherwise tedious modeling problem. Electric circuits consider superconducting devices via macroscopic quantities in a lumped model. Thus, one needs to be able to extract the appropriate resistances and inductances of the device during, or before, the simulation of the quench propagation.

As briefly described, the quench is a very multiphysical problem. The intensity of the coupling of the physics is also very important. For example, the stress distribution originating from nonhomogeneous temperature within the system can be analyzed after the quench simulation is performed [78]. When modeling decisions are taken, including which physics to include and how to represent the device in the modeling software, one should lean on the pessimistic side. The high complexity and cost associated with constructing superconducting devices does not allow one to break them due to design flaws related to quench. Naturally, the design is not the only thing that can go wrong and cause delays in commissioning and major reconstruction [49]. Also, one should pay a lot of attention to experimentally verifying the designs.

It is very important to find the adequate level for modeling. As well as undermodeling, which does not adequately predict what happens during the quench and may result in an overly complicated protection design, overmodeling should be eschewed in every step too. It prolongs the engineering design and therefore incurs additional costs. A model is very important to design because it gives instructions on what must be considered. On the other hand, design, simulation, and especially experiments give feedback to model.

2.3.2. *Design*

The core of design is the layout and drawings of the protection-related components. Design also represents an interface to the other engineers working with the same system. The particular protection system design depends naturally on the other parts of the system, as it is an interfacing duty. Next, we begin from this viewpoint.

The main input to the quench protection system design is the device, including the auxiliary components, such as the support structure. Naturally, the quench analysis can also give feedback to the other parts of the system. For example, once the required current is analyzed from the stability point of view, there might be a new requirement for the amount of stabilizer in the superconductor. The nominal operation conditions play the most important role, but the protection system must be able to protect the device when it quenches at less demanding operation conditions too. For example, a magnet can naturally quench during ramping up. It is not automatic that if a protection system is good enough at high currents, it will be good enough at low currents too. This is especially the case if a quench must be externally induced on a large fraction of the coil. When the temperature margin has increased, the energy of the quench ignition system, such as quench protection heaters, may not be enough.

Design should always reflect the model. Only those things are designed and described in the design report that are considered important. To take a very particular example, let us consider large magnets where quench protection is implemented with quench protection heaters. If one considers, for example, from previous experience, that the adhesive, which has earlier been utilized to attach the heater traces to the coil, does not have an influence on the modeling, one should not pay special attention to it in the design of the device. On the other hand, if the influence of the adhesive on the performance of the protection heaters is studied, one should especially focus on questions related to that: how to spread the glue, how to control its thickness, how to prepare the surfaces, how to cause the pressure contact, etc. This is important in the scientific method. One needs to deliver results that are usable in future engineering designs. However, in many scientific collaborations, overdoing is a risk. One can focus on small details as long as one wants, but that may not be useful from the point of view of the whole project. From a quench engineer's point of view, it is important to identify what is expected and then deliver the solution, which is often the design based on reliable modeling and other references.

In the construction drawings, it is necessary to include everything that is needed to construct the device. This does not mean necessarily that everything is or even could be simulated. Simulations only assist the design and help understand the experiments. In modeling, one necessarily simplifies the design and, additionally, considers whether some parts of the model have such a small influence on the simulation results that they can be neglected. More details mean longer simulation times as well as prolonged construction of the modeling domain. In case of 3D modeling, the modeling domains should be simplified in every possible way but without losing the reliability of the simulation results.

In design, the most crucial choice to be made is between different protection options. Protection methods can be divided into two classes: *active* and *passive*. Active protection needs quench detection and activation of the protection system and, correspondingly, can fail if the quench is not detected or if the protection system triggering fails. Passive protection, on the other hand, works without quench detection. However, passive protection is rarely enough. Passive protection methods typically require fast development of resistance and corresponding current decay. Often, passive and active methods can be combined. For example, subdividing a coil into diodes is a passive method, whereas bypassing the current source with a switch requires quench detection and is an active method.

If external energy extraction via a dump resistor is possible, that is the easiest option to implement quench protection and also the most economical because heat is not dissipated into the cryogenic atmosphere. However, in magnets that have large stored energies (in the order of MJ), this is not typically possible, and active initiation of quench in a large coil volume is needed. One option to ignite such a widespread quench is to utilize quench protection heaters. These are stainless-steel strips that are mounted on top of the coil's surface and powered via a capacitor bank in the case of a quench. They heat up the coil and cause rapid increase in resistance, which reduces the current decay integral equation (2.13) and, consequently, the hot-spot temperature. The most important protection methods and their operational principles are summarized as follows:

- Dump resistor: It extracts the energy to an external dissipative circuit.
- Quench protection heaters: It causes a widespread quench in the magnet and dumps most of the energy in the magnet.
- Quench-back and/or secondary coil: It utilizes a time-varying magnetic field to cause losses either in the magnet or in the support structure to extract energy and to rapidly spread the quenched volume.
- Subdivision: It bypasses the quenching part of the coil from the electric circuit in order to have a smaller, or zero, current there, which reduces the maximum temperature.
- Coupling loss induced quench (CLIQ): It connects an auxiliary circuit to the coil, which causes part of the transport current to oscillate in two or more parts of the coil. These oscillations cause AC losses [27] (typically interfilamentary or interstrand coupling losses in the case of low-temperature superconducting strands or Rutherford cables made of them) that induce a widespread normal zone [45].

In addition to the design of the protection, the quench detection has an important role. Because quench is a process where resistive voltage begins to develop, the most obvious detection method is to monitor the voltage. However, if the quench originates during a current ramp, it can be difficult to detect the quench early enough, especially due to the associated inductive voltage. The contribution of the time between the quench onset and the start of the current decay to the current decay integral can be substantial if the quench detection does not work adequately. Therefore, monitoring only the terminal voltage of the device is not necessarily enough. In case of a magnet, a voltage tap that splits the magnet inductance into two is useful for canceling the inductive voltage component. However, two symmetrically originated simultaneous quenches cannot be detected only by this way.

In the case of a magnet, other options for the quench detection include the so-called quench antenna [40], which monitors the acoustic emission from the coil and the positioning of a pick-up coil array into the coil bore. At the quench onset, the current density distribution changes due to a local increase in the resistivity in the superconducting filaments, which also changes the magnetic flux density and, consequently, the Lorentz force distribution. The first can be detected with pick-up coils via an induced voltage; the second typically causes small vibrations due to the variation in the stress distribution, which can be detected acoustically.

Important times in the quench detection are the *delay time*, i.e. how soon the quench is detected after it happens; the *validation time*, i.e. the time that the system takes to decide whether a quench has actually originated; and the *activation time*, i.e. the time to activate the protection system. The delay time greatly depends on the particular device. The validation time is a design decision. A too small validation time can cause false detections, whereas a too high one increases the hot-spot temperature. The activation time depends on the protection system electronics. With currently available fast semiconductor-based power electronics, it is typically in the order of few ms.

2.3.3. *Simulation*

In brief, simulation is about verifying the feasibility of the design and its optimization based on the model. The simulation part represents the computer-modeled experiments, which should be cheaper and faster to perform than the actual experiments. With simulations, the engineer can justify the design decisions to the other team members before the experiments are conducted. The simulation phase is merely a tool that aids us in understanding the decisions related to the design. Therefore, though essential, simulations are often considered secondary — when the final marketing value of the device is considered. Thus, for an engineer, it is difficult to concentrate only on simulations, but knowledge of design, experiments, and commissioning is valuable too. Next, we discuss simulations by considering how they aid the design.

Various simulations aid the entire design process in different phases. For example, if the simulations indicate too high a hot-spot temperature in a magnet, either the magnet design (typically the amount of copper in the cable) or the protection system must be modified. The link to the model is crucial because the model tells which physics must be simulated. This does not mean that a single simulation solves all the physics, but the problem can be divided into adequate sub-problems that together deliver the contribution of the simulations.

Typically the most tedious work in starting a superconductor simulation is to find the relevant material properties. This is not restricted to properties such as the heat conductivity and capacity; the critical current surface is especially important too. Often, when devices including new materials are designed, a lot of experiments are required to analyze the critical current surface. Furthermore, as the normal operation conditions are the most important for the engineering design, it is sometimes secondary to do the critical current characterization at elevated temperatures. However, this is necessary for reliable quench simulations and must be kept in mind at every phase of each project. Running the simulations may also be time-consuming, especially if a 3D modeling domain is considered and parametric simulations are run. To cope with the tedium of finding the material properties, many laboratories keep their own cryogenic material property libraries, and there are also commercial ones available. CryoComp is a commercial library [15], whereas MATPRO [39] is an example of a documented material property library maintained by an accelerator laboratory in Italy.

First, it is important to know how much stabilizer is needed in the conductor or cable that is utilized to construct the device. Analytical computations, such as the MIITs, help to get to the ballpark, and they also aid the design work in the beginning, but later, numerical simulations may be necessary in order not to leave too much margin for the design. Too much margin, or too conservative modeling, may mean that something is not considered feasible even though, in reality, it could be done. When the design proceeds, the quench protection engineer typically needs to return to the conductor-level analysis to check if the requirements are met again.

From the simulations of the conductors, one proceeds to the simulation of the device and, possibly later, considers the interplay of several entities, for example if different magnets are connected to the same electric circuit. In a simulation at the level of a device, it is impossible to consider all the details. Therefore, one needs to simplify the situation and scrutinize a substitute that is simplified enough to make the modeling feasible but still represents the original device with adequate accuracy.

The model determines which simplifications are made. Common simplifications in the modeled physics of the quench protection system, or in the quench analysis, are

- neglecting cooling by utilizing adiabatic boundary condition,
- modeling cooling with Dirichlet boundary condition (fixed temperature at the boundary),
- neglecting heat conduction,
- neglecting transverse heat conduction,
- neglecting thermal and electric contact resistances,
- neglecting all the structure outside the superconducting assembly, such as the magnet,
- neglecting the discontinuity in the heat capacity at the superconducting–normal-conducting transition,
- neglecting the device's internal structure and considering average volumetric material properties,
- neglecting the current diffusion at the quench frontier from the superconductor to the stabilizer,
- using a scaling factor to consider heat generation in the current-sharing region,
- neglecting all magnetoquasistatic effects,
- modeling magnetoquasistatic effects with analytic formulae,
- neglecting the effect of a stress state on critical current density or assuming constant strain over the winding,
- neglecting the effect of nonlinear magnetic materials, and
- considering homogeneous field distribution in the modeling domain.

Naturally, some of these simplifications are exclusive, and particular typical situations combine some of these assumptions.

Cooling, clearly, plays a very important role in superconducting devices. However, its modeling is non-trivial, whether the modeling includes the mass transfer in the coolant or only a non-homogeneous and nonlinear Neumann boundary condition representing the cooling heat flux. In impregnated windings, it is often reasonable to consider an adiabatic situation. This is naturally not the case if the cooling phase is of special interest. When the performance of quench protection heaters is modeled, the utilization of a Dirichlet boundary condition on the magnet outer impregnation and coolant surface is justifiable because it is a pessimistic estimate.

In the MIITs concept, all the heat conduction is neglected. In the presented 1D MQE simulations, the transverse heat conduction was neglected (see Section 2.1.1.2). Sometimes, a flat cable represents an important entity in the coil, and one neglects the heat conduction from a cable to another but not inside a cable. The greatest benefit in neglecting the heat conduction in a given direction is that it immediately reduces the dimension of the modeling domain. However, this may easily lead to too pessimistic an estimate in terms of the vulnerability to quench or to too optimistic an estimate in terms of the normal zone propagation velocity [69].

If the thermal and electric contact resistances are not neglected, the modeling domain cannot be reduced. Often, these resistances are neglected because their characterization requires tedious experiments, and still, it is difficult to know how well they generalize to other situations than the specific characterization experiment. Especially in superconducting magnets, the coil is under a heavy stress load and very tightly packed. Therefore, the contact resistances tend to be small. Thus, neglecting them is not necessarily a bad estimate. For example, in many quench protection heater simulations, the adhesive between the heater and the coil's surface is neglected. This means that the modeling is slightly optimistic. When this is combined with Dirichlet-type boundary conditions at the coolant–surface interface, which is a pessimistic estimate, the two effects somewhat cancel each other. When several optimistic and pessimistic simplifications are made, it becomes difficult to know the influence of each decision because the simplifications influence on different directions. Thus, even though the necessarily discrete large-scale measurement results can be compared with good accuracy to the corresponding simulation characteristics, it is not legitimate to draw too far-reaching conclusions from those simulation characteristics that are not measured.

Neglecting the structure outside the superconducting assembly means, for example, that one does not consider the heat conduction via the support structure or the eddy current losses due to the decaying magnetic field. Both the given examples are pessimistic. Then, if the design is pushed to the limits, there is actually some margin. In conduction-cooled systems, the superconducting device is typically well interfaced to a high-purity copper cooling path [55]. Then, the role of this thermal conduction path can be very substantial in causing a quench-back: Eddy currents are induced in the copper, which heats up, and the heat flows into the coil, causing a rapidly widely spread quench.

An individual superconducting strand can consist of a hundred thousand filaments embedded in matrix metal. Several strands can be assembled to form a cable. The coil can be wound from this cable, making the structure highly detailed. It is typically impossible to represent such a structure in a CAD program, not to mention its detailed discretization in a numerical modeling software. Consequently, one often neglects the detailed structure and averages the material properties. Sometimes, the insulation is represented between the cables, but the cables or wires are homogenized. These approaches can also be mixed [32]. All this, of course, depends on the specific device under study.

The idea of homogenization is that the specific heat is averaged, and as for the thermal and electrical conductivities, the materials are considered to be in parallel or in series, depending on the direction that is scrutinized. Therefore, some of the material properties are made anisotropic. If the structure is highly irregular, one can model the unit cell and correspondingly derive the effective properties, such as thermal conductivity [37] and permeability [66]. An example of using homogenization in the MQE and normal zone propagation velocity modeling was presented by Stenvall *et al.* [67].

Many effects are time-consuming to simulate with field models, such as the AC losses or the current diffusion at the quench frontier. Therefore, simplified cases can be utilized to solve simpler problems analytically, and based on experiments, the results can be scaled to the more difficult simulation problems. This is especially the case with magnetoquasistatic problems. Sometimes, a magnetostatic problem can also be simplified. If only a single turn of a coil is considered, one may be able to consider it under a homogeneous magnetic flux, though that is not the case in reality. Simulations with such simplifications may still be able to give enough feedback to the design. A similar approach can be taken if an iron yoke is utilized to burst the magnetic field. To control the produced magnetic field via the coil's current, the effect of the yoke is important. However, in quench simulations, the nonlinearity that the yoke causes to the inductance can be neglected if one begins from the same total energy.

In all the simulations, the simulation tools are of interest too. Many general modeling tools can be utilized to simulate basic problems in superconductors, but often, the particular scenarios require extraordinary approaches, and thus, many laboratories have developed tools for particular purposes [52, 58]. If particular purpose tools are implemented, one must ensure the continuity of the development and the frequent usage of the tools to make the economical investment in its development sensible and invest in expanding the user base of the software. Currently, the STEAM community promotes a unique approach to combine commercial tools within a tailored framework [8].

Naturally, the simulation and design can be coupled too in such a way that simulation proposes the design. If the design is formulated as an optimization problem, the simulations suggest a particular design. Thus, here also, the point of view of how model, design, and simulation are related is subjective. Next, we discuss the experiments related to quench protection.

2.3.4. *Experiment*

One can approach experiments from several perspectives. For example, first, the engineer wants to know if the simulations predicted the experiment results adequately. Second, the working group gets information on the successful construction and design. Third, some things cannot be simulated reliably, so necessarily experiments are required. Finally, the experiments, especially those performed on the entire system, contribute greatly to the credibility of the project — especially if a new future technology is studied.

Performing experiments for conductors and cables can be routine-like, but for large devices, such as magnets, specific test plans are of utmost importance. First, the devices are typically made of very important components, and second, their testing is expensive and time-consuming. The most important tests include the performance tests: Does the device perform as designed and modeled? Although quench protection is very important, it is useless if the device does not reach its target values.

Under nominal operation conditions, the superconductor typically carries a very high current density (even in the order of $1000\,\mathrm{A/mm^2}$). Therefore, failures in quench analysis can cause the device to fail in the first test. Thus, in the case of large-scale devices, one must first do many tests before trying to find out when the coil quenches first naturally. The electric integrity test is the first one that must be done. Undesirable short circuits must not exist in the device. Furthermore, the lumped parameters, the inductance and normal state resistance, must be checked. Small currents are adequate for these checks. Then, a relatively rapid current that ramps up to a safe current value and goes down must be measured in order to check the safe ramping of the magnet. A rapid predefined triangular ramp prevents the magnet damage in the case of an unexpected quench. After the triangular ramps are verified to be safe, a hold time for the current must be considered. Development of abnormalities, such as continuous point disturbances, must be monitored.

It is important to test the magnet quench protection system at low currents. The current must be taken to such a value that the quench is safe when only external energy extraction is utilized. A quench must be manually triggered in such a way that the energy extraction begins automatically regardless of the operation of the quench detection system. At the same time, one must monitor if the detection system works. If everything goes as expected, the test is repeated but without the predetermined energy extraction. After enough confidence with low currents is reached, including successful comparison to simulation results, the test progresses toward situations where the quench is not triggered manually. Then, everything relies on the automatized quench detection system. If the simulation results are reproduced in low-current experiments, the confidence in high-current experiments is gained. If this is not the case, one must rerun the simulations and possibly adjust the fitting variables to assure safe quenches in the short-sample limit also.

By positioning multiple potential taps in the coil and by monitoring their changes, the quench origin can be traced to certain parts of the coil. It is important to find out if the coil quenches at the critical point or elsewhere. If the quench occurs elsewhere after the conductor-limited quench plateau is reached, one must scrutinize why. Was the critical point computed erroneously or was some location damaged during the coil manufacturing? Also, often one uses a fit, or the so-called scaling law, for the critical current characteristic. One should check if this really reflects the reality at the operation point where the magnet quenched. In case of anisotropic conductors, such as REBCO-coated conductors, one should ensure that the conductor is from the same batch for which the fit was made. The variation in the critical current's angular dependence of $\mathbf{B}$ is very large between differently doped conductors [65].

The current decay curve can be compared most reliably against the simulation results. Measurement of the maximum temperature in the coil is very difficult due to the delay in temperature sensors. Typically, the readings of the temperature sensors increase even after the current has decayed to zero. Therefore, the temperature measurements only give lower boundaries for the hot-spot temperature. Recently, a new technique based on fluorescent thermal imaging for temperature measurement has been developed [30]. With such an approach, more timely temperature measurements can be performed on the surface of a sample.

Destructive testing is also important during the R&D phase. In a destructive test, one searches for the maximum value of the current decay integral that can be tolerated in a coil. Then, a quench is triggered manually to know the time of origin, and the validation time of the detection system is increased step by step to increase the current decay integral. If the quench detection is based on a voltage tap measurement, the validation time means the time that the voltage must be above a detection threshold before the protection system is activated. This provides a way to delay the quench detection and increase the current decay integral. Between two delayed tests, one checks if the coil still reaches the conductor-limited quench. When this does not occur anymore, the maximum tolerable current decay integral has been reached. This gives the absolute upper bound that a quench simulation can tolerate for the current decay integral. Correspondingly, in the simulation, a maximum tolerable hot-spot temperature can be determined, but this value is very difficult to confirm reliably in an experiment.

Naturally, testing gives input to the simulation either to confirm that the device performs as expected or to discard the simulation result. Also, it confirms, or discards, the expectations for safety margins and the feasibility of the protection system. The feedback goes all the way back to the modeling phase too. Perhaps, the considered physics were too simplified or the same results could have been achieved with a less rigorous approach. Experiments should never be forgotten, especially when new modeling tools are developed, as the benchmarking of the tools is invaluable to all future work.

2.4. Numerical Modeling of a Quench Event

Perhaps, the most tangible task in the simulation phase is that of the modeling of a quench event: *the quench simulation.* In quench simulations, one studies the temperature evolution in the superconducting device before it is fully de-energized. The principal target of the simulation is to validate, by modeling, that the protection system design allows a safe quench. The protection system and circuit model are also included in the simulation. The most important output is the current decay curve, which can be easily compared to the corresponding experiment, and the determination of the hot-spot temperature, which is difficult to investigate experimentally with high accuracy. Other important characteristics can include, depending on the particular scenario, the time to detection, the time to activation of the quench protection system, the amount of externally dissipated energy, and the maximum voltage to ground. Next, we consider a general numerical quench simulation, with emphasis on the instructions required to perform particular simulations. Then, we present a simulation case that was carried out in an HTS accelerator magnet R&D project.

The computer-assisted modeling of the temperature evolution during a quench has traditionally been divided into two very different approaches. Both approaches have been developed originally for LTS-based magnets, where the temperature margins are modest and current densities high. These mean a high normal zone propagation velocity.

The first approach was developed before it was feasible to solve the heat diffusion equation in a 3D modeling domain using a computer with adequate accuracy and time. Wilson [75] proposed to utilize normal zone propagation velocities to study the growth of the normal zone and adiabatic consideration for the hot-spot temperature. In this approach, one assumes that the normal zone begins from a specific location and extends according to the normal zone propagation velocity. However, the heat does not diffuse, but the hot-spot temperature is computed adiabatically. To initiate the simulation, one begins with a normal conducting, i.e. quenched, ellipsoid and, at every time step, adds a new isothermal shell on top of the previous normal zone according to the normal zone propagation velocity. Then, because the current is known, the temperature increase in each shell and, correspondingly, the normal zone resistance can be computed. The heat generation can be tweaked using analytical formulae for the AC losses. Also, the normal zone propagation velocity can be computed analytically from several available expressions [74, 75] or input from known experimental data. Tuning this method for particular cases is relatively easy after some current decay curves are known at low, i.e. safe, currents. Either the normal zone propagation velocities or the heat generation can be scaled. Also, multiple coils can be studied, and coupling with a circuit model is straightforward via self and mutual inductances and normal zone resistance. Correspondingly, multiple quench origins can also be utilized. After Wilson's pioneering work, this work has been continued by others [48].

The alternative approach does not depend on the normal zone propagation velocities, which can be merely post-processed from the simulation result. Then, one solves directly the heat diffusion equation within the magnet volume. Typical numerical approximate solution methods rely on discretizing the volume of the modeling domain in order to represent the unknown information, i.e. the temperature distribution from which everything else can be derived, with a finite amount of information. The most popular solution methods are finite element and difference methods [18, 19, 22, 63, 64, 68, 77]. Some of these utilize commercial software for implementing the simulations, while some include tailored tools.

Next, we focus on a homemade tool, QueST (Finite Element Method based **Que**nch **S**imulation **T**ool), which has been developed to study HTS magnets at Tampere University of Technology, on top of the GMSH mesh generator with `C++` programming language [24, 32]. The tool utilizes FEM for spatial discretization and also includes a module to solve the magnetostatic problem. The latter is important because the critical current depends on the temperature and the magnetic flux density, and it determines the heat generation in the heat diffusion equation. To keep the discussion at a level that is relevant to an engineer working with a commercial general simulation tool, a tailored tool, or even developing a new tool, we present the input that an engineer wants to consider in a simulation and the output she wants to eventually get.

Then, we go a bit into FEM without deriving the weak formulation, but we merely show the equation system and the representation of unknowns that one eventually solves. As it is well known, one cannot solve for functions in a computer, but only real numbers, or floating point numbers — to be precise. Therefore, we consider how the unknown fields are represented in a FEM software. We also consider the temporal discretization that is external to the FEM matrix assembly.

At the end, we consider the quench simulations. First, we discuss how to ignite the quench in a superconducting magnet. The quench simulation in a 3D modeling domain can be very time-consuming. However, typically, only part of this modeling domain is of interest and has notable influence on the output characteristic. Especially, when one simulates HTS magnets, the quench may be very localized, and therefore, one can leave out a large fraction of the modeling domain from the simulations. At the end, we also present a full-fledged quench simulation for an R&D magnet. The simulations presented here are from different R&D phases of an HTS magnet project. Therefore, they cannot be directly compared, but only the overall conclusions from each investigated case should be considered.

2.4.1. *Input and output of a quench simulation*

Input and output should be considered first when starting to describe the problem at hand to the computer. Input refers to things that the engineer needs to address in the modeling. These can be divided into two specific classes: (i) inputs that make this particular simulation different from other ones and (ii) inputs that are common to all the simulations of this device. Item (i) refers to setting operation conditions, whereas item (ii) refers to the equations that are solved. Output, on the other hand, means the feedback that the simulation gives to the design. If one is interested in deriving the normal zone propagation velocities in the post-processing, the tool needs to have this possibility. Therefore, the input and output are related by the selection of the computational tool.

One very important input is the magnetic flux density distribution. First, we discuss that, then operation conditions and other parameters that are often worth studying. Finally, we discuss the post-processing data.

2.4.1.1. *Magnetic flux density distribution*

Magnetic flux density distribution as a function of current is typically required by the heat diffusion equation solver for several purposes: most fundamentally for the critical current but also for the magnetoresistance of copper that has influence on its resistivity and heat conduction [35]. In LTS magnets, where the critical current does not depend on the orientation of $\mathbf{B}$, $||\mathbf{B}||$ suffices, but in HTS magnets made of REBCO-coated conductors, one needs the angular dependence of $\mathbf{B}$ [65].

To solve the magnetic flux density distribution, one solves a magnetostatic problem. In case of a magnet, it is typical to consider the homogeneous current distribution in the magnet cross-section, or in the cables, and neglect the screening currents. If there are no materials that magnetize, such as iron, the permeability is that of vacuum everywhere, and one can solve the magnetic flux density with unit current and, consequently, only scale the solution by the actual current, as shown in Section 2.1.2. Next, we will look at what kind of partial differential equation one solves in the magnetostatic problem. However, the problem can also be approached in other ways: The magnetic flux density can be directly computed from the Biot–Savart law [14]. On the other hand, this direct computation does not necessarily make the task easier.

In a magnetostatic problem, one finds such a magnetic field $\{\mathbf{B}, \mathbf{H}\}$, where $\mathbf{H}$ is the magnetic field intensity, which satisfies the defining properties of magnetostatics: Gauss' law for magnetism

$$\nabla \cdot \mathbf{B} = 0, \tag{2.14}$$

Ampère's law

$$\nabla \times \mathbf{H} = \mathbf{J}, \tag{2.15}$$

where $\mathbf{J}$ is the known current density distribution, and the constitutive law of magnetic field

$$\mathbf{B} = \mu(\mathbf{H} + \mathbf{M}), \tag{2.16}$$

where μ is the permeability and $\mathbf{M}$ the magnetization [14]. Typically, $\mathbf{M}$ refers to the magnetization of the permanent magnets, and in this work, $\mathbf{M} = \mathbf{0}$ holds. However, if additional equations are considered for the screening currents, their influence can be taken into account with $\mathbf{M}$. Other constraints may be posed via boundary conditions. However, boundary conditions can also be utilized to take advantage of discrete symmetries in particular modeling methods or to model a field that vanishes far away enough — to suppress the modeling domain.

When we search for the magnetic vector potential $\mathbf{A}$ that results in $\mathbf{B}$ as

$$\nabla \times \mathbf{A} = \mathbf{B}, \tag{2.17}$$

we can eliminate Equation (2.14) because $\nabla \cdot \nabla \times \mathbf{F} = 0$ holds for all vector fields $\mathbf{F}$. Then, we can substitute the constitutive law in Equation (2.16) with Equation (2.15), and we get a second-order partial differential equation to solve:

$$\nabla \times \mu^{-1} \nabla \times \mathbf{A} = \mathbf{J}. \tag{2.18}$$

This is the $\mathbf{A}$ formulation of the magnetostatic problem.

Curl ($\nabla\times$) is a linear operator, and if μ is linear as well, the problem is linear and can be solved with unit current corresponding to a specific $\mathbf{J}$. One gets for this a $\mathbf{B}_{unit}$ from the solved $\mathbf{A}$. Then, during the simulations, one can just scale the unit magnetic flux density distribution by the operation current as

$$\mathbf{B} = I_{op} \mathbf{B}_{unit}. \tag{2.19}$$

Then, the $\mathbf{B}$ distribution can be considered as an input to the quench modeling problem. Otherwise, if nonlinearities and hysteresis are considered, $\mathbf{B}$ must be solved simultaneously with the heat diffusion equation.

In QueST, one solves $\mathbf{B}_{unit}$ in the modeling domain with FEMor imports it from another software in a GMSH post-processing file format. In the case of coated conductors, one needs the parallel and perpendicular flux density components due to anisotropic critical current. Determining them is not trivial, for example, for accelerator dipoles with flared ends. However, tools specialized in computation of critical currents in these coils have this information already available, and that can be imported to QueST [81].

2.4.1.2. *Operation conditions*

Another very important input considers the operation conditions. Even though quench analysis is the most important at nominal operation current, one cannot forget the low-current protection as well as the quench at I_{max}. At low currents, the temperature margin is higher, and consequently, for example, quench protection heaters may not be adequate to quench a large fraction of the coil in the case of a spontaneous quench. Magnets are typically also tested above nominal operation conditions if possible.

In the case of conduction-cooled R&D magnet systems, one does not necessarily know beforehand the specific operation temperature at which the magnet is tested. Therefore, once the protection system design is finished and its feasibility is verified through simulations at nominal conditions, it is good practice to study other temperatures too. In the case of forced-flow-cooled or non-impregnated magnets, one should study the effect of heat transfer to the coolant. Different heat fluxes can be utilized in the simulations to gain confidence in the design.

Often, in homemade tools, the input data is controlled by specific input files. Many commercial software, on the other hand, allow importing parameter lists via graphical user interfaces. Because quench simulations are often time-consuming, it is important to pay attention to storing the input data with the results in order to be able to later reopen the simulation cases for examination. The downside of commercial software is that often there are some inconsistencies between different versions of the program. That makes it difficult to return to older simulation cases. In the case of homemade codes, it is easier to keep track of the revisions of the code, especially if programming is tracked with a version control system [26, 70].

In QueST, a specific `JSON`-formatted file is given to the command-line-executable solver. This file is used to map the variables directly onto a specific object in the program. Then, these variables can be utilized everywhere in the code. Some variable names are fixed, such as the operation temperature, but some can be freely introduced, such as external lumped parameters. This `JSON`-formatted file also prevents one from compiling the software if only minor modifications to the operation conditions are made, such as when the operation current is changed.

2.4.1.3. *Post-processing data*

The solution is the main reason for the simulations to be performed in the first place. But what is the solution from a quench simulation? The answer to that question is not simple. There are multiple solutions or outputs from which an engineer can draw conclusions.

The most important post-processing data consider global quantities as a function of time. These are, for example, the hot-spot temperature, normal zone resistance, size of the normal zone, and circuit currents. Because simulations are time-consuming, it is also important to have access to these parameters during the simulation. That gives the engineer the option to suspend the simulation in the case of abnormal behavior. For example, if one aims to quench a magnet with a heater, but it does not quench, it is not valuable to continue the simulation. As shown in Section 2.1.1, it is not always easy to determine if the system quenches or not. In simulations, one must also be able to position potential taps in the modeling domain to investigate voltages over specific sectors. These characteristics can then be compared to experiments. The voltage curves can be recorded during a simulation or afterward in the post-processing of the temperature distributions. Post-processing is preferred for wider possibilities to determine what to consider; online derivation, on the other hand, helps to understand the simulation even while it is ongoing.

The evolution of the temperature distribution in the magnet is also an important result. The propagation of the normal zone can be determined from the temperature distribution, and it should be saved at multiple time instants. Furthermore, these distributions can be utilized as an input for thermal stress computation by a solid mechanics engineer. An appropriate interval is 0.1 ms before the maximum temperature is below T_{cs} and about 1–5 ms after that.

Often, the quench simulation is terminated if the hot-spot temperature has risen to a sufficiently high temperature (e.g. 400 K) or if the current has decayed to 10% of the nominal current. The latter means, roughly, that the heat generation is only 1% of the maximum.

During the simulation, QueST displays the operation current, maximum temperature, resistive voltage, and volumetric fraction of the normal zone. All these are written to `.m`files and MATLAB can be used for visualization.

The temperature distributions are saved in GMSH-style post-processing files at specific time steps. The time steps can be determined from the input `JSON`-formatted file. GMSH can be used for post-processing the temperature distributions. In principle, any tweaks can be added to the code by programming and then compiling.

2.4.2. *Spatial and temporal discretization in a FEM based tool*

The finite element solution of a problem relies on representing the field quantity to be solved as an element in a predetermined finite-dimensional vector space where the elements are functions. A mesh is generated in the modeling domain Ω to fix the basis for this space. Then, the task of the problem is to find coefficients for these basis functions. The mesh offers spatial discretization for the solver. A denser mesh means more basis functions than a sparser one. Consequently, the mesh should be made dense in the regions where the solution changes rapidly in space.

Because the heat diffusion problem requires integration in time, temporal discretization is needed too. The nonlinearity of the problem encourages us to utilize an implicit backward differential formula with adaptive time stepping for the time integration. Those kinds of numerical solvers are available for homemade tools in the open-source package, Sundials [53]. However, in case of a quench, the problem is very time-consuming, and if the adequate time step is known beforehand, it can be more efficient to utilize simpler methods, such as Runge–Kutta or a mix of implicit and explicit Euler, where less attention is paid to the computation of error estimates. Even with such an approach, an adequately accurate solution is achievable.

Next, we detail spatial and temporal discretizations from the FEM and QueST perspectives. A further improvement in the quench modeling software is the utilization of adaptivity in spatial mesh as well as local time stepping. With local time stepping, one can get a high resolution of the solution at the quench frontier. On the other hand, a high resolution can be achieved with long time steps in locations where no heat generation occurs. However, these topics are still under development for superconductor-modeling-tailored software [72].

2.4.2.1. *Spatial discretization*

In order to be able to utilize FEM, one needs to discretize Ω in space. This means splitting Ω into polyhedra, which then form the finite element mesh for Ω. QueST relies on simplicial elements,[4] with a slight extension to be able to include other polyhedra.

In FEM, one attaches to first-order simplices one basis function per node — the barycentric basis function N_i. The basis function gets a value of 1 in a given node and 0 in others and linearly varies in between. Thus, if points $\mathbf{r}_j$ determine the nodes of a simplex, $N_i(\mathbf{r}_j) = \delta_{ij}$ holds, where δ_{ij} is the Kronecker's delta. This relation determines uniquely the linear basis functions. In the elements where node i does not belong to, N_i vanishes. The restriction of the support of the basis functions allows sparse system matrices. However, one can also attach more than four basis functions to an element — in the case of a tetrahedron having one basis function per node. In the case of second-order simplices, one basis function is attached to the center of each edge too. Then, $N_i(\mathbf{r}_j) = \delta_{ij}$ holds for all $2(n+1)$ points in an n-dimensional simplex, making the basis functions quadratic.

In the case of an unknown vector field, such as $\mathbf{A}$, one must have basis functions for a vector field. These can be either tuples of N_is or vector fields attached to the edges or faces of the mesh. In the case of $\mathbf{A}$, one can utilize the edge elements [6] to span the approximation of $\mathbf{A}$. Naturally, these concepts generalize to various element shapes (even to isoparametric elements [76]) and higher-order basis functions.

Because the solution to the original partial differential equation lies typically in an infinite-dimensional function space, an exact solution cannot be provided by a computer. Therefore, the solution one gets is an approximation. The FEM finds the solution from the subspace, spanned by the basis functions that are determined by the mesh, in a way that the energy norm of the approximate solution and the actual solution is minimized [12]. Thus, it finds the projection of the actual solution from the subspace when the metric for the projection is induced by the energy of the problem. This is a well-accepted criterion, which has made the FEM very popular for field computations. From now on, we do not make a difference in notation if we are discussing the approximate solution of the field problem or its actual solution. Also, we focus on solving the heat diffusion equation, where the unknown can be represented with scalar functions.

In the heat diffusion equation, we need to span the space for temperature T with functions, i.e. in case of first-order elements, those attached to the

[4] An n-dimensional simplicial element, a simplex, is a polyhedron with $n+1$ nodes. Thus, in 2D, it is a triangle, and in 3D it is a tetrahedron.

nodes of the mesh. Thus, we have

$$T(\mathbf{r}) = \sum_{i=1}^{\#\,\mathrm{DoF}} T_i(\mathbf{r}) N_i, \tag{2.20}$$

where $\mathbf{r}$ refers to a point in the node. Therefore, T_i is also the temperature in node i. Next, we put all the T_is into a tuple $\mathbf{T}$:

$$\mathbf{T} = \left[T_1\ T_2\ \cdots\ T_{\#\,\mathrm{DoF}}\right]^T. \tag{2.21}$$

According to the weak formulation of FEM, it is straightforward to represent Equation (2.1) as

$$M \frac{\partial \mathbf{T}}{\partial t} = A\mathbf{T} + \mathbf{Q}, \tag{2.22}$$

where M is the mass matrix, A is the stiffness matrix, and $\mathbf{Q}$ is the load vector. Here, M, A, and $\mathbf{Q}$ depend on T and $\mathbf{B}$ as well as the operation current. $\mathbf{Q}$ captures the role of heat dissipation and boundary conditions.

In a FEM software, it is straightforward to include domains with different material properties. However, the meshing should be done in a way that the element boundaries also respect the domain boundaries. Therefore, the fine structure of the modeling domain also causes a dense mesh. An important option to bypass this is to utilize effective, almost necessarily anisotropic, material parameters, as in Section 2.1.1. This is also the approach QueST utilizes, with the possibility of including finely structured domains as well.

2.4.2.2. *Temporal discretization*

To solve the ordinary differential equation (2.22), one needs time integration. For a computer solution, temporal discretization is required. In QueST, we use the generalized Euler algorithm [34, p. 205], with parameters θ and Θ that tune how implicit ($\theta = \Theta = 1$) or explicit ($\theta = \Theta = 0$) the method is:

$$\mathbf{T}_{k+1} = \mathbf{T}_k + hf\left(t_k + \theta h, \mathbf{T}_k + \Theta(\mathbf{T}_{k+1} - \mathbf{T}_k)\right), \tag{2.23}$$

where k refers to the time step and h to the step size. f is the function to be integrated over time, i.e. here,

$$f(t, \mathbf{T}) = M^{-1}\left(A\mathbf{T} + \mathbf{Q}\right). \tag{2.24}$$

Naturally, this method does not allow algebraic constraints like, for example, Sundials does [53]. However, to make the computation of M, A, and $\mathbf{Q}$ explicit, we compute those at $\mathbf{T}_k$. Therefore, the numerical integration scheme of QueST is

$$\mathbf{T}_{k+1} = \mathbf{T}_k + hM(\mathbf{T}_k, \mathbf{B}_k)^{-1}(A(\mathbf{T}_k, \mathbf{B}_k)(\mathbf{T}_k + \Theta(\mathbf{T}_{k+1} - \mathbf{T}_k)) + \mathbf{Q}(\mathbf{T}_k, \mathbf{B}_k)), \tag{2.25}$$

which can be explicitly solved for $\mathbf{T}_{k+1}$ with any Θ. We have emphasized here the dependence of the mass and stiffness matrices and the load vector on the values of $\mathbf{T}$ and $\mathbf{B}$ at time step k. Naturally, they also depend on the current at time step k. In QueST, we use Θ equal to 1. QueST also has an option to use Sundials.

2.4.3. *Triggering the quench in the simulation of an HTS magnet*

Whereas in an actual operation situation, the magnet quenches abruptly, in a simulation, the quench must be triggered at a predetermined location. Of course, one can also consider a continuous point disturbance within a magnet volume, but in a detailed quench simulation, the diffusion process for a few seconds can be very time-consuming, and the quenches from continuous disturbances do not develop rapidly. Therefore, triggering is an issue to consider in a quench simulation.

With a method based on normal zone propagation velocities, triggering is straightforward. One has an initial volume, no matter how small, from which the quench begins to propagate. An obvious possibility is to add a heat pulse somewhere in the system or to its surface [20]. Whereas this can be used to quench LTS magnets in a way they also quench in experiments, the quench triggering of an HTS magnet with a high stability margin through this way can make the simulation results completely useless. First, the required triggering energy density for a quench may be so high that it causes the hot-spot temperature to overstep a value that could never be achieved in an experiment. If the artificial heater is made large enough to prevent this, the voltage that it causes may almost immediately exceed the detection threshold limit, which makes it impossible to study how the quench develops. Therefore, with HTS magnets, another option must be found.

A coil quenches from a location where the critical current surface is punctured first. This locally reduced critical current can be caused by a

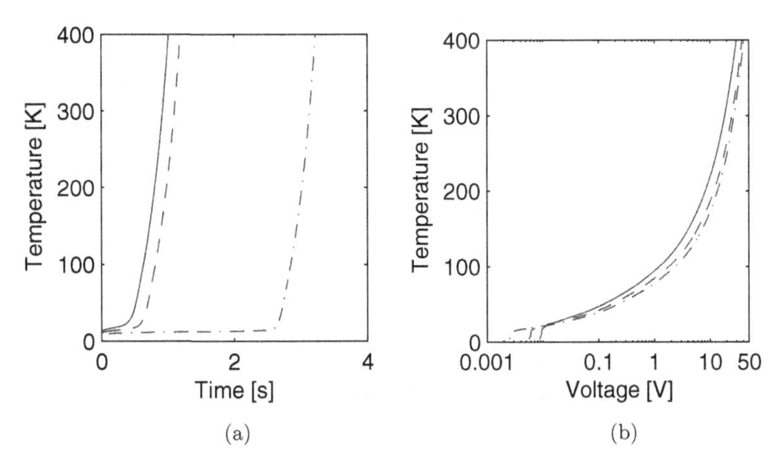

Figure 2.15. Temperature as a (a) function of time and (b) normal zone voltage during a quench simulation of a YBCO-based HTS coil. The solid (—), dashed (– –), and dash-dotted (–·–) lines correspond to reduction of critical current in the hot spot by 100%, 65%, and 30%, respectively.

locally increased temperature as well as a local damage or inhomogeneity [73]. To emulate the situation of reduced critical current, one can artificially create a small volume in the modeling domain where the critical current is degraded. This volume must be so small that the initial voltage is considerably below the quench detection threshold voltage. The threshold voltage depends on the quench protection system, with typical values ranging from highly sensitive 10 mV to 300 mV.

We studied [31] the influence of the current degradation on the development of the hot-spot temperature as a function of time or voltage over the normal zone. The results are shown in Figure 2.15. In that particular case, we studied a YBCO-based HTS magnet that was designed in the framework of an accelerator magnet R&D project. In that case, the protection was neglected, and we only studied how rapidly the magnet's hot spot reaches 400 K when operated at 4.2 K with a coil current density of 250 A/mm^2 in a background field of 13 T. Here, without getting into the details of the considered magnet, we conclude that the more the critical current degrades, the faster the detectable quench propagation begins. However, it seems that only the actual onset is delayed. The relation between the voltage and the temperature is very close to being identical between different situations. Also, one should note the very short time, around 500–600 ms, that is required to reach 400 K. Therefore, the early detection of the quench is of utmost importance in superconducting magnets. Finally, to speed up the quench simulations of HTS magnets, it is justified to consider a small volume inside the magnet and set the critical current in it to zero and to simulate the consequences.

2.4.4. *Reducing modeling domain to speed up quench simulations for HTS magnets*

Due to the wide spectrum of temperature margin in HTS magnets, the normal zone propagation velocities are much slower than in LTS magnets. Therefore, it is typical that the whole magnet, even when it is relatively small, does not go through the resistive transition — or reach T_{cs}. To decrease the duration of a simulation, it seems to be reasonable to reduce the modeling domain and not to solve the heat diffusion equation in the whole magnet. In principle, this means smaller a number of unknowns and faster simulation time — or that one can add more resolution, i.e. make the mesh denser at the interesting locations near the quench origin.

Next, we present an analysis considering the reduction of the modeling domain for an R&D magnet similar to that considered in Section 2.4.3. This analysis is based on our published research results [32]. The basic principles are demonstrated here. For the quench ignition, we utilized a volume with a degraded critical current. First, we describe the modeling domain of interest and then consider how the main simulation characteristics behave when we reduce the modeling domain. Also, in these simulations, we neglected the protection and quench detection and let the simulation run until the hot-spot temperature reached 400 K.

2.4.4.1. *Modeling domain*

The full modeling domain consisted of six stacked racetracks made from a twin YBCO tape. In this twin tape, two $CuBe_2$ and copper stabilized tapes were soldered together. Copper offers thermal stabilization, whereas copper–beryllium is for mechanical rigidity. In addition, both tapes had a 50 μm thick Hastelloy substrate. In total, the Kapton-insulated twin tape had a thickness of 460 μm and a width of 12.06 mm. In the modeling domain, we modeled the twin-tape insulation structure near the quench onset. The material properties within the thin tape were homogenized. Farther away from the location of a degraded twin tape, the structure was modeled with homogeneous material parameters. Figure 2.16 presents the

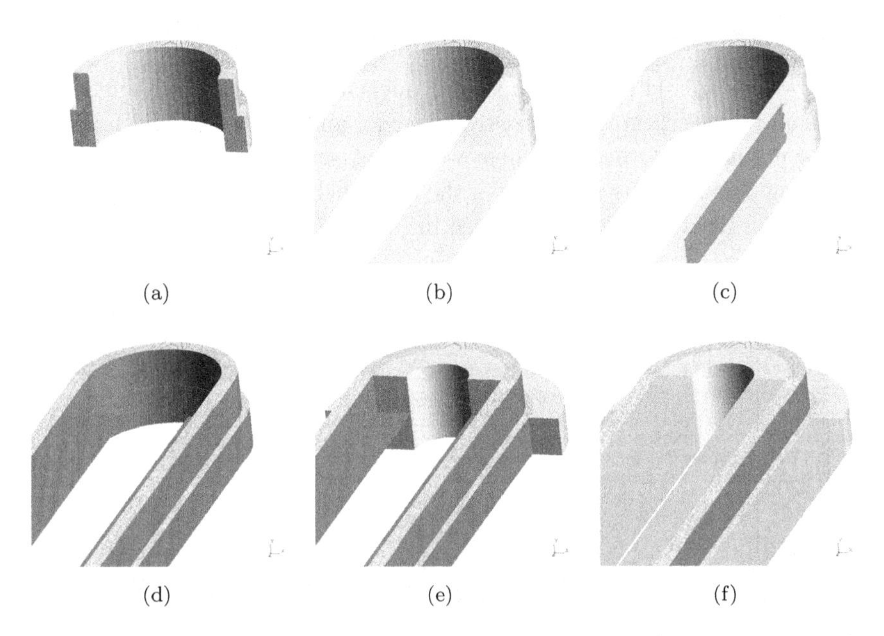

Figure 2.16. The structure of the modeling domain, including the two top-most coils as (new inclusions described) (a) twin tapes at the other end of the coil (17 turns on the topmost coil), (b) twin tapes at the straight parts, (c) volume with reduced critical current, (d) insulation in blue, also between the adjacent twin tapes, (e) rest of the magnet's end modeled with effective homogeneous anisotropic material properties, and (f) straight parts modeled with effective homogeneous anisotropic material properties.

Table 2.2. Relative material proportions of unit cells.

	Coil (%)	**Twin tape (%)**	**Insulator (%)**
Copper	15	17.7	0
$CuBe_2$	42	49.4	0
Hastelloy, buffers, YBCO	28	32.9	0
Kapton	15	0	100

structure of the modeling domain of the two topmost coils. The volumetric fractions of the constituents are shown in Table 2.2.

A number of modeling decisions were made prior to the simulations. The heat generation was computed according to the current-sharing model. For the boundary conditions, adiabatic condition was used. The reduction of the critical current in the quench onset volume was 100%.

2.4.4.2. *Simulation results*

First, we considered how the important quench characteristics, namely the maximum temperature, the resistive voltage, and the volumetric fraction of the normal zone, behave when we neglect some of the coils from the modeling domain. The results from the cases including only the quenching coil and all the six coils are displayed in Figure 2.17.

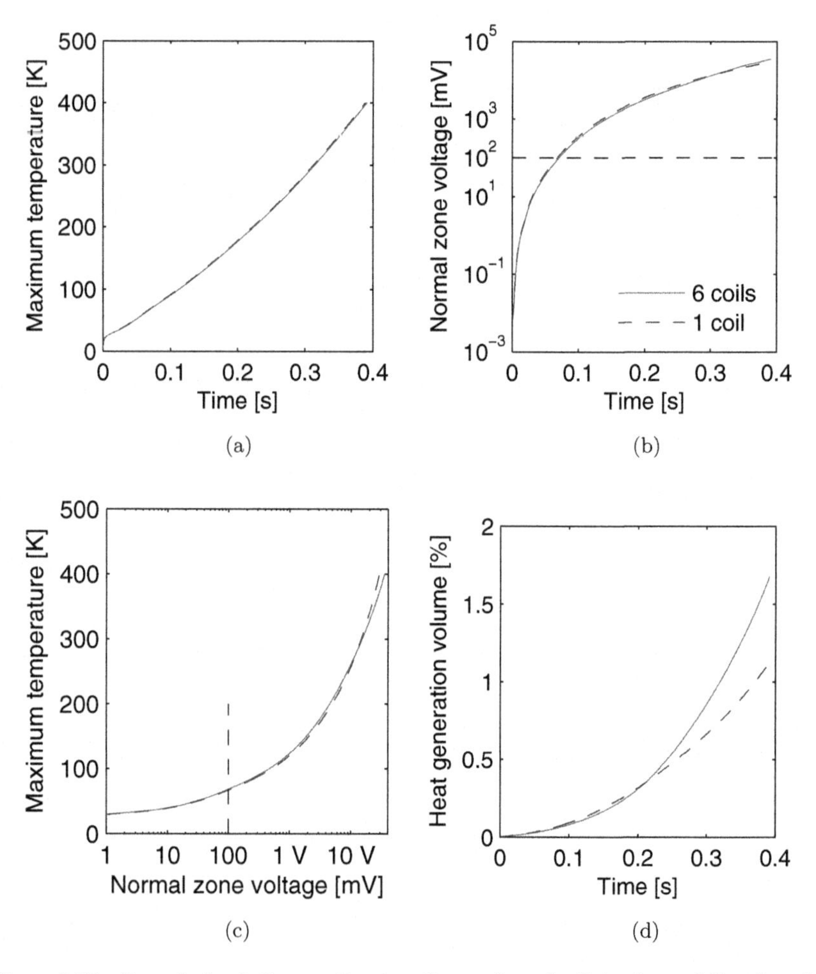

Figure 2.17. Quench simulation results when the number of coils in the modeling domain was varied: (a) hot-spot temperature as a function of time, (b) voltage over the normal zone as a function of time, (c) hot-spot temperature as a function of normal zone voltage, and (d) heat generation volume. The dotted lines in (b) and (c) denote a possible quench detection threshold voltage of 100 mV.

A very important characteristic is the time to reach the quench detection threshold voltage. For the analysis, we selected a conservative value of 100 mV. In the case of six coils, the quench detection time was 71 ms, but the analysis of only the topmost coil resulted in a time only 3 ms shorter. Therefore, the stabilization that the other coils bring about as a heat sink was not important in this case. Also, the variation in the hot-spot temperature at the time of detection was almost negligible. The single coil simulation resulted in 67 K, whereas the whole magnet analysis gave 68.5 K. However, 200 ms after the start of the simulation, the normal zone propagated to the adjacent coil also in the full magnet model, and correspondingly, the normal zone increased more rapidly. Therefore, in order to compute reliably the normal zone resistance, one needs to include more than one coil in the computation.

The normalized computation times for the simulations including different number of coils are shown in Table 2.3. The savings in the computation time when shifting from the six-coil-structure to the modeling domain including only the single coil was fivefold. Therefore, if one needs to simulate what happens before the quench detection, the modeling domain should be reduced considerably — especially if it consists of several coils.

Next, we considered only the topmost coil and studied the stabilization that the transverse heat diffusion offers. This study was performed by reducing the number of twin tapes included in the modeling domain. Figure 2.18 compares the single coil simulation and the simulations with 2, 4, and 10 twin tapes. Whereas the difference in the maximum temperature as a function of time is almost negligible, the maximum temperatures as a function of the normal zone voltage behave very differently.

Table 2.3. Quantitative data from different modeling domains that were considered. Normalization of computation time is done to that of full magnet simulation. The absolute time was two days and eight hours, but this should not be used as a reference value for any other case because the simulations were limited to a single core to isolate them from all other operation system processes.

Simulated coils	# Elements	# DoFs	Computation time (%)
1	157 058	39 622	18
2	386 189	90 127	46
3	631 355	142 418	63
4	820 000	182 935	86
6	940 000	209 009	100

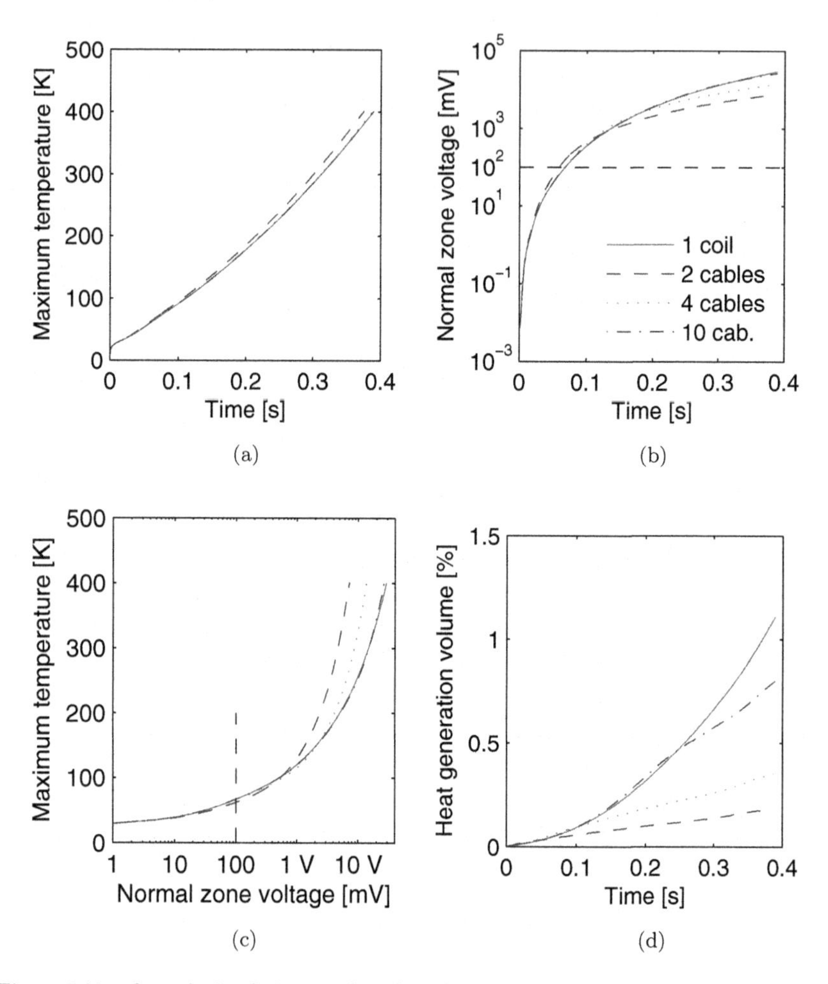

Figure 2.18. Quench simulation results when the number of twin tapes in the modeling domain was varied: (a) hot-spot temperature as a function of time, (b) voltage over the normal zone as a function of time, (c) hot-spot temperature as a function of normal zone voltage, and (d) heat generation volume. The dotted lines in (b) and (c) denote the considered quench detection threshold voltage of 100 mV.

When the normal zone reaches the edge of the modeling domain, the transverse heat diffusion does not offer additional stabilization or a volume that would quench next. Therefore, the voltage increase is limited even though the temperature continues to increase. Also, the temperature increase is bursted but only marginally. The quench detection times for 10 twin tapes and the single coil model differ only by 0.2 ms, and the

Table 2.4. Quantitative data from different modeling domains that were considered in the twin tape analysis. The normalization of the computation time is done to that of the single coil simulation.

Simulated twin tapes	# Elements	# DoFs	Computation time (%)
1	2 476	1 218	1.3
2	14 856	4 283	8.3
4	29 712	7 961	19
10	74 280	18 995	44

hot-spot temperatures at the corresponding time instances differ by 0.1 K only. When less than 10 twin tapes were considered, the results became more unreliable. The simulation times are summarized in Table 2.4.

The lesson to be learned from these simulations is that often in time-consuming simulations, it is possible to consider a reduced modeling domain — not the whole device of interest. However, it is important to study when the characteristics that one really needs converge. Each new case must be studied in detail before general conclusions can be made. Furthermore, the quench onset simulation can be performed with a reduced domain, and after that, one can completely change the approach to consider the quench propagation during the protection system activation phase.

2.4.5. *Quench analysis of an R&D REBCO magnet*

The quench analyses of large and small magnets are very different. An issue unique to each large magnet is the design of the protection system. In small magnets, the most important question is what happens before the detection because the protection can be easily taken care of with an external dump resistor. The amount of stored energy and the operation current are the key parameters that determine whether a magnet is small or large. The stored energy necessarily needs to be dissipated or extracted from the magnet. The operation current limits the terminal voltage and thus the power for extraction. We consider a magnet to be large if the external dump resistor is not adequate for protection and small otherwise. The definition of adequate protection may vary for individual cases: the hot-spot temperature, the voltage to ground or turn-to-turn, as well as the temperature gradient may constrain the protection.

Because the classification of a magnet as small or large is vital for all the phases of the quench analysis, one must first consider the size of the magnet. The MIITs computation is the first approach that one takes. It is

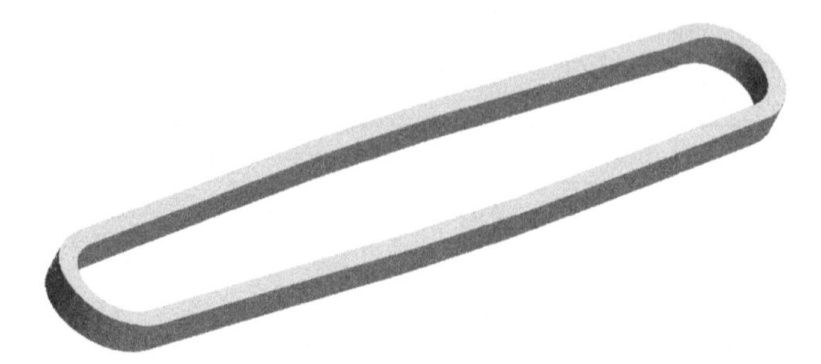

Figure 2.19. Design of prototype coil with the location of the reduced critical current volume at the other end of the coil shown in brown color.

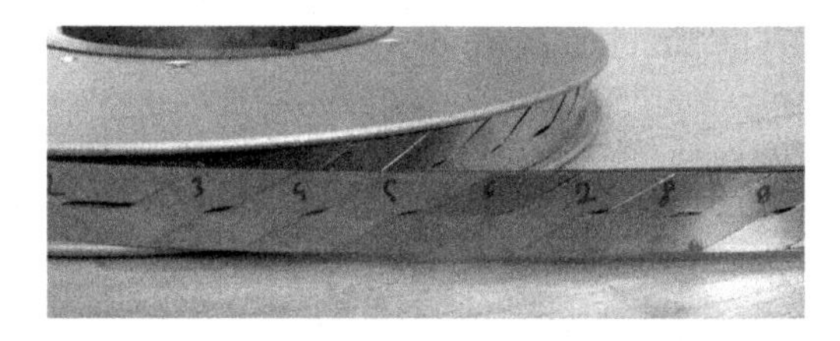

Figure 2.20. Dummy Roebel cable made of stainless steel.

pessimistic in the sense that heat conduction is neglected, and if it implicates that the magnet is small, one can focus first on the quench onset only. The MIITs analysis depends on the inductance of the magnet, the maximum terminal voltage, the current, the current density and the effective coil unit cell properties: its area, heat capacity and normal state resistivity.

Here, we consider a Roebel-cable-based [29] five-turn, racetrack-like prototype coil having a length of 220 mm and a width of 44 mm at its widest cross-section [36]. This coil with the location of the reduced critical current zone is shown in Figure 2.19. The Roebel cable was 12 mm wide and 1 mm thick and made from 15 REBCO tapes (see Figure 2.20 for such a dummy assembly on a spool). Copper plating was added to offer thermal stabilization. The volumetric fractions of copper; Hastelloy, substrates, and REBCO; and the insulation on the coil unit cell are shown in Table 2.5. The insulated cable cross-section area was 12.2 mm^2. The insulation material was Kapton. The aim of constructing this coil was to study the feasibility of the Roebel cable in a coil with flat sides. For example, accelerator dipoles, block or cos-Θ design, have flat sides. When the required magnetic flux densities overstep those that can be achieved with LTS conductors, one needs to change the technology to HTS-based magnets operated at low

Table 2.5. Volumetric fractions of the most important compounds on the coil's cross-section.

Copper	31.5%
Hastelloy, substrate layers, and REBCO	38.5%
Insulation	30%

Table 2.6. Operations conditions considered in the simulations and the most important parameters needed for the simulations. The given operation current has a corresponding temperature. Operation currents correspond to 80% of I_{max} at a given temperature.

Parameter	**Value**
Operation Current	9/5.6/0.9 kA
Operation Temperature	4.2/30/77 K
Copper, RRR value	100
Superconductor, n-value	20

temperatures. The R&D project EUCARD[2] [50] on which this section is partially based was part of such an effort.

The goal of the work package on developing future magnets in the EUCARD[2] project was to produce a 5 T stand-alone accelerator-quality magnet with an aperture for a beam tube. The results were reported by van Nugteren *et al.* [82]. Eventually, the target was to test the magnet also as an insert in an ~13 T background field. For the current intermediate development step, the first prototype coil was characterized at various operation temperatures in order to study quench and the relation between the achievable current and the computed critical current. The operation conditions and the most critical parameters utilized in the simulations are displayed in Table 2.6. For the heat generation, we utilized the power law in superconducting domain model (see Section 2.1.1). Cooling was also neglected because the coil was impregnated. Therefore, adiabatic boundary conditions for temperature were utilized in the simulations.

Based on the most demanding operation conditions, one can deduce whether the magnet is small or large. With a terminal voltage limitation of 1 kV and a coil current of 9 kA, the maximum dump resistor resistance R_d is 0.11 Ω. The terminal voltage limitation and the operation current determine the maximum dump resistor size via Ohm's law. The inductance L of the coil was $10\,\mu$H. When the current decay was determined only by the energy extraction and the coil inductance, one has for the current

$$I(t) = I_{op} \exp\left(-\frac{t}{\tau}\right), \tag{2.26}$$

where the circuit time constant τ is L/R_d. Therefore, the current decay integral equation (2.12) that considers only the energy extraction results in 0.00368 MIITs (only the time after the activation of the protection system is considered), corresponding to a current decay of 99% of the operation current in just 0.41 ms. This short decay time is caused by the very small inductance of the coil. The corresponding temperature increase from 4.2 K is 3.7 K, from 50 K is 0.2 K, from 100 K is 0.2 K, and from 200 K is 0.3 K, according to Equation (2.13). Therefore, this coil is small and thus the temperature increase after the activation of the protection system is negligible. One can then focus only on studying the temperature increase before the activation of the protection system. To handle all the different cases simultaneously, one can study how long it takes for an unprotected coil to reach 400 K, and then, one can analyze the temperature and voltage characteristics.

The main quench simulation results are displayed in Figure 2.21. The most important observation is that when a hot-spot temperature of 100 K is reached, the terminal voltage is around 100 mV. Therefore, the quench is a safe event if the detection and the energy extraction systems work as they are supposed to. However, the temperature increase from the corresponding value at 100 mV to 400 K will last only 43, 102, and 4.3 s for operation currents corresponding to initial temperatures of 4.2, 30, and 77 K, respectively. Thus, the successful detection of the quench, as well as the rapid operation of the quench protection system that bypasses the current source and connects the dump resistor, is very important.

Figure 2.22 presents the temperature profile of the coil at the time instant when the hot spot has reached a temperature of 400 K. In this case, the coil quenched at 4.2 K. As already known from the previous observations, the high-temperature volume is very localized, not even extending through the thickness of the coil, which only consists of five turns. This forces the quench detection system to be such that it considers the whole coil volume. Also, the thermal stresses during the quench can be critical,

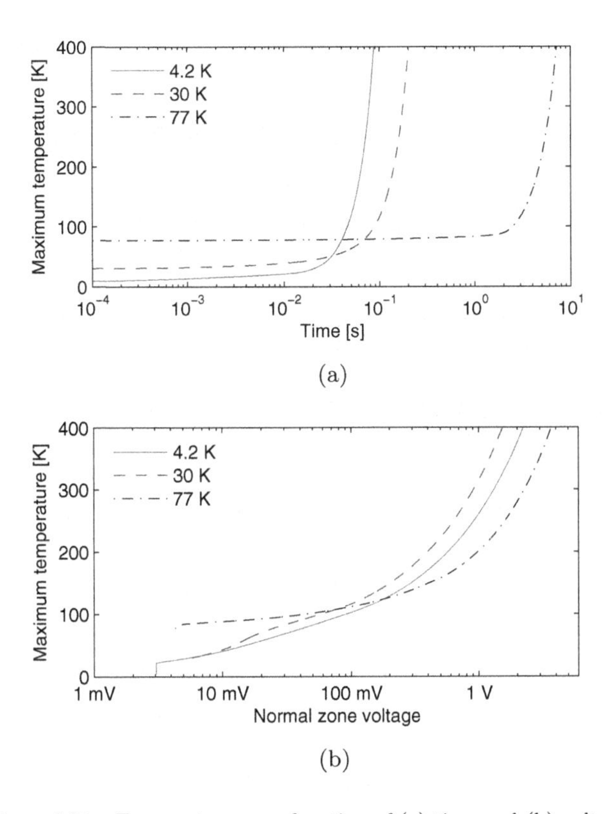

Figure 2.21. Temperature as a function of (a) time and (b) voltage.

especially with REBCO-based coils that are prone to suffer delamination of the tape due to its layered structure [83].

To conclude the quench analysis, the pessimistic analytical computations indicated that the coil is small and that it can be protected with an external dump resistor. The current extraction in the worst case takes less than 0.5 ms. Then, the simulation of the quench event showed that before the safe detection threshold voltage of 100 mV was reached, the coil temperature had risen to around 100 K, slightly depending on the initial temperature. A critical observation was made that the time derivative of the temperature at 100 mV is steep and that, in the worst case, there is only 43 ms for the protection system to react before a temperature of 400 K is reached. It is not certain if 400 K is even a safe limit. One reason for this is the localized hot spot, which can cause high thermal stresses in the winding. The thermal stresses must be considered, and experiments are required

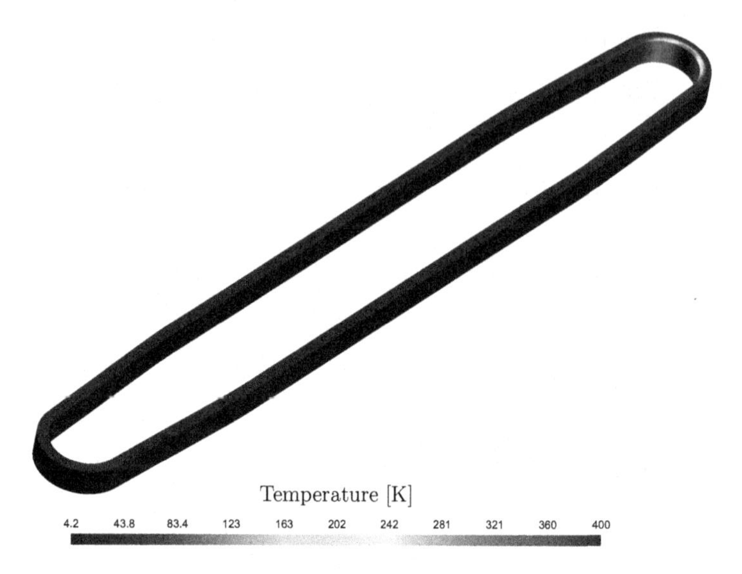

Figure 2.22. Temperature profile on the coil's surface when the hot spot reaches 400 K.

to assure that the quench can be carried through safely. Also, the quench detection system needs special attention. Because the quench is localized, the entire coil must be closely monitored. This is the main output of this kind of analysis for the engineering team to take the quench protection to the level of actually implementing and characterizing the system.

2.5. Design of Quench Protection Heaters for Nb_3Sn Accelerator Magnets

In large accelerator magnets, the external energy extraction with a dump resistor is not enough for protection. Therefore, even in subscale prototyping, alternative protection methods must be investigated. In high-energy accelerator magnets, especially dipoles and quadrupoles, protection has traditionally included quench protection heaters [41, 62]. These heaters cause a widespread quench in the magnet and, correspondingly, rapidly enlarge the heat dissipation volume from the initial hot spot, thus bringing down the operation current. The most important characteristic of a protection heater is the time delay between powering them and quenching the magnet below the heaters.

This section considers an analysis of quench protection heaters, starting from their design via simulations to experiments and, therefore, to the benchmark of the simulation results. Also, the model behind the simulations is discussed. This case study arises from the quench protection of the Nb_3Sn accelerator quadrupole magnets[5] that are being designed for the Large Hadron Collider (LHC) luminosity upgrade [25]. We focus on the quench protection heater design of a subscale-model magnet, called long high-gradient quadrupole (LHQ), that has been designed during the R&D phase. We show an example of how simulations can be used to guide the heater design.

In this section, we first briefly consider the technological development related to Nb_3Sn quadrupoles for the LHC luminosity upgrade. This is important for understanding the target of the heater design. We include the MIIT considerations of the LHQ model magnet design to demonstrate the need for additional quench protection. Then, we consider the heater technology in general. It is especially important to recognize the parameters that can be optimized. This is related to the design phase of the quench protection heaters and its coupling to the simulations. Thus, we consider the optimization of the heater design for the LHQ through simulations. At the end, we also present an experimental characterization of the heaters that is viable for validating the usability of the design and benchmarking the simulation results.

2.5.1. *R&D of Nb_3Sn quadrupole magnet*

The goal of the luminosity upgrade is to develop high-gradient Nb_3Sn quadrupole magnets for the LHC IR in order to achieve, before the collision, a focusing of the particle beams that is stronger than the currently existing ones [11]. The design parameters of the IR magnets (called MQXF) are the following:[6] a 132.6 T/m gradient in a 150 mm aperture with a nominal current of about 16.5 kA. The stored energy will be about 1.17 MJ/m and the peak field in the conductor, 11.4 T [23]. Magnets with two different lengths are considered: 4.2 and 7.15 m. In the preparation of designing and building such magnets, several shorter and smaller aperture prototypes have been designed and tested [4].

[5]A quadrupole magnet consists of four coils that produce such a gradient of magnet flux density to its aperture that a beam of charged particles is focused toward the center line of the magnet [42].
[6]These values date back to 2015, the time of writing of this chapter. Updated values can be found in the following preprint: https://arxiv.org/abs/2203.06723v1.

The prototypes' sizes and complexities have increased based on the experience with the behavior of the materials and fabrication methods. Thus, the model–design–simulation–test procedure can be seen in the magnet development too, on a scale of several years. At the time of writing this chapter, the closest to the final magnet was a 3.3 m long, 120 mm aperture magnet: the LHQ.[7] The operation current of the LHQ is about 15 kA and the peak field in the conductor, about 12 T. Figure 2.23 presents the coil configuration and the support structure related to the designs of the MQXF and the high-gradient quadrupole (HQ), which has the same cross-section as the LHQ.

The quench protection of these compact high-field and high-energy magnets will be very challenging due to the high current density in the copper stabilizer after a quench has originated. Using the MIITs concept, one can compute that in the LHQ, a current integral of 19.3 MA^2s corresponds to a 350 K hot-spot temperature. Based on the gained experience with the Nb_3Sn magnet R&D, the MIITs corresponding to temperature increase of 350 K is the upper limit for a quench that does not degrade a magnet [3].

In the MIITs calculation, the insulated cable area was 23.6 mm^2, the Cu fraction was 40%, the Nb_3Sn fraction was 33%, the G10 (insulation) fraction was 27%, RRR was 100, and the magnetic flux density for magnetoresistivity was 12 T. The material properties were taken from the MATPRO database [39]. At constant operation current of 15 kA, the considered 19.3 MIITs correspond to 85 ms. Therefore, the magnet energy must be rapidly discharged.

The MIITs calculation can be used to derive the first estimation of the quench protection requirements. Assuming operation at 15 kA and 20 ms for the quench detection, validation, and activation of the protection system, the MIITs left for the current decay phase are about 15. When exponential current decay is considered, the circuit time constant (L/R) can be up to a maximum of 130 ms. The inductance of the magnet is 7 mH/m. Thus, for the 3 m long prototype, the total inductance is 21 mH. To get the needed circuit time constant, a circuit resistance of at least 160 mΩ is required. If this was achieved with only an external dump resistor, its voltage would be about 2.4 kV at nominal operation current. The same voltage would be across the magnet's terminals and beyond the insulation limit (1 kV). Therefore, an external dump resistor is not a solution; the

[7] Finally, the full LHQ magnet was not built. One LHQ coil was built and tested individually.

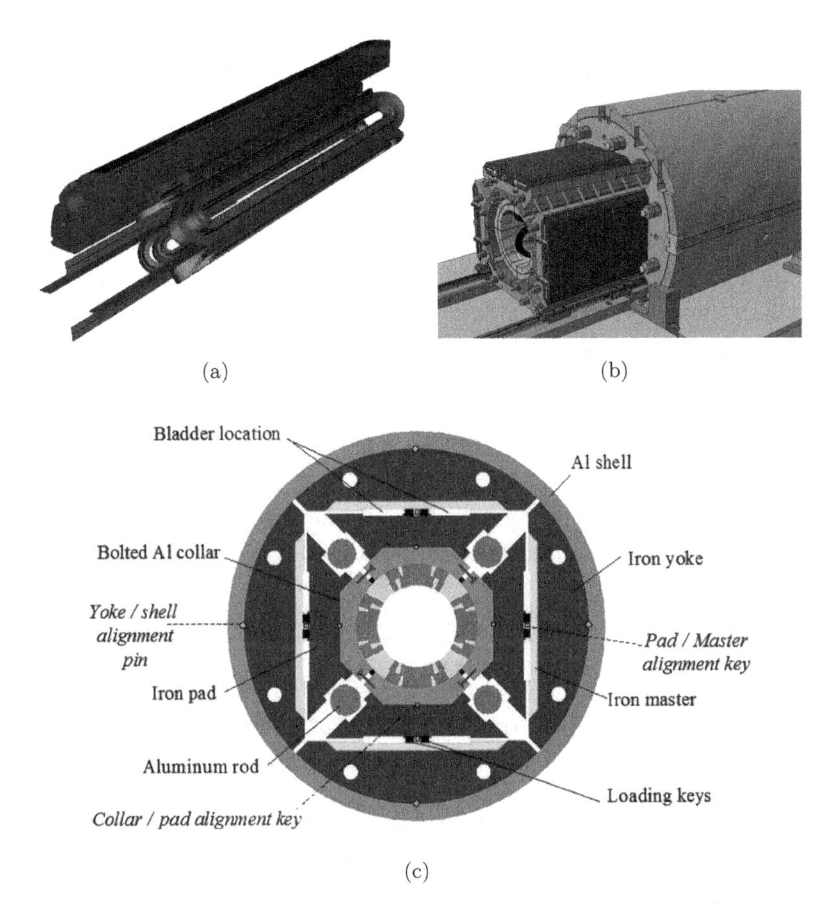

Figure 2.23. (a) Coil configuration in a small-sized model of MQXF, (b) the support structure envisioned for MQXF, and (c) the cross-section of HQ magnet structure.
Source: (a) courtesy of S. Izquierdo Bermudez, CERN, see Ref. [85] for more details; (b) see Ref. [85]; (c) see Ref. [86].

magnet protection must rely on internal absorption of the energy, and the resistance must come from the resistive cable segments. Also, in the case of a longer magnet (final target is 7 m), the inductance of the magnet and the required resistance would be even larger, making a protection system relying entirely on a dump resistor even less practical. However, due to the difficulty of protecting the MQXF, the advantage of using a small dump resistor to extract part of the energy is being investigated separately from the heater design study.

In order to increase the resistance of the magnet assembly, protection heaters are activated upon quench detection to rapidly induce a widely spread normal zone in the coils. The advantage of a large normal zone with respect to hot-spot temperature is also evident when noting that the conversion of the stored magnetic energy to heat occurs only through Ohmic losses in the normal conducting cables. The dissipation of the energy into a larger volume leads to a more uniform temperature distribution and thus a lower peak temperature compared to a case where all the energy is absorbed by only a small quenched cable volume. Therefore, the efficiency of the heaters can be characterized by the heater delay, i.e. the time it takes to initiate quench below the heater, and by the fraction of the coil that is quenched by the heaters.

As the design of the new LHC magnet is an iterative process, so is its protection heater design. R&D magnets, such as the LHQ, are used to test different protection heater layouts and other quench protection design concepts. Here, we describe a heater design study that was done for the LHQ magnet. This case provides an example of how the model–design–simulation–experiment principle applies to a quench analysis task, including the heater design.

This particular study investigates the option of protecting the magnet only with the heaters: dump resistor and potential help from AC losses are neglected in this analysis [45]. The goal of this heater design was to quench a volume as large as possible as fast as possible within certain constraints related to powering the heaters. Though successful protection at operation currents below the nominal one is interesting too, in this study, we focus on the quench at operation current only.

2.5.2. *Heater technology and target variables for optimization*

The heater technology that is considered here is based on the so-called trace technology [21]. The traces are like flexible printed circuits that consist of 25 μm thick stainless-steel heater strips glued onto a polyimide layer. The traces are impregnated on the coils' surfaces. The polyimide layer is needed for electrical insulation. A pre-charged capacitor bank is used to deliver power to the heaters. Therefore, the heater's power is not constant during its operation but decays according to the RC time constant of the circuits, where R is the resistance of the heater circuit and C is the capacitance of the capacitor bank. R changes slightly during the operation because of the temperature increase; however, a constant R typically describes the power decay in the heaters sufficiently well.

Ideally, to heat a coil fast, one would use heaters with very high power, covering the entire coil surface and use a very thin insulation layer to minimize the thermal barrier between the coil and the heater. However, there are several constraints to the heater design set by the available technology and other magnet design aspects. The limitations to consider in this particular case are summarized as follows:

(1) Insulation thickness: A too thin insulation may lead to an electric breakdown and cause damage to the heater or to the coil. A too thick insulation increases the heater delay. Here, the design uses 50 μm thickness because empiric experience suggests that it is the thinnest safe thickness to consider.[8]
(2) Strip location: In wind-and-react cos-$n\Theta$ Nb_3Sn coils [42], the heaters are applied on the coil's inner and/or outer surfaces after heat treatment. In previous R&D models, the heaters on the coil's inner layer surface have shown signs of degraded contact with the coil [2]. Therefore, in this design, the heaters are placed only on the coil's outer surface. Moreover, the innermost and outermost coil turns must be left free for other instrumentation or for mechanical purposes.
(3) Strip length: In order to avoid additional power wires from the middle of the magnet, a single heater strip must extend over the entire coil length.
(4) Heater voltage: The total voltage across a heater strip can be limited either by the available heater power supply or the electrical insulation's withstand voltage (taking into account that the coil's voltage may be of opposite polarity and add to the heater powering voltage). In this design, the heater's terminal voltage is limited to 400 V.
(5) Heater temperature: In this design, the upper limit of the heater strip's temperature is 350 K. Higher temperatures might cause thermal stresses that can damage the heater.

Due to these limitations, most of the heat in the heater is deposited at the so-called *heating stations*. The limitations on the heater's temperature

[8]In recent years, several voltage integrity tests have been done with different insulation thicknesses at Lawrence Berkeley National Laboratory, Berkeley, California, USA. In these tests, a 25 μm thick insulation has failed several times, and 50 μm has been considered the safe option.

determine the energy that can be deposited per heating station. The limitation on the voltage then constraints the number of heating stations one can have. In long coils, the heater strips must also be long. Therefore, the requirements for homogeneous power dissipation can make the heater design inefficient, and it is more effective to deposit the energy in specifically designed heating stations to have a rapid quench onset, i.e. short heater delay. This can be achieved by reducing the total resistance of the heater strip by having lower resistance paths between the periodically placed heating stations. The coil quenches under the heating stations due to the temperature increase in the heaters and the consequent heat conduction to the coil. Then, the quench propagates naturally, filling the space between the heating stations.

From several analyses [21, 56, 57] performed during the development of the R&D model coils, it is known that both longer heating stations and higher power reduce the heater delay. However, their impact saturates at some point so that on increasing their values above a certain point, the decrease in the delay is not noteworthy. Thus, during the heater design phase, the particular design must be selected carefully: The higher the power and the longer the heating stations, the fewer they can be because the total energy and voltage are limited. When the distance between the heating stations is increased, i.e. the heater's period is increased, the time it takes for a quench to propagate between the heating stations increases. The design of the strip geometry and the optimization of the heating stations' length and spacing are the goals of this design study.

There are different ways to make a heater strip with heating stations. One can utilize straight stainless-steel strips on the coil outer layer. The low-resistance path between the heating stations is achieved with copper cladding [46, 47]. This technology was utilized in the LHC Nb–Ti magnets. In the so-called Nb_3Sn-based LQ magnet (also a prototype coil in the MQXF development program) only stainless steel was utilized and the heating stations were narrow segments [21]. The low-resistance segments were implemented with wider stainless-steel paths. The heater design for the LHQ was based on the same technology as the LQ heater and used stainless steel only. The shape of the strip, however, has not been investigated earlier; see Figure 2.24 for both the LQ and LHQ heater shapes. The novelty of the LHQ heater design concept, when compared to existing concepts, is that it has longer heating stations and their length can be optimized based on the field region under them. In particular, more heat can be directed toward the lower field area by using a longer heating station.

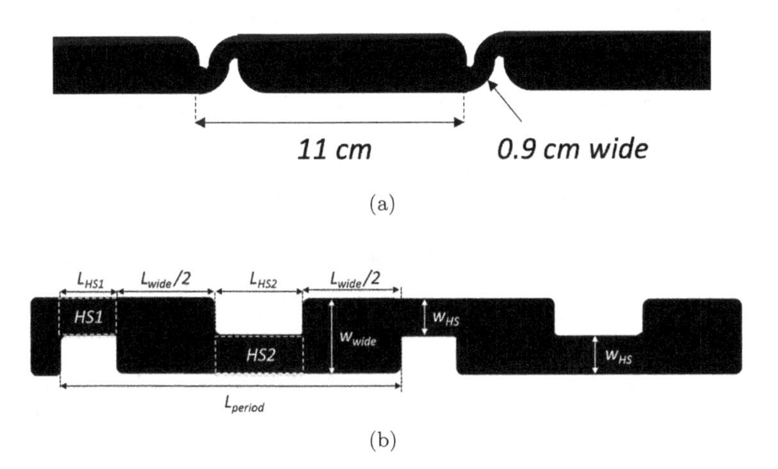

Figure 2.24. (a) Schematic layout of LQ heater. *Source*: Ref. [21]. (b) Schematic layout of LHQ heater strip.
Source: Reproduced from Ref. [59]. Both figures are not to scale.

The low-field region has a higher thermal margin than that near the coil's critical point. Therefore, more heat is required there to cause a quench.

In the LHQ, each coil outer layer of the magnet has four strips, two on each side of the pole. Figure 2.25 shows the magnetic flux density distribution in the cross-section of one half of the coil (an octant of the magnet's cross-section) and the heater strip locations at the high-field (HF) block close to the central pole piece and the low-field (LF) block near the magnetic mid-plane.

The outer layer consists of two cable blocks separated by a copper wedge. One heater strip is used to cover each block. The widths of the strips are based on the dimensions of the block. The heating stations are half of the strip width. The HF block has an arc length of about 24.4 mm. In order to leave the pole turn and the copper wedge uncovered, the heater strip width was determined to be 21 mm. The two heating stations (HS1 and HS2 in Figure 2.24) were decided to be 10.5 mm wide in order to cover all the turns below the heater with heating stations. The arc length of the LF block was about 20.8 mm. To leave one turn uncovered near the mid-plane, the heater strip width was fixed as 18 mm, with two 9 mm wide heating stations. The lengths (L_{HS1} and L_{HS2} in Figure 2.24) and the distances of the heating stations (i.e. the L_{period} in Figure 2.24) have been optimized by simulating the efficiencies of the different heaters in both HF and LF blocks.

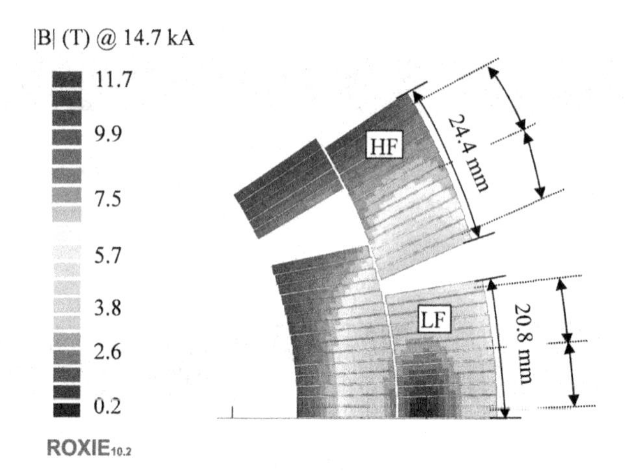

Figure 2.25. Half of the LHQ coil with magnetic flux density distribution and the location of the two heater strips with their azimuthal dimensions based on the coil block dimensions. The shorter arrows show the locations of the heating stations in the heater that occupy both halves of the heater's width.

2.5.3. *Modeling the heater's efficiency*

The interest of the quench heater delay simulation is limited to the time between the heater's activation and the coil's transition above the current-sharing temperature. The quench analysis engineer is also interested in the time it takes for a quench to propagate between the heating stations. It is typically assumed that the magnet operates at constant current until a heater causes a quench. For each turn of the coil, the model for the heater efficiency and cable normal transition is composed of only two parts:

(1) the quench provoked by the heater under the heating stations, and
(2) the quench propagation between the heating stations.

The modeling of these two parts is separated in this analysis. Moreover, since the turn-to-turn thermal propagation is neglected, each turn of the coil is considered as an individual thermal system. The error arising out of this assumption is not important when the adjacent turns are also heated in a similar manner [60]. The heater delay is modeled based on the heat diffusion from the heater to the cable. The longitudinal quench propagation, on the other hand, is computed simply by assuming a known constant quench propagation velocity for the known distance between two heating stations. The total delay is the sum of the heater delay and the quench propagation time. The purpose of the heater delay simulation is to solve the heat diffusion from the heater to the cable in order to know the cable's temperature evolution and its transition from the superconducting to the normal conducting state.

The heater-coil structure always includes several thin layers of different materials with highly nonlinear material properties. In the simulations, high thermal gradients are present on very short time scales (sometimes μs). Therefore, utilizing a computational tool based on numerically solving the heat diffusion equation in 3D leads to a computationally very expensive simulation. Furthermore, because one often simulates several cables with several different heaters, the individual simulation time tends to multiply. Therefore, several simplifications and reductions have to be adopted in the simulations.

The physics of the heat transfer problem can be limited to solving the heat diffusion equation (2.1) within one cable, its insulation, and the heater trace. Considering each coil turn separately and isolated from the other turns allows reducing the modeling domain to 2D, representing the cable longitudinally. The domain further reduces to half of the heater period in the case of periodic heater geometry (with the heating station at the center of the period). A schematic layout of a heater on the coil's surface and the modeling domain are shown in Figure 2.26. The assumptions in the heater delay simulation are as follows:

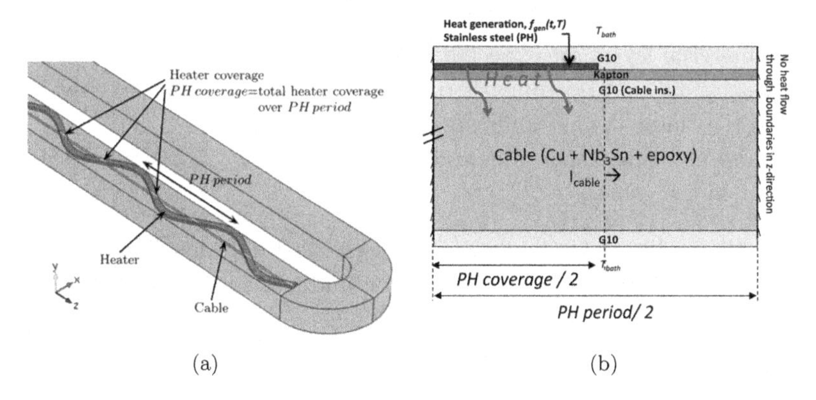

Figure 2.26. (a) The schematic layout of the heater and (b) that of the modeling domain for the 2D heat diffusion simulation that solves for the heater delay. Note that the dimensions are not to scale.

Source: Reproduced from Ref. [58].

- The heat transfer from the heater to the cable occurs through conduction only.
- There is a uniform power distribution within the heater's cross-section in both the heating station and the wide segment. The computation is based on the heater current and the heater cross-sectional area.
- Each turn of the coil is considered an individual thermal system.
- The modeling domain consists of half of the heater period in one cable. The included materials, in addition to the heater and bare cable, are the cable insulation on the top and bottom of the cable and the heater insulation. Above the heater is the coil ground insulation, which insulates the coil from the support structure i.e., the collar. Below the cable is the insulation to the inner coil layers.
- The insulation on the cable's wide sides is neglected.
- The cold components in the immediate surroundings of the top and bottom of the system (metallic collars around the coil or the other coil layer) are modeled as heat sinks at constant temperature (the temperature of the cryogenic bath).
- The time taken for the transition to the normal state is based on the cable's maximum temperature increasing above T_{cs}.
- The magnetic field is considered constant across a cable's cross-section.
- During the whole simulation, the operation current remains constant. This allows pre-computation for T_{cs}.
- The critical current as a function of the field and temperature is computed at a single stress state.
- The material layers are expected to have a constant thickness.
- Homogeneous and isotropic thermal properties are utilized for the bare cable (i.e. individual strands are not modeled).
- Because the simulation is considered only until T_{cs}, no Joule heating occurs in the cable.

The thermal model is characterized by the heat diffusion equation (2.1). Heat generation occurs only in the heater, and it is computed as

$$Q(t,T) = \rho(T)J_{ss}(t)^2, \tag{2.27}$$

where $J_{ss}(t)$ is the current density in the heater strip. Its time dependence is emphasized, as the heaters are energized from a preloaded capacitor bank, and the current decays during the heater operation.

An important modeling decision is the criterion for the quench onset induced by the heater. The possibility of validating the simulations with experiments has been kept in mind. In the measurement, the quench onset is typically associated with a detectable monotonously increasing resistive voltage rise in the coil. That is why we defined the quench onset to be the time instant when the maximum temperature in the cable reaches T_{cs}.

A dedicated simulation tool, based on the finite difference method [13] with explicit time integration [7], has been developed to solve the described heater delay problem [58]. The following results have been obtained with this tool, whose detailed implementation is discussed by Salmi [59].

2.5.4. *Guidelines for parametric optimization of heaters*

Because the parameter space for a heater strip is limited and the computation routine has been simplified and optimized, it is possible to use parametric analysis to find the optimum heater design. In the analysis, both the HF and LF (see Section 2.5.2 and Figure 2.25) strips are considered separately. The goal of optimization is to minimize the maximum sum of the heater delay and quench propagation time between two heating stations among all the turns that are covered by the heater strip. The quench propagation between the heating stations is directly known from the heater's period and the predetermined, or given, normal zone propagation velocity. One should also note that another criterion for the optimization could be set. For example, the average delay could be minimized. The currently selected method minimizes the maximum delay, but it may do so at the expense of increasing the very short delay times somewhere in the coil.

The currently utilized routine is as follows:

(1) Compute the heater delay in each heating station as a function of set heater coverages and powers. The power decay time constant, i.e. the time variation of the current density of the heater strip, for each peak power is defined such that the temperature in the heater does not exceed 350 K in adiabatic computation (this computation is equivalent to the MIITs of a superconductor).
(2) List all possible heater geometries (power, coverage, and period) that are within the design limits.
(3) Compute quench propagation time between the heating stations and, consequently, the longest heater delay in each turn of the coil for every heater geometry as the sum of the heater delay and propagation time.
(4) Choose the final design: The heater geometry that led to the smallest time until all the turns under the heater have been quenched.

In order to define the heater periods, the heating station lengths for the HF and LF blocks must be set. The heater period for a given combination of HS lengths can be defined from the heater's terminal voltage limitation. The period should be the smallest possible that keeps the heater resistance below the limiting value. This computation is based on approximating the heater's resistance by considering it to consist of rectangular blocks, emulating the heating stations and the wider segments in between, which are connected in series.

When computing the delay for each magnetic field region as a function of heating segment length and power, the heater period is assumed to be sufficiently long so that it does not impact the result. Then, each period does not require a separate heater delay computation. The heating provided by the wide segment has been ignored in the simulations, though its influence on the resistance, which determines the current, is considered. Therefore, the period only influences the quench propagation time between the heating stations. To go through all the possible heater layouts, the predefined heating station lengths are combined in all the possible ways. These fix the periods so that the heater's terminal voltage does not exceed the limiting value, i.e. 400 V here.

2.5.5. *Simulations for the LHQ heater design*

The presented modeling approach and the heater design concepts were applied to the LHQ magnet heater design. The strip length was 3 m, and a 0.3 Ω margin was left for the connections to the capacitor bank. The magnet's parameters are detailed in Table 2.7. The optimization was done for a magnet current of 15.4 kA, which is 80% of the short-sample limit. The magnetic flux density on the coil's outer surface was calculated using ROXIE [52], and the lowest value of the magnetic flux density under each heating station was used to determine T_{cs} and, consequently, the heater delay. On the HF block, the magnetic flux densities for HS1 and HS2 were fixed at 8.1 and 5.8 T, respectively, with the current-sharing temperatures of 9.2 and 11.0 K, respectively. On the LF block, the corresponding values were 5.6 and 5.2 T and 11.2 and 11.4 K for HS1 and HS2, respectively. These magnetic flux density values are the maximum ones in the cables that have the lowest maximums below the corresponding heating stations. However, one could also utilize the simulated magnetic flux density distribution, which is shown in Figure 2.25, its minimum, or the average value. Because the delay in experiments is often associated with the first observation of a resistive signal, the utilization of the maximum value is

Table 2.7. LHQ cable and coil parameters for heater simulation.

Parameter	Value
Number of strands per cable	35
Strand diameter	0.79 mm
RRR	90
Strand Cu/non-Cu ratio	1.20
Cable width	15 mm
Epoxy fraction	0.18
Insulation material	G10
Cable insulation thickness	0.1 mm
Ground insulation around coil	0.38 mm
Insulation between coil layers	0.5 mm

justified. Furthermore, with simulations, one does not aim at tackling all the design issues, but prototyping also gives important input. The critical surface utilized in the determination of T_{cs} was based on the Godeke fit [28] and the HQ coil number 15 therein. The initial temperature in the simulations was 4.2 K. The normal zone propagation velocity was assumed to be 12 m/s in the HF block and 7 m/s in the LF block.

The temperature and magnetic flux density dependencies were considered in all the material properties. In the simulations, the effective material properties of the uninsulated cable were utilized in the cable volume. The properties of copper were taken from CryoComp [15], the specific heat of Nb_3Sn was based on a fit proposed by Manfreda [38], and epoxy specific heat was based on CryoComp [15] (below 4.4 K, a linear extrapolation is used with an assumption that the specific heat of epoxy is 0 J/K/kg at 0 K). The thermal conductivities of Nb_3Sn and epoxy were assumed to be negligible when compared to the thermal conductivity of copper. The G10 properties were taken from Ref. [43]. The heater's polyimide insulation was based on the properties of Kapton and taken from Ref. [43] (with an extrapolation presented by Manfreda [38] below 4.3 K). The specific heat of stainless-steel and its thermal conductivity were taken from Ref. [43]. The resistivity of stainless-steel was based on the work of Prestemon [44].

The heater delays were evaluated for heater peak powers from 50 to 300 W/cm^2 with a step size of 25 W/cm^2 in the HF heater and 28 W/cm^2 in the LF heater.[9] Consequently, with the stainless-steel strip thickness of

[9]The power per area is a useful parameter when the efficiency of different heaters are compared. Naturally, it does not represent the heat flux entering the coil because some heat flows into the coolant too.

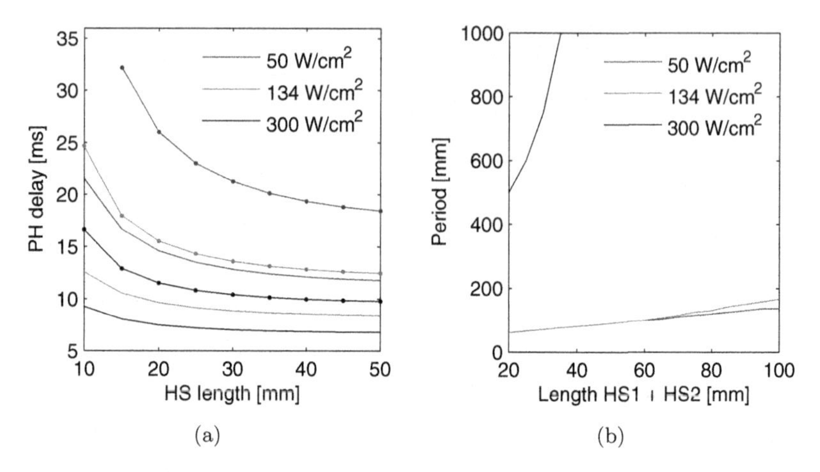

Figure 2.27. (a) LHQ HF heater delay as a function of HS length with peak powers of 50, 134, and 300 W/cm^2: solid lines represent HS1 and lines with markers, HS2. (b) The minimum period that can be used as a function of the sum of HS lengths.

25 μm, the circuit's time constant for the heater's current varied from 73 ms (with 50 W/cm^2) to 12 ms (with 300 W/cm^2). Figure 2.26(a) shows the dependence of the HF heater delays on the heater's longitudinal coverage (HS length) with powers of 50, 134, and 300 W/cm^2. For HS1, the impact of the heating station length started to level off at 25 mm for the powers of 300 W/cm^2 and 134 W/cm^2. The corresponding leveling off starts at 30 mm for the power of 50 W/cm^2. For HS2, the corresponding values were 30 and 35 mm. At the leveling-off length, less than 10% change occurs when the heater length is further extended to 50 mm.

After the heater delays as a function for heater coverage and power were known at each field, all combinations of HS1 and HS2 lengths were considered. The HF heater period length for different HS length combinations is shown in Figure 2.27(b). At 300 W/cm^2, the periods were always larger than 0.5 m and even larger than 1 m if the sum of the heating station lengths was larger than 50 mm. This, combined with only 2–3 ms improvement in the delay compared to the case when the heater power was 134 W/cm^2, suggests that this power is too high for the heater. For the powers of 134 and 50 W/cm^2, the periods were between 60 and 170 mm, which is in the range of practical interest. In this power range, the optimal period depended negligibly on the heater power. This is because there is a need for a minimum wide segment length (the width of the strip at its widest point, i.e. 42 mm) between the transitions of HS1 and HS2 to make it at least a square.

Table 2.8. Results from the heater optimization procedure. The delay times refer to the longest delay times within all the turns the heater covers.

Heater location	HS1 length [mm]	HS2 length [mm]	Period [mm]	Power [W/cm^2]	Time constant [ms]	Heater delay [ms]	Total delay [ms]
HF block	15	50	107	134	27	12.5	14.9
LF block	30	35	100	125	37	14.7	19.3

For both heaters, the layouts that gave the shortest total delays are detailed in Table 2.8. It is clear that these delay times represent a significant quench load for the hot-spot temperature. This heater layout was also drawn and tested in a real coil.

2.5.6. *Testing the designed heater layout*

The designed heater was fabricated and implemented in an LHQ coil that was tested in the so-called mirror structure. This configuration allows testing individual coils without cooling all four coils that are required for a quadrupole magnet [9]. However, the utilization of the mirror structure also brings about some uncertainty in the magnetic flux density distribution.

In the experiment, the outer layer trace had on its one side (including both the LF and HF blocks) the new heater design and on its other side a scale-up of the LQ-style design for comparison. The LF and HF heaters of the new heater design were connected together at the coil end. The reason was that the grooves at the coil's ends were already designed for only one connection, and the coil fabrication schedule did not allow the redesign of the end parts. This was not expected to impact the peak power of the heaters since the voltage will be the same across both strips, i.e. when the LF and HF heaters are in parallel, although the optimal time constants (capacitance values) could not be selected. A more significant disadvantage was that, in this way, the strips could not be tested individually, and the advantage of added redundancy, which is tolerable in experiments but not in the final magnet, was lost. Photos of the implemented heaters are shown in Figure 2.28.

The measurement of the heater's resistances, for both the LHQ and LQ style heaters, and the comparison with calculations highlight some shortcomings of the heater modeling. Table 2.9 lists the computed heater resistances in the design and of the implemented strips using the same analytical resistance calculation as in the design phase and strip resistances

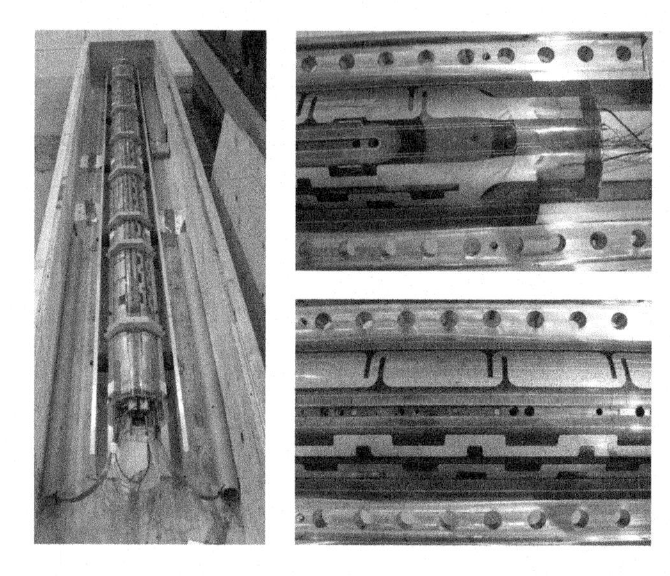

Figure 2.28. Photo of protection heaters impregnated on the outer surface of an LHQ coil, developed as part of the of U.S. LHC Accelerator Research Program (LARP). In this coil, two different heater designs were tested.

Source: Photos by Jesse Schmalzle (Brookhaven National Laboratory) and U.S. LARP.

Table 2.9. Comparison of the strips' resistances (at 4.5 K).

Heater design	**Designed strips in parallel (incl. margin) [Ω]**	**Simple computation for implemented strips [Ω]**	**FEM computation for implemented strips [Ω]**	**Measured from coil in cryostat [Ω]**	**Measured when connected to capacitor [Ω]**
New	2.4	2.6	2.8	3.1	3.5
LQ-style	—	3.7	4.4	4.8	5.2

computation with an FEM-based tool that solves the resistance from a current stationary problem and two measurements. The comparison shows that the analytical model for resistance calculation resulted in about 10% smaller resistance than the FEM modeling, which were computed only afterward because the currents in the heaters were lower than expected. Therefore, the experiment gave feedback to the model too. One cannot directly compare the measured and simulated resistances because the simulations only include the area of the heating stations, whereas in the measurement, the straight end part of the heaters and some wiring are also considered. Of course, one should note that the comparison with the measurement also includes the uncertainty related to the used literature value of stainless-steel resistivity and the fact that the thickness of the implemented heater had uncertainty as well. Also, the connections between the heater strip and the capacitor bank were non-negligible ($0.4\,\Omega$), though in the design, they were not considered. The measured resistance reduced the heaters' power almost by a factor of two from the design value. In future modeling, one should pay special attention to the resistances to get a better idea of the power that can be applied to the heating stations.

The designed heater was powered at 400 V, leading to a power of about $72\,\mathrm{W/cm^2}$ in the HF heater. The LQ-style heater was powered at 280 V in order to obtain the same power. The measured circuit time constants governing the current decay of the heaters, i.e. their power as a function of time, were 25 ms in the LQ-style heater and 17 ms in the new heater. The simulations of the heaters in these conditions were performed in order to find the first delays for both heater layouts that could be compared with measurements. The impact of the wide part was also accounted for in these simulations. It is noteworthy that the measurements can catch only the first delay in the coil. The experimental results and the new simulations with the measured lumped parameters are shown in Figure 2.29. The simulated delays for the new heater agreed very well with the measurement, except at 14.6 kA, where the measured delay was considerably shorter.

Simulation of the LQ-style heater gave delays systematically longer than the measured ones. One reason for this can be the shape of the heating station: It touches only four strands of a cable, and the power is not evenly distributed on the heating station surface. Therefore, the assumption related to the homogeneous current distribution in the heater strip was not adequate in that case — see also Refs. [59, p. 97] and [84].

The goal of inducing a faster current decay with the new heater design was also characterized. The results are summarized in Figure 2.30. According to the results, when the heaters were powered with the same power density, the new heater design was more effective in reducing the MIITs, but when the LQ heater was used at its maximum power at the highest test current, it resulted in slightly smaller MIITs.

The feedback that goes from this experiment back to the heater design process is that for the new heater design style, the simulated delays are

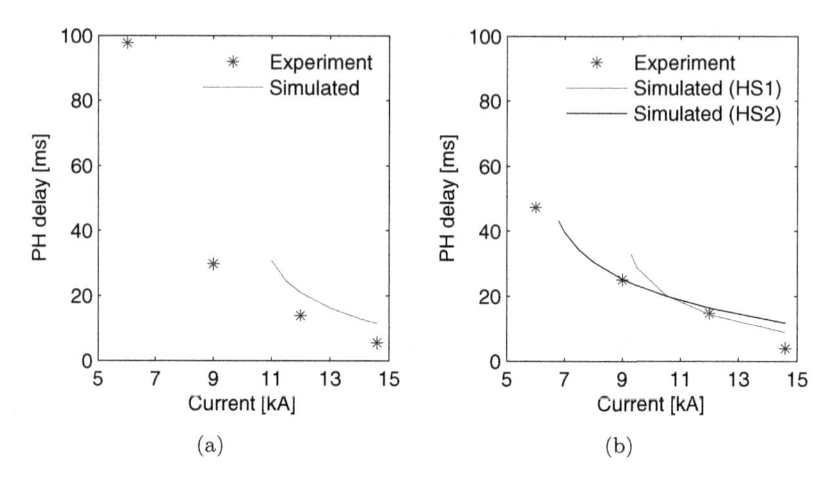

Figure 2.29. Measured heater delays in LHQ compared with simulations for (a) LQ-style heater and (b) new heater design. Delays for both the heating stations of the new heater design at the HF block are shown. The shortest delay times are the ones that should be compared with the measurements.

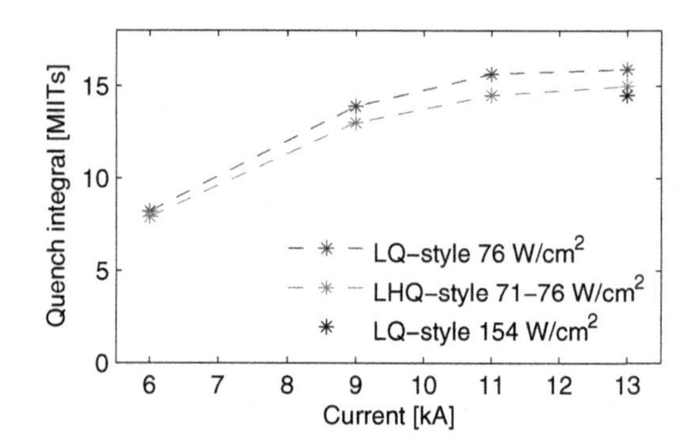

Figure 2.30. Measured LHQ quench load (i.e. the current decay integral in Equation (2.12)) after inducing a quench with the different heater designs.

quite reliable. However, the quench propagation velocity remains uncertain because it could not be characterized in these tests. In addition, the resistance computation for the heaters must be reconsidered, and special attention must be paid to the heaters' connection to the capacitor bank.

Acknowledgements

The research eventually enabling us to contribute this chapter was supported by Stability Analysis of Superconducting Hybrid Magnets (Academy of Finland, #250652) and by Enhanced European Coordination for Accelerator Research & Development (EuCARD-2), which is co-funded by the partners and the European Commission under the Capacities 7th Framework Programme, Grant Agreement 312453. The quench protection heater design study that was presented here for Nb_3Sn magnets was enabled by the US LHC Accelerator Research Program (LARP).

References

[1] Ahoranta M, Lehtonen J, Tarhasaari T and Weiss K. 2008. Modelling of local strain and stress relaxation in bronze processed Nb_3Sn wires. *Supercond. Sci. Technol.* 21, 025005.

[2] Ambrosio G. *et al.* 2011. Test results of the first 3.7 m long Nb_3Sn quadrupole by LARP and future plans. *IEEE Trans. Appl. Supercond.* 21, 1858.

[3] Ambrosio G. 2014. Maximum allowable temperature during quench in Nb_3Sn accelerator magnets. *CERN Yellow Report* CERN-2013-006, 43. doi: http://dx.doi.org/10.5170/CERN-2013-006.43.

[4] Ambrosio G. 2015. Nb_3Sn high field magnets for the high luminosity LHC upgrade project. *IEEE Trans. Appl. Supercond.* 25, 4002107.

[5] Baccaglioni G, Blau B, Cartegni GC, Horvath IL, Neuenschwander J, Pedrini D, Rossi L and Volpini G. 2002. Production and qualification of 40 km of Al-stabilized NbTi cable for the ATLAS experiment at CERN. *IEEE Trans. Appl. Supercond.* 12, 1215.

[6] Bíró O, Preis K and Richter K. 1996. On the use of the magnetic vector potential in the nodal and edge finite element analysis of 3D magnetostatic problems. *IEEE Trans. Mag.* 32, 651.

[7] Blomberg T. 1996. Heat conduction in two and three dimensions: Computer modelling of building physics applications. PhD Thesis at Lund University, Lund, Sweden.

[8] Bortot L. *et al.* 2018. STEAM: A hierarchical cosimulation framework for superconducting accelerator magnet circuits. *IEEE Trans. Appl. Supercond.* 28.

[9] Bossert R. *et al.* 2012. Optimization and test of 120 mm LARP Nb_3Sn quadrupole coils using magnetic mirror structure. *IEEE Trans. Appl. Supercond.* 13, 1297.

[10] Bottura L. 1998. Quench propagation and protection of cable-in-conduit superconductors. In Seeber B. (ed.) *Handbook of Applied Superconductivity,* Institute of Physics Publishing, Bristol, Vol. C3.

[11] Bottura L, de Rijk G, Rossi L and Todesco E. 2012. Advanced accelerator magnets for upgrading the LHC. *IEEE Trans. Appl. Supercond.* 22, 4002008.

[12] Brenner S and Scott R. 2008. *The Mathematical Theory of Finite Element Methods,* Springer-Verlag, New York.

[13] Çengel Y. 2003. *Heat Transfer a Practical Approach,* New York: McGraw-Hill Education.

[14] Cheng D K. 1993. *Fundamentals of Engineering Electromagnetics,* New York: Prentice Hall.

[15] CryoComp material property database of materials often utilized in cryogenic temperatures Eckels Enginering Inc. http://www.eckelsengineering .com/.

[16] Devred A. 1992. Quench origins. *AIP Conf. Proc.* 249, 1262.

[17] Ekin J W. 1987. Effect of transverse compressive stress on the critical current and upper critical field of Nb_3Sn. *J. Appl. Phys.* 62, 4829.

[18] Eyssa Y and Markiewicz W D. 1995. Quench simulation and thermal diffusion in epoxy-impregnated magnet system. *IEEE Trans. Appl. Supercond.* 5, 487.

[19] Eyssa Y, Markiewicz W D and Miller J. 1997. Quench, thermal, and magnetic analysis computer code for superconducting solenoids. *IEEE Trans. Appl. Supercond.* 7, 159.

[20] Fazilleau P, Devaux M, Durante M, Lecrevisse T and Rey J-M. 2013. Protection of the 13 T Nb_3Sn Fresca II dipole. *CERN Yellow Report* CERN-2013-006, 65.

[21] Felice H. *et al.* 2009. Instrumentation and quench protection for LARP Nb_3Sn magnets. *IEEE Trans. Appl. Supercond.* 19, 2458.

[22] Felice H. 2014. Quench protection analysis in accelerator magnets, a review of the tools. *CERN Yellow Report* CERN-2013-006, 17.

[23] Ferracin P. *et al.* 2016. Development of MQXF: The Nb_3Sn low-β quadrupole for the HiLumi LHC. *IEEE Trans. Appl. Supercond.* 26, 400207.

[24] Geuzaine C and Remacle J F. 2009. Gmsh: A three-dimensional finite element mesh generator with built-in pre- and post-processing facilities. *Int. J. Num. Meth. Eng.* 79, 1309.

[25] Gianotti F. *et al.* 2005. Physics potential and experimental challenges of the LHC luminosity upgrade. *Eur. Phys. J. C* 39, 293.

[26] Git — a version control system. https://git-scm.com/.

[27] Grilli F, Pardo E, Stenvall A, Nguyen D N, Yuan W and Gömöry F. 2013. Computation of losses in HTS under the action of varying magnetic fields. and currents. *IEEE Trans. Appl. Supercond.* 24, 8200433.

[28] Godeke A, Chlachidze G, Dietderich D R, Ghosh A K, Marchevsky M, Mentink M G T and Sabbi G L. 2013. A review of conductor performance for the LARP high-gradient quadrupole magnets. *Supercond. Sci. Technol.* 26, 095015.

[29] Goldacker W, Grilli G, Pardo E, Kario A, Schlachter S I and Vojenčiak M. 2014. Roebel cables from REBCO coated conductors: A one-century-old concept for the superconductivity of the future. *Supercond. Sci. Technol.* 27, 093001.

[30] Gyuráki R, Sirois F and Grilli F. 2018. High-speed fluorescent thermal imaging of quench propagation in high temperature superconductor tapes. *Supercond. Sci. Technol.* 31, 034003.

[31] Härö E. *et al.* 2013. Quench considerations and protection scheme of a high field HTS dipole insert coil. *IEEE Trans. Appl. Supercond.* 23, 4600104.

[32] Härö E and Stenvall A. 2014. Reducing modeling domain to speed-up quench simulations of HTS coils. *IEEE Trans. Appl. Supercond.* 24, 4900705.

[33] Härö E, Stenvall A, van Nugteren J and Kirby G. 2015. Modelling of minimum energy required to quench an HTS magnet with a strip heater. *IEEE Trans. Appl. Supercond.* 25, 4701505.

[34] Hairer E, Nørsett S P and Wanner G. 1993. *Solving Ordinary Differential Equations: Nonstiff Problems.* Berlin: Springer-Verlag.

[35] Kim S-W. 2000. Material properties for quench simulation (Cu, NbTi and Nb_3Sn). *Fermilab TD Note 00-041.*

[36] Kirby G. *et al.* 2014. Accelerator quality HTS dipole magnet demonstrator designs for the EuCARD-2, 5 Tesla 40 mm clear aperture magnet. *IEEE Trans. Appl. Supercond.* 25, 4000805.

[37] Lehtonen J, Mikkonen R and Paasi J. 2000. Effective thermal conductivity in HTS coils. *Cryogenics* 40, 245.

[38] Manfreda G. 2011. Review of ROXIE's material database for quench simulations. *CERN Internal Note 2011-24* EDMS Nr. 1178007.

[39] Manfreda G, Rossi L and Sorbi M. 2012. MATPRO an upgraded version 2012: A computer library of material property at cryogenic temperature. *Instituto Nazionale di Fisica Nucleare, Italy, Report* INFN-12-04/MI.

[40] Marchevsky M. *et al.* 2012. Quench performance of HQ01, a 120 mm bore LARP quadrupole for the LHC upgrade. *IEEE Trans. Appl. Supercond.* 22, 4702005.

[41] Martin P S. 1986. Design and operation of the quench protection system for the Fermilab Tevatron. *Fermilab TM Note 1398.*

[42] Mess K-H, Schmüser P and Wolff S. 1996. *Superconducting Accelerator Magnets.* World Scientific, Singapore.

[43] National Institute of Standards and Technology, Material Measurement Laboratory, Cryogenic Technologies Group, Material Properties at cryogenic temperatures. http://cryogenics.nist.gov/MPropsMAY/material/%20 properties.htm.

[44] Prestemon S. *Lawrence Berkeley National Laboratory, Berkeley, California, USA.* Private communication with Salmi T.

[45] Ravaioli E, Datskov V I, Giloux C, Kirby G, Ten Kate H H J and Verweij A P. 2014. New, coupling loss induced, quench protection system for superconducting accelerator magnets. *IEEE Trans. Appl. Supercond.* 24, 0500905.

[46] Rodrigues-Mateos F, Pugnat P, Sanfilippo S, Schmidt S, Siemko A and Sonnemann F. 2000. Quench heater experiments on the LHC main superconducting magnets. *Proc. EPAC 2000* 2154.

[47] Rodriguez-Mateos F and Sonnemann F. 2001. Quench heater studies for the LHC magnets. *Proc. PAC 2001* 53451.

[48] Rossi L and Sorbi M. 2004. QLASA: A computer code for quench simulation in adiabatic multicoil superconducting windings. *INFN/TC-04/13.*

[49] Rossi L. 2010. Superconductivity: Its role, its success and its setbacks in the Large Hadron Collider of CERN. *Supercond. Sci. Technol.* 23, 034001.

[50] Rossi L. *et al.* 2015. The EuCARD-2 future magnets European collaboration for accelerator-quality HTS magnets. *IEEE Trans. Appl. Supercond.* 25, 4001007.

[51] Rostila L, Lehtonen J, Mikkonen R, Šouc J, Seiler E, Melíšek T and Vojenčiak M. 2007. How to determine critical current density in YBCO tapes from voltage-current measurements at low magnetic fields. *Supercond. Sci. Technol.* 20, 1097.

[52] Roxie, A computational tool for thermal and electromagnetic simulation and optimization of accelerator magnets. https://espace.cern.ch/roxie/default.aspx.

[53] SUite for Nonlinear and DIfferential/ALgebraic Equation Solvers (SUNDIALS). https://computation.llnl.gov/casc/sundials/.

[54] Sabbi G L. (for the LARP collaboration) 2011. Progress in high field accelerator magnet development by the US LHC Accelerator Research Program. *CERN Yellow Report* CERN-2011-003, 30. http://arxiv.org/abs/arXiv:1108.1625v1.

[55] Sætre F, Hiltunen I, Runde M, Magnusson N, Järvelä J, Bjerkli J and Engebrethsen E. 2011. Winding, cooling and initial testing of a 10 H superconducting MgB_2 coil for an induction heater. *Supercond. Sci. Technol.* 24, 035010 (5pp).

[56] Salmi T. *et al.* 2012. Quench protection challenges in long Nb_3Sn accelerator magnets. *AIP Cond. Proc.* 1434, 656.

[57] Salmi T. *et al.* 2014. Protection heater delay time optimization for high-field Nb_3Sn accelerator magnets. *IEEE Trans. Appl. Supercond.* 24, 4701305.

[58] Salmi T, Arbelaez D, Caspi S, Felice H, Mentink M G T, Prestemon S, Stenvall A and ten Kate H H J. 2014. A novel computer code for modeling quench protection heaters in high-field Nb_3Sn accelerator magnets. *IEEE Trans. Appl. Supercond.* 24, 4701810.

[59] Salmi T. 2015. Optimization of quench protection heaters performance in high-field accelerator magnets through computational and experimental analysis. PhD Thesis at Tampere University of Technology, Tampere, Finland.

[60] Salmi T, Chlachidze G, Marchevsky M, Bajas H, Felice H and Stenvall A. 2015. Analysis of uncertainties in protection heater delay time measurements and simulations in Nb_3Sn high-field accelerator magnets. *IEEE Trans. Appl. Supercond.* 25, 4004212.

[61] Salmi T, Prioli M, Stenvall A and Verweij A P. 2019. Quench protection of the 16 T Nb3Sn dipole magnets designed for the future circular collider. *IEEE Trans. Appl. Supercond.* 29, 4700905.

[62] Schmidt R. *et al.* 2006. Protection of the CERN large Hadron Collider. *New J. Phys.* 8, 290.

[63] Schwerg N, Auchmann B and Russenschuck S. 2008. Quench simulation in an integrated design environment for superconducting magnets. *IEEE Trans. Magn.* 44, 934.

[64] Schwerg N, Auchmann B and Russenschuck S. 2009. Challenges in the thermal modeling of quenches with ROXIE. *IEEE Trans. Appl. Supercond.* 19, 1271.

[65] Selvamanickam V, Yao Y, Chen Y, Shi T, Liu Y, Khatri N D, Liu J, Lei C, Galstyan E and Majkic G. 2012. The low-temperature, high-magnetic-field critical current characteristics of Zr-added (Gd,Y)$Ba_2Cu_3O_x$ superconducting tapes. *Supercond. Sci. Technol.* 25, 125013.
[66] Stenvall A, Korpela A, Mikkonen R and Kováč P. 2006. Critical current of an MgB_2 coil with a ferromagnetic matrix. *Supercond. Sci. Technol.* 19, 32.
[67] Stenvall A, Korpela A, Mikkonen R and Grasso G. 2006. Stability considerations of multifilamentary MgB_2 tape. *Supercond. Sci. Technol.* 19, 184.
[68] Stenvall A, Korpela A, Mikkonen R and Grasso G. 2006. Quench analysis of MgB_2 coils with a ferromagnetic matrix. *Supercond. Sci. Technol.* 19, 581.
[69] Stenvall A, Mikkonen R and Kováč P. 2010. Comparison of 1D, 2D and 3D quench onset simulations. *Physica C* 470, 2047.
[70] Subversion — version control system. https://subversion.apache.org/.
[71] Todesco E. 2014. Quench limits in the next generation of magnets. *CERN Yellow Report* CERN-2013-006, 10. doi: http://dx.doi.org/10.5170/CERN-2013-006.10.
[72] Wan A. 2014. Adaptive space-time finite element method in high temperature superconductivity. PhD Thesis at École Polytechnique de Montréal, Montréal, Canada.
[73] Wang Y, Lu Y, Xu X, Dai S, Hui D, Xiao L and Lin L. 2007. Detecting and describing the inhomogeneity of critical current in practical long HTS tapes using contact-free method. *Cryogenics* 47, 225.
[74] Whetstone, C N and Roos C E. 1968. Thermal phase transitions in superconducting Nb-Zr alloys. *J. Appl. Phys.* 36, 783.
[75] Wilson M N. 1983. *Superconducting Magnets.* Oxford University Press, Oxford.
[76] Wilson E L. 1998. *Three Dimensional Static and Dynamic Analysis of Structures.* Computers and Structures Inc., Berkeley.
[77] Yamada R, Marscin E, Lee A, Wake M and Rey J-M. 2003. 2-D/3-D quench simulation using ANSYS for epoxy impregnated Nb_3Sn high field magnets. *IEEE Trans. Appl. Supercond.* 13, 1696.
[78] Zhao J, Stenvall A, Salmi T, Gao Y and Lorin C. 2017. Mechanical behavior of a 16-T FCC dipole magnet during a quench. *IEEE Trans. Appl. Supercond.* 27, 4004407.
[79] Zhao J, Prioli M, Stenval A, Salmi T, Gao Y, Caiffi B, Lorin C, Marinozzi V, Farinon S and Sorbi M. 2018. Mechanical stress analysis during a quench in CLIQ protected 16 T dipole magnets designed for the future circular collider. *Physica C* 550, 27.
[80] van Lanen E and Nijhuis A. 2011. Simulation of interstrand coupling loss in cable-in-conduit conductors with jackPot-AC. *IEEE Trans. Appl. Supercond.* 21, 1926.
[81] van Nugteren J. 2011. Software development for the science and design behind superconducting magnet systems. *CERN Internship Report.*
[82] van Nugteren J. *et al.* 2018. Powering of an HTS dipole insert-magnet operated standalone in helium gas between 5 and 85 K. *Supercond. Sci. Technol.* 31, 065002.

[83] van der Laan D C, Ekin J W, Clickner C C and Stauffer T C. 2007. Delamination strength of YBCO coated conductors under transverse tensile stress. *Supercond. Sci. Technol.* 20, 765.
[84] Marchevsky M, Turqueti M, Cheng D W, Felice H, Sabbi G, Salmi T, Stenvall A, Chlachidze G, Ambrosio G, Ferracin P, Izquierdo Bermudez S, Perez J C and Todesco E. 2016. Protection heater design validation for the LARP magnets using thermal imaging. *IEEE Trans. Appl. Supercond.* 26, 4003605.
[85] Izquierdo Bermudez S, Ambrosio G, Bossert R, Cheng D, Ferracin P, Krave S T, Perez J C, Schmalzle J and Yu M. 2016. Coil end optimization of the Nb3Sn quadrupole for the high luminosity LHC. *IEEE Trans. Appl. Supercond.* 25, 4001504.
[86] Felice H, Ambrosio G, Anerella M, Bossert R, Caspi S, Cheng D W, Dietderich D R, Ferracin P, Ghosh A K, Hafalia R, Hannaford C R, Kashikhin V, Schmalze J, Prestemon S, Sabbi G L, Wanderer P and Zlobin A V. 2009. Design of HQ—A high field large bore Nb_3Sn quadrupole magnet for LARP. *IEEE Trans. Appl. Supercond.* 19, 1235.

Chapter 3

Finite Element Structural Modeling

Nathaniel C. Allen and Luisa Chiesa
Mechanical Engineering Department,
Tufts University, Medford, MA 02155, USA

3.1. Introduction

In practical applications, high-temperature superconductors (HTS) are subject to stress from fabrication, thermal contraction, and operation. This stress can at times be substantial, producing high levels of strain, which contribute to a critical current degradation of the superconductor. The performance of superconductors under stress and strain is in a large part governed by their mechanical properties.

The main types of stress that have the most significant influence on HTS, particularly for magnet applications, are bending, torsion, uniaxial and transverse loads. Among other sources, bending stress occurs from winding of magnet coils, torsion stress from twisting during fabrication to minimize AC loss, uniaxial stress from different thermal contraction rates, and transverse stress from electromagnetic Lorentz forces.

Understanding the critical current behavior of HTS under stress and characterizing their mechanical properties is an important research area being investigated. Structural finite element analysis (FEA) is a powerful computational method being used to study this electromechanical behavior of superconductors.

FEA uses a numerical technique called the finite element method, where an object or system is discretized into simple geometric shapes called finite elements, which are interconnected at nodes. By assembling and solving a system of governing equations for each finite element (comprising constitutive relationships, material properties, degrees of freedom, etc.), the behavior over the domain of the entire system can be approximated. The advantage of FEA as a numerical tool is its ability to approximate complex systems with no closed-form analytical solution.

3.2. HTS Tapes and Cables

Second-generation rare-earth-barium-copper-oxide (REBCO)-coated conductors have a thin rectangular tape geometry and a multilayered composite architecture, as illustrated by the schematic layout in Figure 3.1. These HTS tapes have a strong structural material for a substrate, on which ceramic oxide buffer layers and a superconducting layer are deposited by chemical or physical means. A silver cap is applied on top of the REBCO layer followed by a coating of copper for stability. These HTS tapes are available from a variety of manufacturers, each with its own unique characteristics, including materials, thicknesses, layers, and deposition techniques [1–6].

HTS such as these REBCO-coated conductors have great mechanical properties, which are governed by the strength of the substrate material and

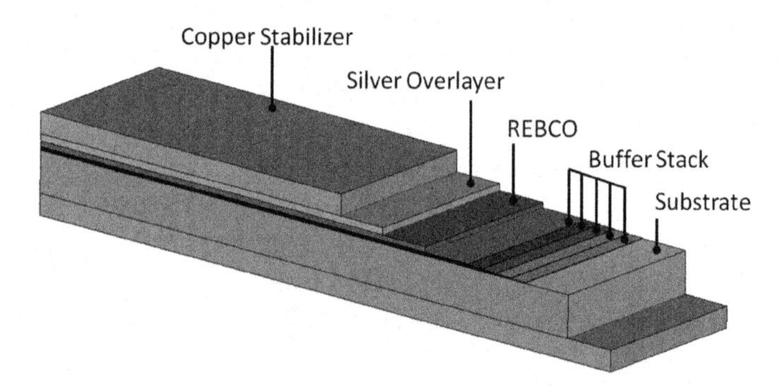

Figure 3.1. Schematic layout of the multilayered composite architecture of REBCO-coated conductors (HTS tapes).

Note: Layers are not to scale.

Source: M. Takayasu, J. V. Minervini and L. Bromberg, Massachusetts Institute of Technology.

the thickness of the copper stabilizer. They also have excellent high-current capabilities at high magnetic fields, making them very promising conductors for applications, such as accelerator and fusion magnets. The main challenge of REBCO-coated conductors is their flat tape architecture, which causes mechanical and field-orientation anisotropies.

Various applications require HTS-cabled conductors with low AC loss and high current density. Several novel cabling methods for REBCO-coated conductors are under development. Examples of the five predominant cabling techniques are shown in Figure 3.2. Each cabling method has differences in assembly, tape utilization, and transposition and mechanical properties, among other things. The Roebel assembled coated conductor (RACC) punches HTS tapes with a zigzag pattern and then assembles them into a Roebel bar configuration [7]. The conductor on round core (CORC) cables tightly winds multiple coated conductors in a helical fashion on a small round former [8]. Twisted stacked-tape cables (TSTC) stack multiple tapes into a rectangular configuration and then twist them along the stack axis [9]. A variation of the stacked-tape cable is round strands made of twisted stacks, in which the stack of coated conductors is placed between two half-round copper profiles [10]. These cables can also be used as the base conductor for larger multistage Rutherford or cable-in-conduit

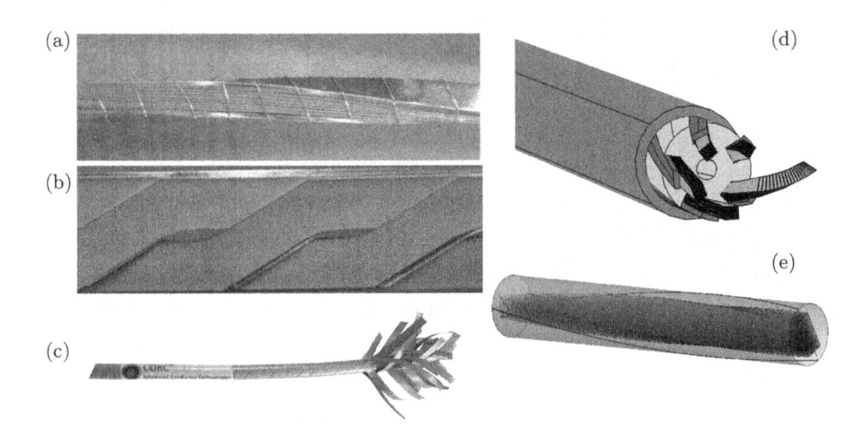

Figure 3.2. HTS cabling methods: (a) twisted stacked-tape cable, (b) Roebel assembled coated conductor, (c) conductor on round core cables, (d) HTS slotted-core CICC, and (e) round strands made of twisted stacks.

Source: (a) M. Takayasu, J. V. Minervini and L. Bromberg, Massachusetts Institute of Technology; (b) A. Kario, Karlsruhe Institute of Technology, Germany; (c) Advanced Conductor Technologies; (d) G. De Marzi and G. Celentano, ENEA, Italy; (e) N. Bykovskiy, Paul Scherrer Institute, Switzerland. More details in Refs. [11–13].

conductors (CICC) made with multiple cables. Another cabling technique, designed particularly for fusion applications, is the HTS slotted-core CICC, where stacks of tape are placed in five helical slots of an aluminum core [11]. The cable also has an external jacket and a central cooling channel for liquid cryogen.

To improve cable processing, limit fabrication degradations, and maximize operational performance, it is essential to characterize the mechanical properties of the HTS tapes and cables and determine their electromechanical behavior under stress. Structural FEA is one numerical technique being used for this investigation.

3.3. FEA Research Areas

Various institutions are utilizing FEA as an insightful tool to investigate the electromechanical behavior and mechanical properties of HTS. Finite element simulations are being used in conjunction with experiments to predict and validate results. Numerical modeling is also being done to better understand the degradation mechanisms of HTS tapes under stress. A variety of methods are being used for modeling the stress and strain states of these tapes under mechanical loads with both simple two-dimensional (2D) and more complex three-dimensional (3D) simulations. Single REBCO-coated conductors have been investigated as well as the behavior of HTS cables.

The structural FEA modeling being conducted focuses on two main aspects: mechanical characteristics and electromechanical behavior. Mechanical characteristics investigate stress–strain response, forces, and torques, as well as yield strength and elastic–plastic behavior. These mechanical characteristics are typically a good way of validating the numerical model when compared with experiments. The electromechanical behavior analyzes how the critical current changes with applied stress or strain to determine irreversible limits under various loads. The critical current can be predicted from structural models by identifying the longitudinal strain state in the superconductor.

This chapter briefly describes the type of numerical FEA simulations being done on single tapes and cables. Their corresponding results are described in Section 3.6.

3.3.1. *Single-tape simulations*

Single HTS tapes under a variety of loading conditions (including tension, torsion, transverse, and thermal) have been analyzed using 3D structural

FEA models [12–18]. The simulations have been conducted on tapes from a variety of manufacturers and using a mixture of numerical modeling techniques.

Single tapes under uniaxial tension have been investigated a great deal [12, 14–17]. Structural finite element modeling has been used to both characterize the stress–strain curves of the coated conductors and predict their critical current behavior under tensile stress. Simulating the mechanical stress–strain behavior is typically done to identify the yield strength and modulus of elasticity of the composite tapes and also to validate the numerical model being used. Predicting the critical current with structural models is normally done to identify the critical load that causes irreversible degradation in the superconductor. The critical current behavior can be predicted by identifying the strain state in the superconducting layer under stress. Specific to Roebel cables, the uniaxial tensile characteristics of single zigzag tapes have also been analyzed to determine the ideal punched shape to minimize stress concentrations [17].

Single tapes under pure axial torsion are also an important topic being explored [12, 13, 18]. Numerically, FEA has been used to examine the strain state of the coated conductors under torsion to predict their critical current behavior and ultimately to identify a minimum twist pitch achievable before irreversible degradation. The required mechanical torque to twist these samples was also analyzed for cable fabrication purposes.

Single tapes under a combined loading case of tension–torsion were also studied [12–16]. This loading case is of particular interest for twisted cables used in magnet coils because the twisted tapes within the cable will experience a tensile hoop stress, leading to a combined tension–torsion load. Structural FEA is being used to determine the change in strain of a twisted conductor as tension is applied and how that change affects the predicted critical current behavior. Tensile stress was applied to tapes with various twist pitches to determine the influence of the degree of twist. The change in mechanical torque as tension is applied to the twisted tape was also analyzed.

Single tapes under transverse compressive stress were investigated with FEA [15, 16]. Numerical simulations were used to identify the strain in the superconductor to determine the critical load corresponding to irreversible critical current degradation. The mechanism of degradation for transverse load was studied to discover any correlation with uniaxial tension. The influence of the copper stabilizer thickness on the coated conductor under transverse load was also analyzed.

The intrinsic strain state of single REBCO tape after the production process was analyzed by including the temperature variations during the tape production into the structural model [15, 16]. The FEA modeling was done to determine the intrinsic residual strain in the superconducting REBCO layer after the production process. Structural simulations were run using thermal (temperature variation) loads along with temperature-dependent material properties. The strain effect from cool down to cryogenic operation temperature was also studied.

3.3.2. *Cable simulations*

Building on single tape simulations, numerous structural models of cables have been developed. Focusing on the implementation of HTS cables in magnet applications, the bending behavior and electromagnetic transverse load characteristics have been investigated.

The electromechanical behavior of TSTC, HTS slotted-core CICC, and CORC cables under bending has been investigated a great deal [12, 16, 19]. Numerically, FEA has been used to study the overall cable and individual tape performances under pure bending loads. Full 3D simulations of each cable were required to capture the appropriate bending characteristics. The influence of friction on the mechanical and electrical performances of the cable was also investigated. In addition to the mechanical characteristics of the cable under bending, the structural simulations were used to determine the bending strain in each tape, which was used to predict the critical current behavior. Identifying the minimum bending diameter before irreversible critical current degradation was the primary focus of the analysis.

Transverse compression stress on CORC and TSTC conductors has also been explored [20, 21]. Numerically, simple 2D FEA models of the cable cross-section were used to limit the computational time and complexity. The transverse stress equivalent to that anticipated from electromagnetic loads in operation was simulated. FEA was used to determine the maximum transverse stress on an individual tape to identify any potential critical current degradation. The simulations were also used to identify any cross-sectional deformation under load to investigate the best support mechanism for high-current HTS cables.

Similar 2D simulations of the HTS slotted-core CICC cross-section were run to analyze the radial compaction caused by external jacketing of the CICC [10]. The radial shrinkage of the HTS CICC cross-section during cool down to cryogenic operating temperature was also simulated numerically.

The potential strain produced on the coated conductors within an HTS cable after cool down to 4.2 or 77 K was also analyzed [16]. The HTS tapes and cable support structure, unavoidably, have different rates of thermal contraction, which could lead to stress accumulation in the tapes, depending on their end constraints and their ability to slide freely.

3.4. Modeling Techniques for Single Tapes

Single-tape models of coated conductors have been analyzed using 3D structural FEA under tension, torsion, transverse, and thermal loads. Modeling and loading a thin rectangular tape is relatively straightforward, but the challenge in modeling coated conductors is accurately defining their internal layered structure and the appropriate elastic–plastic material properties. This chapter addresses the numerical techniques and approaches used to modeling single HTS tapes under various loads.

3.4.1. *Finite element software and settings*

FEA is a powerful numerical tool for modeling the mechanical behavior of superconductors. The main commercial finite element software packages being used are ANSYS®, ABAQUS®, and COMSOL® Multiphysics. ANSYS® and ABAQUS® are the primary programs being used for structural modeling of HTS tapes and cables. COMSOL® is more widely used for thermal and fluids simulations and employs a different approach to finite element modeling and therefore will not be discussed here.

ANSYS®, and ABAQUS® are alike in the manner in which they approach FEA modeling. They both have a diverse library of finite elements, each built from governing equations and degrees of freedom. The modeler needs only to choose the appropriate element type for the physics and geometry being modeled, and the software will compile and calculate the system of governing equations automatically. For clarity, terminology relevant to both software packages will be used throughout the chapter.

The first step in modeling is preprocessing, where the user must build or input the system geometry being modeled. During this step, the modeler must: apply appropriate material properties, choose a suitable element type and discretize the model geometry, apply contact relations and nodal connectivity, and prescribe degree-of-freedom constraints and loads.

The processing or analysis step allows the modeler to choose the desired analysis settings before the simulation is computed. The vast majority of structural modeling for HTS is run as static simulations. Linear analyses are used to model simple elastic behavior, while more commonly, nonlinear analyses are used to capture nonlinear structural behavior arising from plastic material properties, geometric nonlinearities, or irregular boundary conditions and loads. Large deformation and enhanced strain formulations are typically used in conjunction with nonlinear simulations. Various mathematical formulations, including sparse direct and iterative matrix solution techniques as well as Newton integration methods for nonlinear problems, are commonly used in practice.

The final step in finite element modeling is postprocessing, where the user evaluates and displays the simulation results. For structural modeling, the stress and strain are of particular interest, as well as the resulting forces and moments. Additional data manipulation steps are also done. One example of this is taking the numerical strain results from the FEA simulations and relating them to an expected critical current using experimental data, as described later.

3.4.2. *REBCO-coated conductor architecture*

HTS tapes have a layered structure comprising a thick substrate material, a thin buffer and REBCO layers, and a copper coating, as shown in Figure 3.1. In most circumstances when modeling the mechanical behavior of these coated conductors, the strain in the REBCO layer is of interest, in addition to the overall behavior of the composite tape. It has been established that the mechanical characteristics of these layered tapes are primarily governed by the strength and thickness of their substrate and copper layers. Consequently, the numerical models of coated conductors must include the substrate, copper stabilizer, and superconducting REBCO layers as parts of their composite architecture. Determining how to appropriately model this layered structure can be the most challenging, yet most important, aspect of modeling.

There are two main approaches being used to model the composite architecture of the REBCO-coated conductors. The first approach models the layered structure with physical geometric volumes, each having its own material properties and finite element mesh. The second approach models the entire composite tape as one single uniform volume with its layered structure defined within the finite elements used to mesh it. A depiction of both approaches is provided in Figure 3.3.

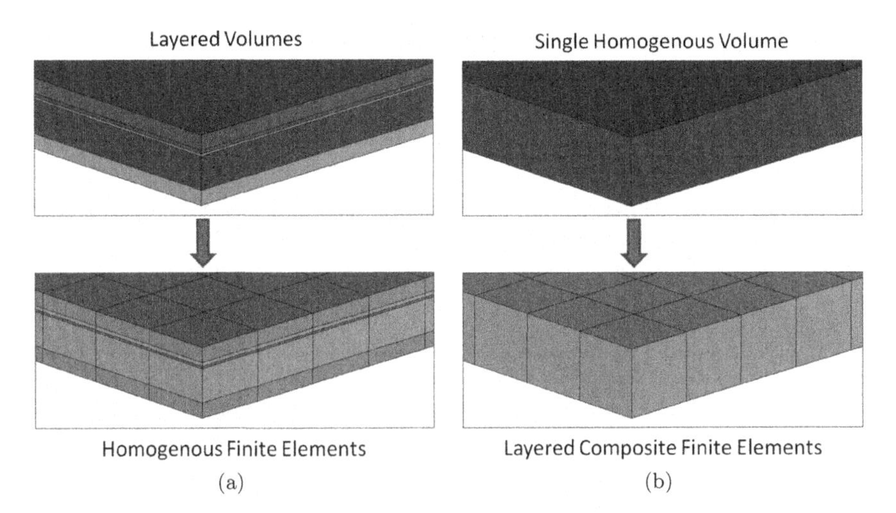

Figure 3.3. Depiction of the two approaches used to model the layered architecture of HTS tapes: (a) homogenous and (b) layered composite elements.

In the first approach, the entire internal layered structure of the coated conductor is included in a 3D geometric model. Every layer is created as a separate physical volume with its own thickness and material properties. The geometric model consists of a stack of volumes representing each layer in the coated conductor. In this approach, the geometric model must be meshed with a minimum of one element through the thickness of each layer or volume. The thicker layers (substrate and copper) are typically meshed with multiple elements through their thickness, while the thinner layers (buffer, REBCO, and silver) are meshed with one element to produce a more uniform element size.

The second approach uses a novel technique to simplify the 3D geometric model by defining the internal layered structure of the coated conductor within composite finite elements. In this approach, the entire tape is modeled simply as one homogenous geometric volume. The homogenous volume must be discretized with exactly one finite element through its thickness. The order, thickness, and material properties of each layer are then defined within the finite elements used to mesh the homogenous volume. Even though one element is used through the entire tape thickness, the results for each individual layer can be selected during postprocessing.

The second approach not only greatly reduces the 3D geometric model complexity but also eliminates the challenge of meshing neighboring layers with very different thicknesses (e.g. substrate and buffer layers). Additionally, the second approach greatly reduces the total number of nodes and elements and therefore tends to reduce the overall computation time. This makes the second approach particularly well suited for modeling full-scale HTS cables composed of many tapes. Defining the layered structure within the finite elements also makes customizing the model to the exact composite architecture from each manufacturer easier.

One limitation of the second approach is that though interlayer failure can be predicted by stress and strain levels, physical interlayer failures, such as delimitation and cracking, cannot be simulated. The first approach, with the addition of appropriate contact relations between every layer, could be used to study these physical interlayer phenomena. The first approach, with many elements through the thickness of each layer, is also a more advantageous way to model stress through the thickness of the tape for loads such as transverse compression.

3.4.3. *Element types*

The two main types of finite elements that are typically used to model coated conductors in 3D are structural solid (continuum) and solid-shell hexahedron "brick" elements. The names of the equivalent solid and solid-shell elements in ANSYS® and ABAQUS® are listed in Table 3.1. The continuum and solid-shell elements are designed to model both uniform and layered structural solids and have three translational degrees of freedom at each node (x, y, and z directions). Both element types also have linear elastic and full nonlinear capabilities, including plasticity, stress stiffening, large deflection, and large strain.

Brick elements are almost exclusively used because they are the most tolerant element geometry to high aspect ratios (the ratio of element thickness to width and length), making them ideal for modeling the thin rectangular geometry of the HTS tapes. Additionally, the similarity between the brick

Table 3.1. Finite elements used to mesh coated conductors.

Element Type	ANSYS®	ABAQUS®	Nodes	Geometry
3D structural solid (continuum)	SOLID185 SOLID186	C3D8 C3D20	8 20	Hexahedron (brick)
3D structural solid-shell (continuum shell)	SOLSH190	SC8R	8	Hexahedron (brick)

element and the tape geometry permits the use of mapped (structured) meshing, allowing the modeler to define a more uniform and repeatable patterned mesh.

Typically, linear eight-node brick elements with sufficient mesh density are adequate for most applications, while certain circumstances can necessitate the use of higher-order 20-node hexahedron elements with quadratic shape functions. Higher-order elements typically produce more accurate results for the same element density but also require longer computation times because of the higher node count.

Solid-shell elements are structural solid elements with built-in shell element capabilities. They have full 3D continuum element topology and are specifically designed to analyze thin shell-like structures. The elements have an internal suite of special kinematic formulations to produce additional shell-like characteristics. Solid-shell elements are, as a result, more suitable for modeling thin geometries and produce more accurate results compared to standard solid elements.

Solid and solid-shell finite elements both have the ability to be defined as either homogenous elements or layered composite elements, as illustrated in Figure 3.4. Therefore, both element types can be used to model coated conductors in either approach described in Section 3.4.2. For composite elements, the layered structure, including the number, order, thickness, and material properties, are all defined within the finite elements.

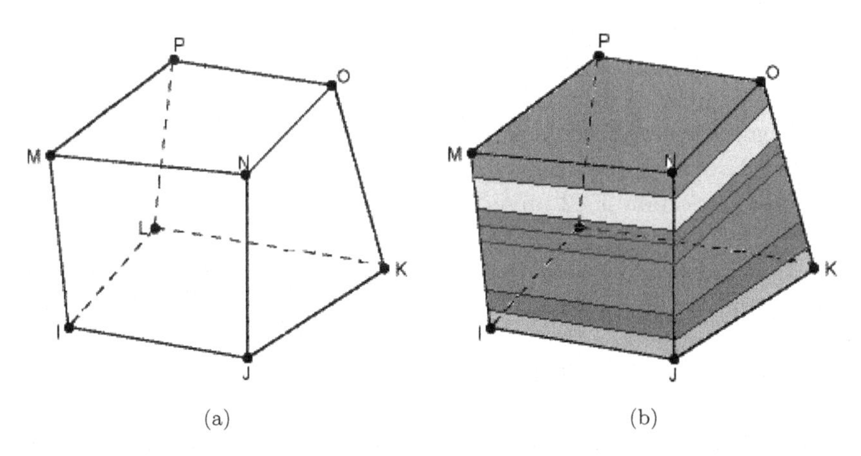

Figure 3.4. Schematics of a solid or solid-shell brick element having (a) homogenous and (b) layered composite characteristics.

3.4.4. *Meshing*

The overall number of elements (mesh density) as well as the element type and geometry influence the results of numerical FEA simulations. Generally, the greater the mesh density, the more accurate the results. At some point, the analysis results will converge to a unique solution, and any further increase in the mesh density will result in greater computation times with negligible improvement in accuracy. Mesh analysis should be performed to determine the optimal mesh density for accurate results with reasonable computation times.

Coated conductors are commonly meshed using hexahedron "brick" elements, as discussed in Section 3.4.3. Brick elements are favored because they can tolerate the high aspect ratios required to mesh the relatively long and thin rectangular geometry of HTS tapes. Brick topology is also advantageous because it permits mapped (structured) meshing, which provides the modeler with the most control over the finite element mesh. Mapped meshes are used to produce very uniform and repeatable patterned meshes. A mapped mesh can easily be defined by prescribing the number of elements through the thickness, width, and length of the coated conductor.

For accurate strain results, sufficient mesh density in each dimension was chosen based on the particular load being modeled. Transverse load requires a higher mesh density through the thickness of the tape to accurately model the perpendicular stress in each layer. Torsion loads necessitate a higher mesh density through the width of the tape to capture the resulting parabolic strain distribution. Finally, bending-type loads need a finer mesh through the length of the tape to ensure uniformity along the entire tape. Certain situations may also require locally higher mesh densities on top of the uniform mapped mesh.

The mesh through the thickness of the tape not only depends on the load type but also on the modeling approach being used. The second approach described in Section 3.4.2 prescribes the internal layered structure of the coated conductor within the finite elements. This approach requires the tape to be modeled as a single homogenous volume meshed with only one element through its thickness, as shown in Figure 3.5(b). In this method, the mesh density in the thickness dimension cannot increase, but the number of integration points within each internal layer of the finite element can be increased to improve the accuracy.

The first approach described in Section 3.4.2 models every layer of the coated conductor as a unique geometric volume that can be meshed

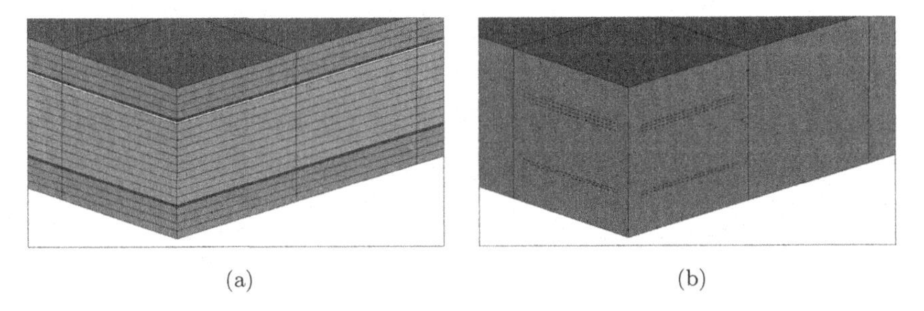

(a) (b)

Figure 3.5. Mesh through the thickness of a REBCO-coated conductor using (a) 21 homogenous elements and (b) single-layered composite element with the internal structure indicated by dashed lines.

independently. At least one element must be meshed through the thickness of every layer. For a standard tape with five layers, a minimum of five finite elements through the thickness are needed. The thinner layers (REBCO and silver) are exclusively meshed with only one element, while the thicker compositional layers (substrate and copper) are typically meshed with multiple elements, as displayed in Figure 3.5(a). The thicker layers have been meshed with up to 10 elements per layer in some instances. This is done to produce a more uniform element size throughout the tape thickness, thus improving the mesh quality, rate of convergence, and accuracy of results.

The mesh through the width of an HTS tape depends both on the load being analyzed and on the length of the sample being modeled. A very long tape may be meshed with fewer elements through its width to limit the computational burden, while a shorter tape may utilize additional elements to capture a more detailed strain distribution through the tape width. A standard 4 mm wide coated conductor has been meshed with as few as 10 to as many as 40 elements through the width of the tape. For most load applications, an average of 20 elements through the width has been found to provide a good balance between accuracy and computation time.

The mesh density along the length of the tape is highly dependent on the mesh density through the width and the length of the sample being modeled as well as the load being applied. The element width-to-length aspect ratio should, in most circumstances, be kept less than two to avoid any element shape warnings and potentially invalid results. For example, a standard HTS tape, as shown in Figure 3.6, can be meshed with 20 elements through its 4 mm width and at least 50 elements over its 20 mm length.

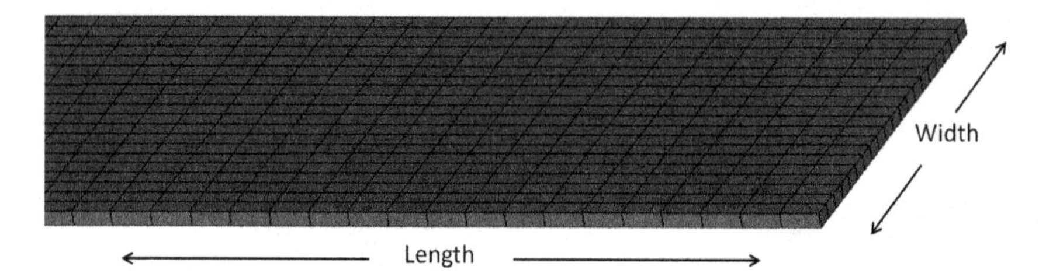

Figure 3.6. Discretization of an HTS tape with 20 elements through its width and 50 elements along its length.

3.4.5. *Material properties*

Material properties are one of the most important components of structural FEA because of their direct influence on the resulting mechanical behavior of the system. The wrong choice of material model or properties could be the difference between an accurate and an erroneous solution. HTS tapes and cables are most often tested in liquid nitrogen at 77 K or in liquid helium at 4.2 K, requiring their material properties to be known at these temperatures. When available, material properties should always be taken from direct stress–strain measurements, manufacturer's data, or published values in literature.

HTS tapes are expected to experience both elastic and plastic deformations under fabrication and operational loads. When modeling only the elastic response of the system (i.e. small strains and loads sufficiently below the yield strength of all constituent layers), linear elastic isotropic material models are adequate. However, in most circumstances, it is more common to model the full elastic–plastic behavior of the coated conductors, typically under large strain and high loads. To do this, it is necessary to use nonlinear rate-independent inelastic constitutive material models for plasticity.

The essential components of nonlinear plasticity models are the yield criterion, flow rule, and hardening. The yield criterion defines the material's transition from elastic to plastic behavior, the flow rule determines the rate of inelastic plastic deformation, and the hardening governs the change in the yield criterion following plastic deformation.

Isotropic (work) hardening and kinematic hardening are two common types of hardening rules used in plasticity models, in addition to perfectly plastic or no hardening. Isotropic hardening causes a uniform increase in the yield criterion under plastic loading (i.e. an increase in the yield strength) and is suitable for modeling the behavior of monotonic loading and elastic unloading. Kinematic hardening causes a shift in the yield criterion under plastic deformation and is designed for modeling the inelastic behavior of cyclic loads. The von Mises yield criterion is typically used for determining the onset of plastic deformation in these hardening models.

The effective stress and strain behavior of a material is generally represented in plasticity models by two main cases: bilinear and multilinear curves. An example of a bilinear and a multilinear curve used to define elastic-plastic material properties is shown in Figure 3.7. A multilinear material curve is a piecewise linear approximation of a true stress–strain curve defined by a set of positive stress and strain values. A bilinear material curve is an approximation of a true stress–strain curve using two straight lines defined by two slopes (modulus of elasticity and tangent modulus) and a yield stress.

Multilinear plasticity models more accurately represent the stress–strain curve of a material and in return tend to simulate more realistic elastic–plastic behavior compared to bilinear models. That said, multilinear models require a complete stress–strain curve, while bilinear plasticity models only need a modulus of elasticity, yield strength, and tangent modulus. Therefore, bilinear material models are more advantageous to model plastic deformation when full stress–strain data are not available. Additionally, since bilinear curves are parameterized, they are easily modifiable and can be adjusted if the exact material parameters are unknown. The material properties at 77 K used in these plasticity models are provided in Table 3.2.

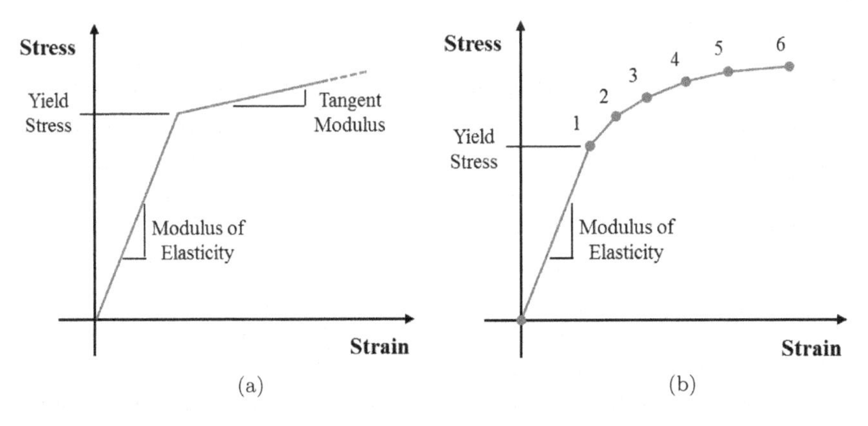

Figure 3.7. Classical elastic–plastic material behavior characterized by (a) a bilinear and (b) a multilinear curve.

Table 3.2. Range of structural properties used for coated conductor materials at 77 K.

Material	Modulus of Elasticity	Yield Stress	Refs.
Processed Substrate			
Hastelloy® C-276	180–200	1150–1300	[15, 22, 23]
Stainless Steel	150–200	850–1050	[23, 24]
Ni-5AT.%W	125–130	250–260	[25, 26]
Copper Stabilizer			
Electroplated	85–100	340–400	[15, 22]
Thin film	95–110	300–350	[27, 28]
Bulk	120–135	375–400	[29, 30]
Silver overlayer	60–85	200–275	[27, 28, 31]
REBCO/Buffer	145–160	—	[31–33]

REBCO-coated conductors are layered tapes composed of many materials, as mentioned earlier. The exact materials, layered architectures, and even manufacturing techniques are unique to each tape manufacturer. Yet, all tape types include the same basic structure composed of a substrate, buffer, REBCO, silver, and copper layers. Together, each material contributes to the overall composite performance of the tape. Mechanically, the substrate and copper stabilizing layers comprise the majority of the tape volume and, as a result, govern the overall mechanical behavior of the tape. Therefore, accurate material properties for these layers are of the utmost importance for structural modeling. Table 3.2 provides a list of constituent materials used in coated conductors and their corresponding mechanical properties at 77 K for reference [15, 22–33].

The choice of substrate material is the reason for one of the biggest structural differences between coated conductors from different manufacturers. Most manufacturers are using either Hastelloy® C-276, nonmagnetic stainless steel or a nickel–tungsten alloy. Hastelloy® is a superalloy with the greatest mechanical strength of the three materials, followed by stainless steel and then nickel–tungsten, as indicated by their modulus of elasticity and yield stress listed in Table 3.2. The variance in mechanical strength of the substrate materials is generally compensated by their thickness. For example, nickel–tungsten is the weakest material and, as a result, requires a thick substrate.

The substrate materials in their initial polished condition (before deposition) are extremely strong. The buffer and REBCO deposition processes on the polished substrate occur at elevated temperatures that tend to anneal the substrate material, reducing its mechanical strength. When defining material properties for the substrate layer of the coated conductor, it is important to use those for the processed substrate (after deposition), as done in Table 3.2. The stress–strain behavior of individual processed substrate materials has been well studied experimentally [22–27].

The stabilizing layer or coating of an HTS tape is another area of disparity between tape manufacturers. Typical stabilizers are made of pure copper or a copper alloy and are applied via electroplating or lamination by means of soldering. The mechanical properties of electroplated copper are different from that of thin-film copper, which are both different from standard bulk copper properties. The differences in copper properties, however, are small when compared to the mechanical strength of the substrate. In general, copper has a significantly lower yield strength and modulus of elasticity than the substrate materials, as seen in Table 3.2. The low mechanical strength of the copper leads to early plasticity, which tends to lower the strength of the composite tape [22].

Identifying accurate material properties for the stabilizing material and specific fabrication techniques can be challenging. Some stress–strain data are available for copper stabilizers (both electroplated and films), but experimentation is challenging and limited [22, 27–30]. In lieu of experimental data, the rule of mixtures has been used to calculate the material properties of copper, knowing the properties of the substrate material and of the entire composite tape [15].

The number, order, thickness, and composition of the buffer layers, as well as the thickness and type of "rare-earth" compound used for the REBCO film, vary for each tape manufacturer. The buffer and REBCO layers are the most important layers electrically, but structurally, they have a negligible influence on the mechanical strength of the tape. The buffer and REBCO layers are both brittle ceramics, which tend to crack and fracture before plastically yielding. Therefore, the buffer and REBCO layers are generally modeled together as a single layer with linear elastic isotropic properties. Table 3.2 provides a range of modulus of elasticity used for REBCO and buffer materials [31–33]. Fracturing of the REBCO layer is generally not simulated but instead inferred from the magnitude of strain in the REBCO layer.

The mechanical strength of the thin silver overlayers also has minor influence on the elastic–plastic behavior of the entire tape. Complete stress–strain curves are limited, but the available material properties of thin-film silver show similarity to those of copper, as seen in Table 3.2 [27, 28]. The inclusion of the silver layers in coated conductor models is primarily for completeness. Some models ignore the silver layers and incorporate their thickness into the copper stabilizers, considering the comparable mechanical behavior.

For simulation of the internal residual strain incurred by the manufacturing process of the tape and the cool down to operational temperatures, the temperature dependence of the material properties must be taken into account. This requires the elastic–plastic properties of each layer to be known for multiple temperatures, typically the REBCO deposition temperature (>1000 K), room temperature (298 K), and operating temperature (77 or 4.2 K). A linear dependence is classically used to define the stress–strain curve of the materials in between the mentioned temperatures. Thermal expansion coefficients for the materials are generally assumed to be constant with temperature.

3.4.6. *Boundary conditions and loads*

Tension can be applied to the ends of the coated conductor as a displacement or as a force. A tensile displacement can be applied via translational degrees of freedom at the ends of the tape. A tensile force load can be applied on the ends of the tape as an outward-facing surface pressure or as a uniformly distributed force. For both load cases, one end of the tape is generally fixed, while the axial force or displacement is applied to the other. When applying a displacement load, you are defining tensile strain in the tape and are computing the resulting stress, while by applying a force load, you are prescribing a tensile stress in the tape and are interested in the resulting strain.

Torsion is generally applied via rotational degrees of freedom at the ends of the tape geometry. Rotation is applied to one end of the tape, while the opposite end is prescribed to be fixed. The end of the tape with the applied rotation is also permitted to move freely in the axial direction, allowing the tape to naturally shrink while being twisted. This axial freedom guarantees pure torsion by removing the possibility of unwanted axial tension.

Combined tension–torsion loading refers to the tension of a twisted tape producing a combined tensile and torsional strain condition. Combined tension–torsion load can be applied as a combination of pure tension and pure torsion loads. Rotation is typically applied first to the ends of the coated conductor, producing a desired twist pitch. Next, axial tension is applied to the twisted tape via a displacement or a force load. The resulting loading condition should also be the same if the loads were reversed and tension was applied first followed by rotation.

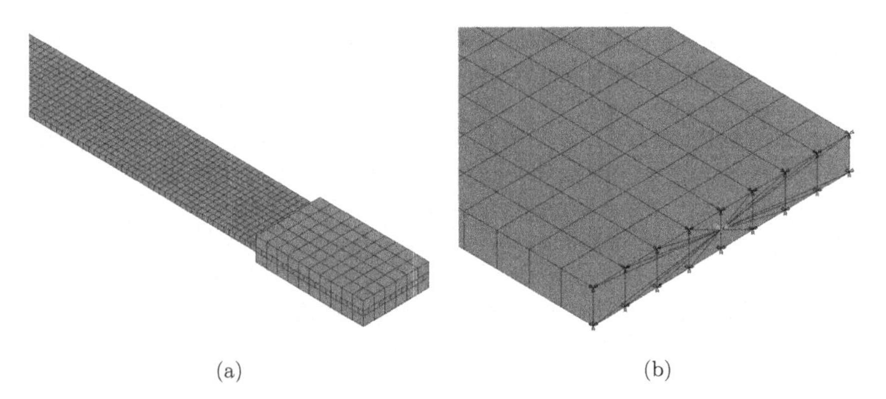

(a) (b)

Figure 3.8. Tensile, torsion, and combined tension–torsion loads applied via (a) physical clamps and (b) pilot node constraints on the end surface.

One modeling technique for prescribing tensile, torsion, and combined tension–torsion loads is using physical clamps at the ends of the tape, as shown in Figure 3.8(a). This method is most similar to what is done experimentally; the clamps hold the tape, and the corresponding displacement, rotation, and force loads are applied directly to them. The clamps are generally modeled as rigid bodies that cannot deform, allowing for direction translation of the applied loads to the tape. One advantage of this method is the ability to model the interaction (clamping force and friction) between the clamps and the tape; however, in most cases, the influence of the clamps is negligible and therefore not included.

Instead, another technique for modeling tensile, torsion, and combined tension–torsion is to apply the loads directly to the tape geometry. Pressure and force loads can be defined directly on the surfaces at the ends of the tape. Translation and rotation loads are normally applied to the ends of the coated conductors using a multipoint kinematic coupling (pilot node) constraint. A pilot node or coupling constraint is when the motion of a collection of nodes on a surface is constrained to the rigid body motion of a reference or pilot node. The pilot node can be a node on the surface of coupled nodes or a node at any arbitrary location. Its location is only important when rotation is applied because the rotation occurs about the pilot node.

Pilot node constraints can be defined at each end of the tape, rigidly coupling every node on the end surfaces to a pilot node located in the center of the tape (along its axis of rotation), as depicted in Figure 3.8(b). The rotations and displacements at the ends of the tape can then be prescribed on the pilot nodes. Pilot node coupling constraints are particularly well suited for modeling pure torsion because they only constrain the degrees of freedom in the rotation plane and therefore allow the ends of the tape to move freely in the axial direction.

For tension, torsion, and combined tension–torsion loads, the strain along the length of the tape is constant. This uniformity along the length of the tape allows shorter models to be simulated using symmetry-style end conditions. Symmetric surfaces must remain planar but be allowed to move axially. Pilot node constraints are ideal for applying loads to symmetry end conditions without creating end effects. When applicable, additional planes of symmetry can be used to further reduce the model size and computation time.

Transverse compression can be modeled as a surface pressure or a uniformly distributed load applied directly to the tape. In experiments, however, the exact loading surface or contact area made by the pushing head is unknown due to the slight thickness nonuniformities (dog bone shape) of the coated conductor. Similarly, the load distribution through the pushing head is also unclear in experiments. These uncertainties can necessitate the inclusion of the complete load fixture in the model, as shown in Figure 3.9. The load fixture, including the pushing head and support anvil, can be modeled as a rigid body. The contact between the tape and the load fixture must be defined. The load can be modeled by applying either a force or a displacement to the pushing head, while the anvil is prescribed with a fixed replicating realistic transverse compression on the tape.

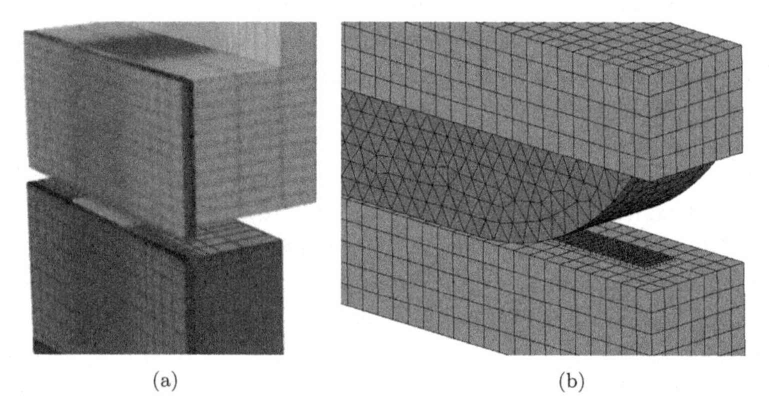

(a) (b)

Figure 3.9. Transverse compression stress models highlighting two different load structures [15]. (a) Simplified load distribution using two flat surfaces. (b) More realistic load distribution applied during the measurements (with a rounded anvil).

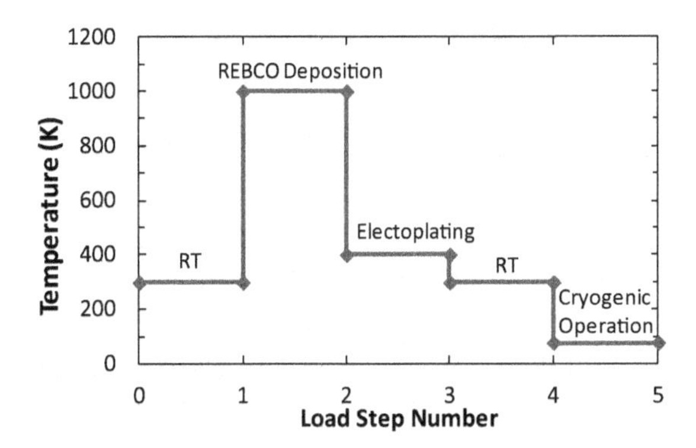

Figure 3.10. Example of temperature load steps applied to structural model to analyze internal residual strain in the coated conductor after processing and cool down to operational temperature.

The temperature load can be defined in a structural model as a simple change in temperature, assuming that appropriate temperature-dependent material properties have been defined. The change in temperature from the tape production processes to the cryogenic operating environment can be prescribed using a set of temperature load steps, as illustrated in Figure 3.10. The simulation will calculate the steady state mechanical effects (residual strain) caused by each relative step.

3.5. Modeling Techniques for Cables

HTS cable simulations have focused primarily on bending loads and electromagnetic transverse loads to support their future implementation in magnet applications. The resulting strain in the coated conductors within a cable after residual cool down to cryogenic temperatures was also analyzed. The structural FEA cable models employ many of the same modeling techniques used for single-tape simulations; however, modeling full-scale cables also has its own set of challenges, which require the use of additional modeling techniques. This chapter describes these modeling aspects which are unique to cables.

3.5.1. *Model simplifications*

One of the main challenges of modeling HTS cables can at times be the size of the numerical model. Cables consist of many coated conductors (upwards of 100 tapes for certain cables), in addition to a support structure or former. Considering this, full 3D models can easily become very large, requiring a significant amount of computational power and time. Using 2D models when suitable is one of the biggest simplifications that can be made to reduce the model complexity and computation time.

The dimensionality of models depends on the type of cable being modeled as well as the structural loads being analyzed. Simulating bending characteristics, for example, is one circumstance that requires the use of 3D models to capture the full electromechanical behavior of the cable. On the other hand, electromagnetic transverse compressive loads are one instance that can generally be modeled accurately in 2D.

When 3D models must be used, one of the biggest simplifications that can be made is to remove the internal layered architecture of the coated conductor. This can accurately be done by simply modeling the layered tapes as a single uniform volume defined with mechanical properties equivalent to those of the composite tape. Removing the layered structure of the coated conductors greatly reduces the size and complexity of the model and simplifies the matrix calculations, leading to faster convergence and shorter simulation times.

Another simplification that can be used with full-scale 3D models is to reduce the length of the cable being modeled. The length of the model directly relates to the size and thus the computation time of the simulation. Models with half the cable length will typically take half the time to run. Although length reduction can cause a significant decrease in computation time, it can also make applying boundary conditions at the end of the cables more challenging. Considering this, the length of the cable should be chosen based on the twist pitch of the cable. If less than one full twist pitch is modeled, special end constraints must be used.

The last major simplification for 3D modeling of cables is to eliminate a part or all of the cable support structure. In certain circumstances, the support structure of the cable can be represented in the model by comparable boundary conditions or constraints. This technique, however, is only suitable for certain types of cables and loads and, as a result, has limited applicability.

3.5.2. *Element types*

The coated conductors used in HTS cables are generally modeled in the same manner as done for single-tape simulations. Their internal layered structure can be modeled using either approach described in Section 3.4.2.

Alternatively, for the simplification of the cable models, the layered structure of the tapes can be neglected and instead modeled as a uniform tape with equivalent composite properties. The tapes should be meshed with either solid continuum or solid-shell hexahedron elements.

In addition to the coated conductors, most cables have some form of constituent structural support component. The cable structure varies depending on the kind of cable but is typically some type of central former used to wrap and wind the tapes around or some sort of external support used to protect and reinforce the tapes. No matter the type, the support structure can be meshed with solid continuum elements. The chosen solid element geometry and resulting mesh will be unique to each cable type.

In most cases, if the geometry of the cable support is suitable, solid brick elements are used because of their ability to produce uniform mapped meshes along the cable length. Brick elements are also advantageous because they can be elongated along the length of the cable, reducing the number of elements without a loss of accuracy. Another element topology commonly used with solid elements is tetrahedral. Tetrahedron geometry is well suited to model irregular and curved structures, giving it an advantage for certain cable support structures. The tetrahedral elements commonly used in finite element software are listed in Table 3.3.

For 2D cable models, planar structural solid elements with translational degrees of freedom are used. Planar solid elements can be defined as either plane stress or plane strain and can have linear or quadratic shape functions. Generally, higher-order elements are chosen to increase the accuracy of 2D simulations. A list of planar elements used in ANSYS® and ABAQUS® are also provided in Table 3.3.

Planar elements can take on quadrilateral or triangular geometry. Similar to the advantages of using hexahedron elements in 3D, quadrilateral

Table 3.3. Finite elements used to mesh 3D cables and 2D cable cross-sections.

Element Type	ANSYS®	ABAQUS®	Planar Type	Nodes	Geometry
3D Structural solid (continuum)	SOLID285	C3D4	N/A	4	Tetrahedron
	SOLID187	C3D10	N/A	10	
2D Structural solid (continuum)	PLANE183	CPE8/CPS8	Plane Strain/ Stress	8	Quadrilateral
		CPE6/CPS6	Plane Strain/ Stress	6	Triangular

elements are most commonly used for the rectangular cross-section of coated conductors. Triangular elements are particularly well suited to model the generally circular or irregular cross-sectional shape of the cable support structure.

3.5.3. *Meshing*

The mesh density again plays an important role in the accuracy of the results of cable simulations. That being said, the size of the numerical model (total number of elements) is one of the biggest challenges in simulating cables. Full 3D models, for example, consist of many coated conductors along with a cable support structure and can have potentially long lengths, which as a result require significant amount of computational power and time. Finding the right balance between computational resources and simulation accuracy is needed and is generally determined from mesh studies.

Cable models can generally be meshed in stages. Typically, the coated conductors are meshed first, followed by the cable support and then any additional load structures when applicable. As done in single-tape models, the coated conductors can be meshed with a uniform patterned mesh of brick elements. The mesh density of the individual tapes is usually less than that in single-tape simulations to help reduce the computation time of full-scale cable models. The element topology and mesh of the support structure depend directly on the type of cable being modeled. Brick elements are used whenever possible so that a uniform mapped mesh with elongated elements can be used, which greatly reduces the overall number of elements along the cable. Tetrahedral elements can also be used to mesh very irregular support structures but must be used cautiously since they tend to require more elements. If applicable, the load structure can be modeled with either brick or tetrahedral elements depending on the structure's geometry.

The discretized models of two HTS cables used for 3D bending simulations are shown in Figure 3.11. The cutaway model of the HTS CICC demonstrates the uniform meshing of the cable support structure (slotted core, spacer, and external jacket) with brick elements. Similarly, the CORC cable model highlights the use of mapped pattern meshing with elongated brick elements for each individual coated conductor. The meshes shown in Figure 3.11 are meant to be illustrative and are not necessarily the final discretizations. Mesh analyses should be conducted to determine

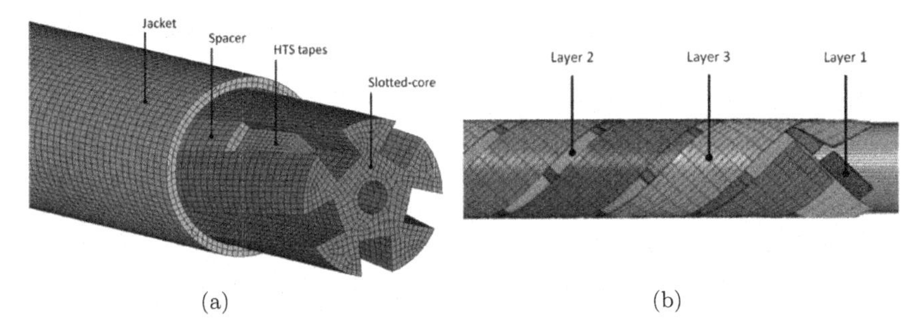

Figure 3.11. Discretized 3D cable models of (a) HTS CICC and (b) CORC cables [12, 16].

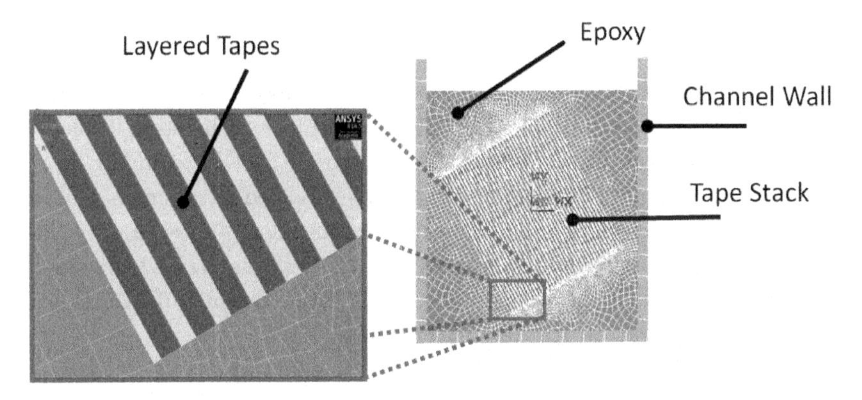

Figure 3.12. Discretized 2D cross-section of a TSTC conductor in epoxy-filled channel [21].

the optimal mesh density (number of nodes and elements) for each specific cable structure.

The size of the mesh used for 3D cable models varies but is generally on the order of hundreds of thousands of elements, which result in long numerical computations. Planar models of cable cross-sections are used when applicable because the size of the mesh for 2D models is typically one order of magnitude smaller (tens of thousands of elements). The lower element count, in addition to the simpler matrix calculations, results in faster computation times. A 2D discretized model of a TSTC conductor cross-section in an epoxy-filled channel used for electromagnetic transverse load simulations is shown in Figure 3.12. The model shows the layered quadrilateral mesh used for the composite tapes, the mixed triangular and quadrilateral mesh used for the epoxy region, and the uniform quadrilateral mesh used for the channel walls. The meshes for 2D models need to be optimized for the unique cable cross-section and load being analyzed.

3.5.4. *Material properties*

Like coated conductors, HTS cables are frequently tested at 77 K in liquid nitrogen or in liquid helium at 4.2 K, requiring the material properties to be defined at those temperatures. The cables are expected to experience both elastic and plastic deformations during cable fabrication and magnet operation. If only small strains are modeled and the loads are below the yield strength, linear elastic isotropic material models are suitable; otherwise, nonlinear plasticity models defined by bilinear or multilinear stress–strain curves should be used.

The layered composition of HTS tapes can be modeled with same nonlinear properties for the substrate, stabilizer, and silver, as listed in Table 3.2, as well as the same elastic properties for the REBCO and buffer layers. The coated conductors can alternatively be modeled as a single uniform volume with no layers, in which case the elastic–plastic material properties of the entire composite tape should be used. The stress–strain behaviors of the composite tapes are commonly available from the manufacturer or in the published literature [22, 23, 26, 36].

The cable support structures differ between cable types but are typically made of aluminum or copper. CORC conductors have cylindrical central formers made up of either a solid aluminum rod or a bundle of copper strands. The HTS CICC uses a twisted and slotted extruded aluminum core along with copper spacers and a cylindrical aluminum jacket. The round strands made by twisted stacks are made with a cylindrical copper rod, which is cut into two pieces and machined in the center to fit the stack of tapes. TSTC and ROEBEL cables, by themselves, do not have any support structure but are frequently reinforced by sheathing or placing in machined channels that can be solder filled or epoxy impregnated.

The appropriate elastic–plastic properties of each support material (aluminum, copper, solder, or epoxy) should be defined using nonlinear plasticity models, as done for the coated conductor materials [29, 30, 37–39]. When available, their material properties should be taken at operating temperatures from direct stress–strain measurements of the exact support material. This should be done because bulk aluminum properties in the literature are different from those of extruded aluminum, which are also different from those of slotted and twisted aluminum [29, 37]. Similarly, copper strands have different properties from copper rods, and both show variation in their respective properties depending on if they are hardened or annealed [29, 30]. Solder properties also have very different properties depending on the composition, temperature, and use [38, 39]. When direct measurements are not available, manufacturer's data or published literature values are favored.

Material properties are highly temperature dependent and therefore should be defined for the particular cryogenic environment being modeled. To simulate the cool down of the cables, material properties should be prescribed at both room temperature and the desired operating temperature, 77 or 4.2 K. The thermal expansion coefficients for each material must also be defined in addition to the nonlinear stress–strain properties.

3.5.5. *Contact relationships*

One of the most influential aspects governing the behavior of HTS cables is their mechanical contact interactions. Many different contacts exist within a cable, and the specific type of interaction depends primarily on the type of load being analyzed. For example, bending loads may be of more interest in frictional sliding contacts and the accumulation of shear stress along the length of the cable, while transverse compression loads may be of more concern with penetration contacts and how loads are transmitted through the contacts.

One of the biggest contact interactions occurs between neighboring tapes. The relative motion of these adjacent tapes has a major influence on the mechanical and electrical performances of the cable. The characteristics and behavior of these coated conductor interactions are highly dependent on friction and the type of cable being modeled. TSTC conductors stack tapes and twist them along their length, creating a single contact area over their entire length. CORC cables helically wind layers of coated conductors around a central former in alternating directions, creating many small periodic contact areas as different tapes obliquely cross over each other.

Another major contact interaction that influences the cable is that of the tapes and remaining cable structure. The type of contact depends on the kind of cable and whether the support is internal, external, uniform, slotted, twisted, etc. As a result, the interaction of the structure with the tapes could potentially play a large role in the overall behavior of the cable. The use of solder or epoxy impregnation within the cable support will also greatly influence the contact interactions and the overall cable characteristics.

Accurately modeling these interactions is essential to developing realistic cable simulations. The interactions are generally modeled using surface-to-surface contact pairs, which are defined as having a target (master) surface and a contact (slave) surface that share a set of specified characteristics or settings. These contact surfaces are generated on existing solid geometries and are solely used to model the interaction of the solid volumes. Similar contact pairs also exist for 2D planar models, where the edges of areas are selected instead of the surfaces of volumes as in 3D.

In ABAQUS®, the model surfaces or edges are simply selected to define the contact pair, while in ANSYS®, the surfaces and edges are instead overlaid with a mesh of specific target and contact elements that create the contact pair. The meshed contact surfaces have the same geometric characteristics and element topology as the underlying solid or planar geometry. Table 3.4 provides a list of the target and contact elements specific to ANSYS®.

Although the methods of generating the contact pairs are slightly different in the two programs, they both have equivalent contact settings and properties that can be defined to control the characteristics and overall behavior of the contact interaction. The mechanical contact properties usually specified include friction coefficients and contact damping, as well as the allowable contact pressure and penetration. Individual contact pairs for every interaction should be used so that the unique contact settings representing the specific characteristics of each interaction can be defined. The properties and settings can also be adjusted to help improve the convergence of the model.

Table 3.4. Target and contact elements used for contact pairs in ANSYS®.

Element Type	Surface type	ANSYS®	Nodes	Geometry
3D surface-to-surface contact pair	Target (master)	TARGE170	Variable	Node, line, triangle, or quadrilateral
	Contact (slave)	CONTA173 CONTA174	4 or 3 8 or 6	Quadrilateral or triangle
2D surface-to-surface contact pair	Target (master)	TARGE169	Variable	Node, line, or arc
	Contact (slave)	CONTA171 CONTA172	2 3	Line or arc

Each FEA package has specific built-in contact algorithms or formulations, which may include hard and soft contacts, small and finite sliding, rough and frictionless contacts, as well as bonded and no-separation contacts. In many cases, the default settings associated with these contact algorithms may be appropriate; however, for more advanced interactions, the following settings can also be changed. Penalty and augmented Lagrange methods are typically chosen for contact constrain enforcement. The contact detection points are generally specified either at the finite element nodes or at the Gaussian integration points. Automatic contact surface adjustment can be selected to move the contact points with an initial gap (clearance) or penetration (overclosure) onto the target (master) surface. Similarly, the effects of the initial geometric penetration or overclosure can be included or excluded. Contact smoothing techniques can also be applied to the contact surfaces if necessary.

3.5.6. *Boundary conditions and loads*

Pure bending of HTS cables can be applied using a variety of modeling techniques. Pure bending requires the cable to be bent into a circular arc with a uniform bending diameter, producing a constant bending moment throughout and no shear forces.

One method to apply pure bending would be to bend the cable around a circular object with a fixed diameter, which is the technique typically used for experiments. For this method, a rigid circular body must be created in the model with appropriate contact settings so that no additional stress is applied to the cable. Bending can be generated by applying a uniform distributed load along the cable, forcing it to conform around the circular object. One drawback of this technique is that a new circular body must be created for every desired bending diameter, which can make modeling cumbersome. Another issue is determining how much force is required to deform the cable around the circular object without adding additional transverse compression.

Another method by which pure bending can be achieved is to apply two equal but opposite moments to the ends of the cable. The moment load must be prescribed on the ends of the cable using pilot node kinematic coupling constraints. This is necessary because the solid continuum elements used to discretize the cable do not directly support rotational degrees of freedom. One disadvantage of this method is determining what

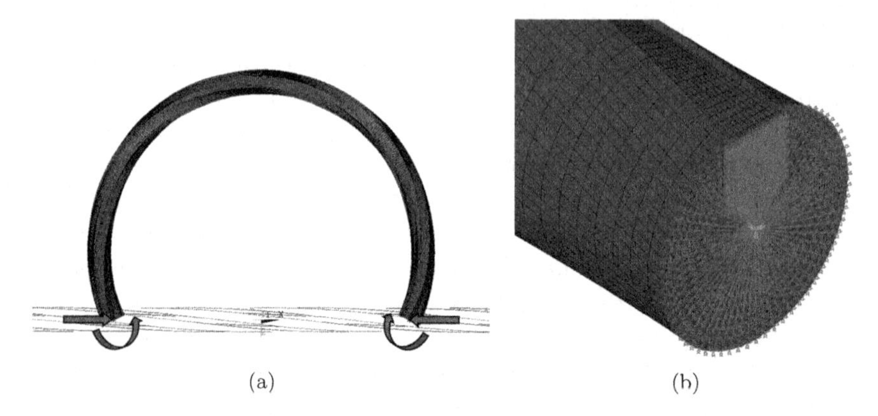

Figure 3.13. (a) Depiction of pure bending load application by equal and opposite rotation and displacement of the cable ends. (b) Illustration of pilot node coupling constraint used to apply bending loads on cable ends [12].

moment value correlates to what bending diameter, which will vary with cable type.

The last approach that is typically used to generate pure bending loads is to apply a coupled rotation and displacement to the ends of the cable, as illustrated in Figure 3.13(a). Pure bending motion consists of moving the ends of the cable closer together as they are rotated in opposite directions relative to one another. The exact coupling between the amounts of displacement and rotation can be solved analytically using a set of constitutive equations. No matter the cable type, the same rotational and translational degrees of freedom can be used to produce pure bending. The coupled rotation and displacement must similarly be applied using pilot node constraints due to the lack of rotational degrees of freedom in solid elements.

Pilot node constraints couple the motion of a collection of nodes on a surface to the rigid body motion of the reference pilot node. In the case of HTS cables, the ends of every tape as well as the ends of the corresponding cable structure all get constrained to a central pilot node, as shown in Figure 3.13(b). The pilot nodes should be defined on the end surfaces of the cable and in the exact center of its cross-section. The position of the pilot nodes is important, as it should be located on the neutral axis under pure bending loads. Rotation and translation, as well as moment loads, can be defined directly on the pilot nodes.

Transverse compression stress on 2D HTS cable cross-sections has been modeled for two main cases: externally applied mechanical transverse compressive loads and internal distributed electromagnetic Lorentz loads.

In the first case, the numerical cable models typically include the load fixture, as done for the single tapes described in Section 3.4.5 and shown in Figure 3.9. The loading structure is generally included to more accurately reproduce the experimental conditions by removing the uncertainties of not knowing the exact load surface or distribution. Using this technique, the pushing head and bottom anvil are modeled as rigid bodies that contact the deformable cable.

The degrees of freedom of the bottom anvil are prescribed to be fixed (no translation or rotation), and the transverse compression is applied to the cable cross-section via the pushing head. The load can be applied either as a force or a displacement to the pushing head. When a displacement is used, the resulting reaction force can be determined along with the contact area to determine the transverse stress applied. Similarly, when a force is applied, the resulting transverse strain and deformation can be found.

In the second case of transverse stress, the electromagnetic load is generally applied in the model through one of two methods. Both methods produce an internal distributed load that acts on the REBCO layer of each individual tape within a cable. A distributed Lorentz load acting on the cross-section of a TSTC conductor in a solder-filled tube is depicted by arrows in Figure 3.14.

The first method to model an electromagnetic load in a structural simulation is to simply apply an equivalent load as a uniformly distributed

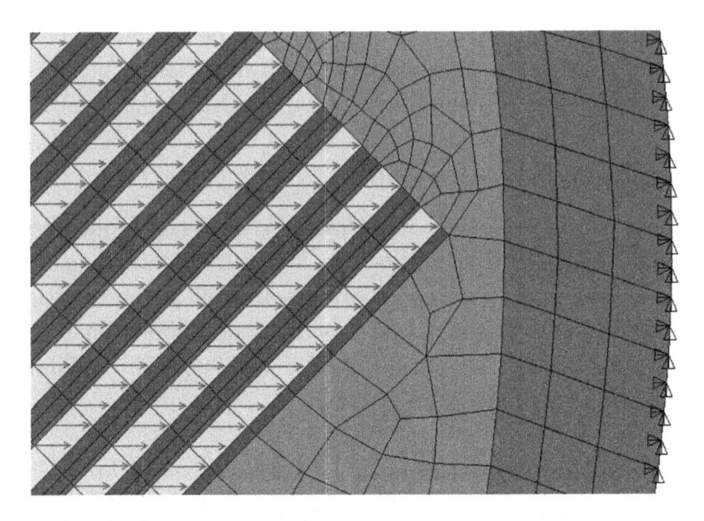

Figure 3.14. Internal distributed electromagnetic load acting on tapes of a TSTC conductor in a solder-filled copper tube [21].

force. The equivalent Lorentz load can be calculated from the anticipated transport current and background magnetic field. The load can be applied to the cable cross-section on an edge or distributed evenly over the REBCO area of the coated conductors, as shown in Figure 3.14.

The second method is to couple simulations from two different physics. First, create an electromagnetic model with the electrical current and background magnetic field to numerically compute the resulting Lorentz load. Then, import that load as nodal forces into a structural model to analyze the stress and strain in the cable. This approach provides a more accurate electromagnetic load distribution; however, it also requires additional steps and simulations.

No matter which method is used to define the electromagnetic load, constraints must be added to the cable cross-section to resist the load. Properly defining accurate constraints without over-restricting the model can be challenging depending on the cable type being modeled. For some cable types, such as the HTS CICC or the round strands made with twisted stacks, the outer edge of their external jacket or former can be used to fix the displacement without influencing the internal tapes. For cable types such as the TSTC, which have tapes on the outside, it is more challenging to apply fixed degrees of freedom without over-constraining the natural movement of the tapes. For these cables, it is typically best to model their external support structure as depicted in Figure 3.14.

The internal strain that may develop in HTS cables after cool down to cryogenic environments can be simulated by defining a change in temperature, as done for single-tape models. All material properties must be adequately defined for every prescribed temperature being analyzed.

3.6. Postprocessing and Results

The final step in finite element modeling is to display, output, record, and then evaluate the numerical results. For structural simulations, the stress and strain results are of particular interest, as well as the resulting forces and moments. The FEA simulations are inherently approximations; therefore, the output results should always be compared and validated with experiments or trusted analytical models.

3.6.1. *Simulation output results*

One of the main benefits of numerical FEA is the amount of output results available from a single simulation. For structural simulations, the reactions, deformations, and stress and strain results can all be output, as well as the contact and failure results, among others. Furthermore, each of these main outputs also includes many additional subcategories of results. For example, the total strain can be broken down into elastic, plastic, creep, thermal, and swelling strains if needed.

The numerical results for both the nodal and elemental solutions are available depending on the desired data. Data for every finite element and node in the model can be displayed and output; however, in most cases, only the results from a small subset of the model in an area of interest are needed. If results for points in between nodes are required, they can be found through interpolation of the nodal data.

Another major benefit of structural FEA is the ability to visualize the numerical results. The deformed shape of the physical geometry can be viewed, on which colored contour plots of the stress and strain results can be displayed. Visualization of the results in this manner provides the modeler with a better understanding of how the system is responding to loads and gives them the ability to identify any problems or troublesome locations that may need further investigation. The ability to visualize the numerical results is one of the most useful aspects of FEA modeling, even if it is only qualitative.

For coated conductors, the main output results of interest tend to be the nodal stress and strain data as well as the reaction forces and deformations. For HTS cables, the same output results are of interest, with the addition of the contact results, particularly with respect to frictional sliding and geometric penetration. The normalized critical current of an HTS tape or cable can be estimated from the structural models using the strain state in the coated conductor along with an analytical model, as described in the following section. The exact output results from each model depend on the tape or cable being simulated and the type of load being analyzed.

The output results from the FEA program are typically compared with experimental data when available to validate the numerical model. Comparison of the numerical model with experiments allows for a better understanding of the electromechanical behavior of coated conductors and HTS cables. It is common practice to start by validating the technique used for modeling the layered composite tapes by comparison with experimental stress–strain data. The structural models can also be validated using measured experimental data for reaction forces and moments as well as deformations (elongation, rotation, and bending) depending on the type of loads being modeled. The numerical strain results can also be used to estimate the normalized critical current behavior; therefore, the structural models can indirectly validate the experimental critical current dependence on stress and strain, as mentioned in the following section.

3.6.2. *Critical current prediction*

The critical current of an HTS tape is known to be a function of axial strain; the larger the tensile or compressive strain, the lower the critical current of the tape. This strain dependence of the critical current has been well documented experimentally and can be characterized using higher-order polynomials fit to the experimental data.

Structural FEA models produce detailed results of the axial strain in the superconducting layer of REBCO-coated conductors under various mechanical loads. These numerical strain results can be used to accurately predict the normalized critical current of the tapes using an analytical model and experimental data as outlined in Ref. [12].

The analytical model calculates the predicted critical current by integrating the critical current densities over the cross-section of the tape. The critical current densities are calculated from the numerical strain data using a polynomial fit to the experimental critical current data. The integration through the width of the tape is evaluated using Gaussian integration [34, 35].

For coated conductors with uniform strain along their length (such as single tapes under tension and torsion), the integration can be conducted at any location because the critical current will not vary along the length of the tape. However, for tapes with strain variation along their length (such as the tapes in HTS cables under bending), the integration must be evaluated at several points along the length so that the total voltage criterion of the tape can be calculated from a summation to identify the normalized critical current for the entire tape.

Predicting the critical current from the numerical strain results in this manner is a simple and easy technique, which has been shown to have relatively good agreement with experimental results. The main advantage of this method is that it is applied during the postprocessing step and can therefore be used on any structural model of coated conductors. Caution should be taken, as this is only a prediction of critical current and not a full electromechanical model and therefore does not include things such as filament breakage or variations with temperature and magnetic field.

3.6.3. *Single-tape results*

Axial tension is one of the main loads used to characterize the electromechanical behavior of REBCO-coated conductors. Modeling a thin tape under axial tension is a straightforward load to simulate and thus is typically one of the first to be analyzed numerically. More importantly, tension is typically modeled first as a means to validate the accuracy of the simulations and the technique used to model the layered composite architecture of the coated conductors. Experimental stress–strain curves are generally used for validation since they describe the mechanical strength and behavior of the composite tape.

Figure 3.15 shows the stress–strain behavior of two types of HTS tapes at 77 K. The experimental data are shown as solid and dashed lines, while the numerical results are plotted with unfilled markers. The numerical stress and strain values are gathered from the loads and deformations in the FEA model. The axial elongation in the model is used along with the initial tape length to calculate the numerical strain in the sample at every load step. Likewise, the axial tensile force on the ends of the tape and the cross-sectional area were used to determine the applied tensile stress. The results were then combined to build the stress–strain curves shown in Figure 3.15. The REBCO-coated conductors were modeled numerically using solid-shell elements with bilinear material properties [14]. The difference in the stress–strain behavior of the two tapes is an example of how

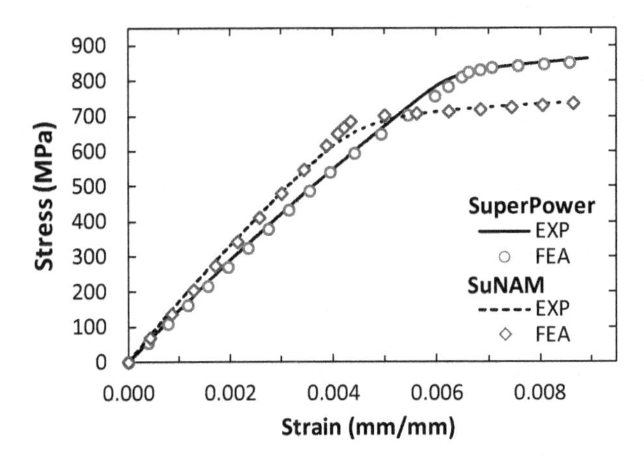

Figure 3.15. Stress–strain behavior of two coated conductors at 77 K determined from experimentation (EXP) and FEA simulations [12, 14].

their layered architecture and choice of substrate governs their composite performance.

Coated conductors are commonly twisted to reduce their AC loss, as demonstrated in a number of HTS cable designs. This is the main motivation behind analyzing tapes under pure torsion and tension–torsion loads. The torsion of a thin rectangular tape generates a parabolic distribution of axial strain through its width. This axial strain distribution can be nicely visualized as a contour plot of the deformed shape of the twisted tape, as shown in Figure 3.16(a). Under pure torsion, the outer edges of the tape will have a high tensile strain, while the center of the tape will experience a slight compressive strain.

The magnitude of the strain distribution under torsion varies with the degree of twist applied to the sample. The amount of twist is typically denoted by the twist pitch, which is the length over which one full rotation of the tape occurs. Figure 3.16(b) shows the axial strain distribution through the width of the coated conductor for three common twist pitches. The strain profile for a 62.5 mm twist pitch is also compared with an analytical model for validation. The plot clearly indicates that the peak strains produced under torsion can quickly become large as the twist pitch length is reduced. These high strains, particularly at the ends of the tape, will lead to a reduction in the critical current performance of the tape.

The strain through the width of the tape, as displayed in Figure 3.16(b), is an example of the type of numerical data that can be used to estimate the critical current capacity of the coated conductor. An example of the predicted critical current behavior under combined tension–torsion load is

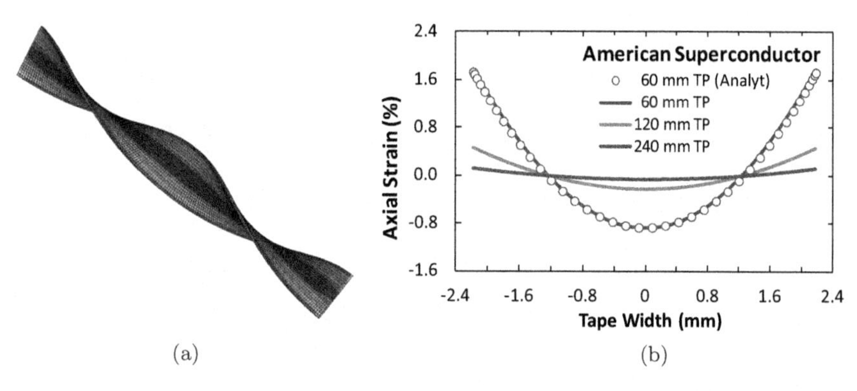

Figure 3.16. Axial strain under pure torsion: (a) strain contour on twisted tape and (b) strain distribution through width of tape for multiple twist pitches [12].

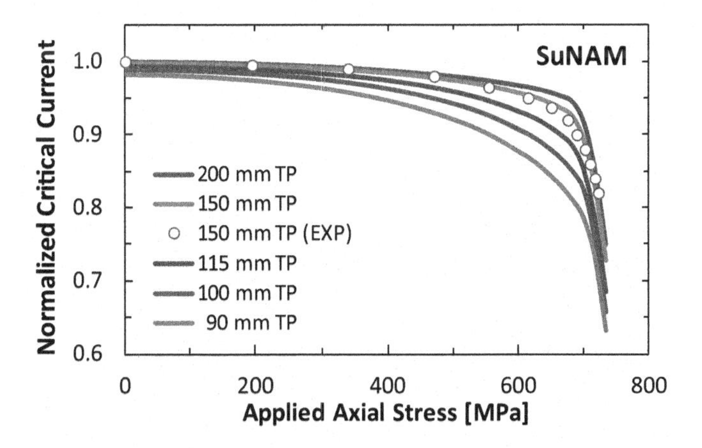

Figure 3.17. Combined tension–torsion critical current behavior predicted from numerical strain results. Experimental data plotted for validation [12, 14].

displayed in Figure 3.17. The critical current was estimated for five different twist pitches over a wide range of axial tensile stress using the method outlined in the preceding section. The calculated results from numerical strain data are in very good agreement with the experimental data for a tape with a twist pitch of 150 mm. The numerical results indicate that the tapes with shorter twist pitch lengths tend to have earlier and more gradual critical current degradations.

For torsion and combined tension–torsion loads, the moment (torque) on the sample is based on the architecture and strength of the tape and is an important mechanical characteristic for the fabrication and operation of twisted cables. Torque can be applied to generate twist or a rotation can be applied, and the resulting moment can be determined. For torsion loads, the simulations can be used to identify the required torque to twist a tape to a certain degree. For tension–torsion loads, the simulations can be used to determine the change in torque on a twisted tape under axial tension. The moment or torque is another mechanical characteristic that is typically used to validate the FEA model against experiments.

Transverse compression on the wide face of coated conductors is another important load that can be analyzed to investigate the resilience of the tape. FEA simulations are uniquely well suited to identify the contact surface and load distribution on the sample, as well as its deformation under the load, as displayed in Figure 3.18. Exaggerating the deformation, as done in the figure, can be particularly useful to identify the behavior of the tape at

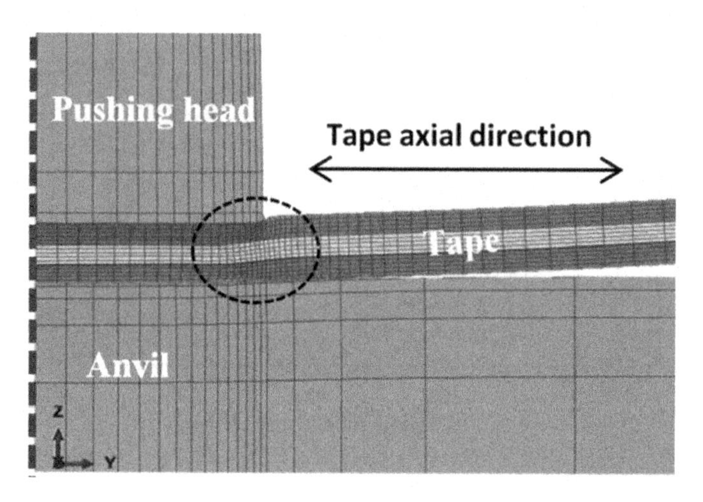

Figure 3.18. Deformed shape of coated conductor under transverse compressive stress [15].

regions of interest, such as the edge of the pushing head where potential load concentrations may exist.

3.6.4. *Cable results*

The adoption of HTS cables for magnet applications requires knowledge of their performance under bending. CORC and HTS slotted-core CICC cables have both been analyzed with FEA to simulate their respective bending characteristics. The deformed shape of both cables under bending are presented in Figure 3.19. Their electromechanical behaviors under bending loads are very different simply due to the method used to cable the coated conductors. The number of tapes, winding method, support structure, and use of external jacket mechanically influence the cables' bending stiffness and electrically influence the cables' critical current performance.

The critical current of the cables under bending loads are dependent on the strain state of the individual coated conductors, as shown by the strain contours in Figure 3.19. The tapes within a CORC cable experience many local regions of high and low strains as they wrap around the central former, while the tapes in an HTS CICC are exposed to the high- and low-strain regions only once over one twist pitch but for a larger region along the tape.

An example of the numerical critical current data predicted from the analytical model described in Section 3.6.2 for the HTS CICC cable over a

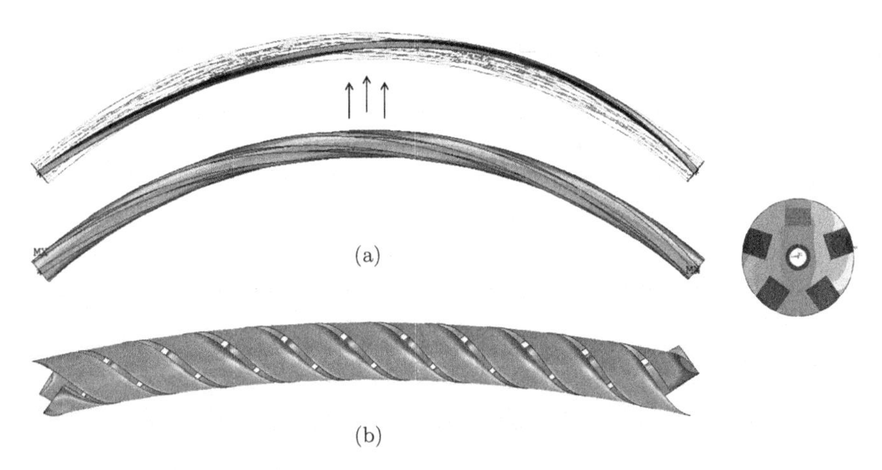

Figure 3.19. Deformed shape of two HTS cables under bending showing strain contours: (a) HTS CICC, highlighting one tape stack; (b) CORC cable, showing tapes from one layer [16].

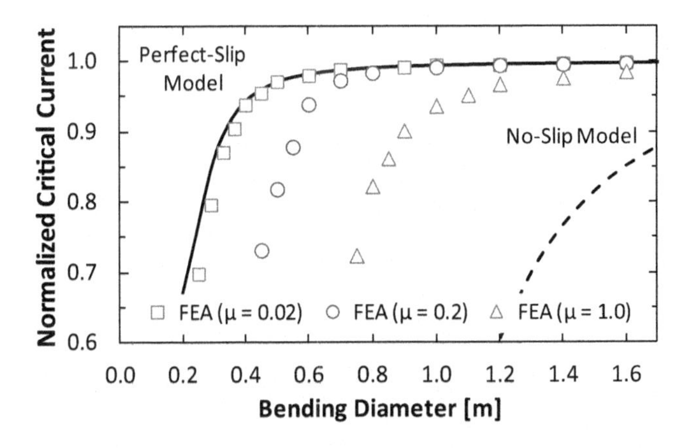

Figure 3.20. Predicted critical current behavior of HTS CICC under bending [19]. Numerical data for various friction coefficients compared with analytical models [35].

range of bending diameters is shown in Figure 3.20. The strain through the width and along the length of each individual tape was used to estimate the critical current performance for each bending diameter. The plot shows the numerical results for three different coefficients of friction, highlighting its effect on the critical current performance due to strain accumulation along the cable. The numerical results are plotted alongside two analytical predictions under perfect-slip and no-slip conditions. Experimental results were found to match the analytical perfect-slip case and the numerical model with the lowest friction.

Simulating transverse load characteristics of HTS cables is a way to investigate the electromagnetic load effects on these cables under operation in magnet applications. In magnets, currents in the cable mixed with the background field produce an electromagnetic load on the cable, which generally acts as a transverse compression. The critical or maximum compressive load on cables before irreversible degradation has been investigated for externally applied transverse mechanical compression, as shown in Figure 3.21. The figure shows a contour plot of the compressive stress distribution through the cross-section of a single CORC cable. The maximum stress is located on the outer layer of the tapes where it contacts the load structure.

External applied loads are important to study and can be validated easily with experiments; however, they do not accurately replicate the real electromagnetic Lorentz load that a cable will experience in a magnet. For that reason, simulations of the internal distributed Lorentz loads acting on each coated conductor in a cable need to be conducted. Figure 3.22 displays

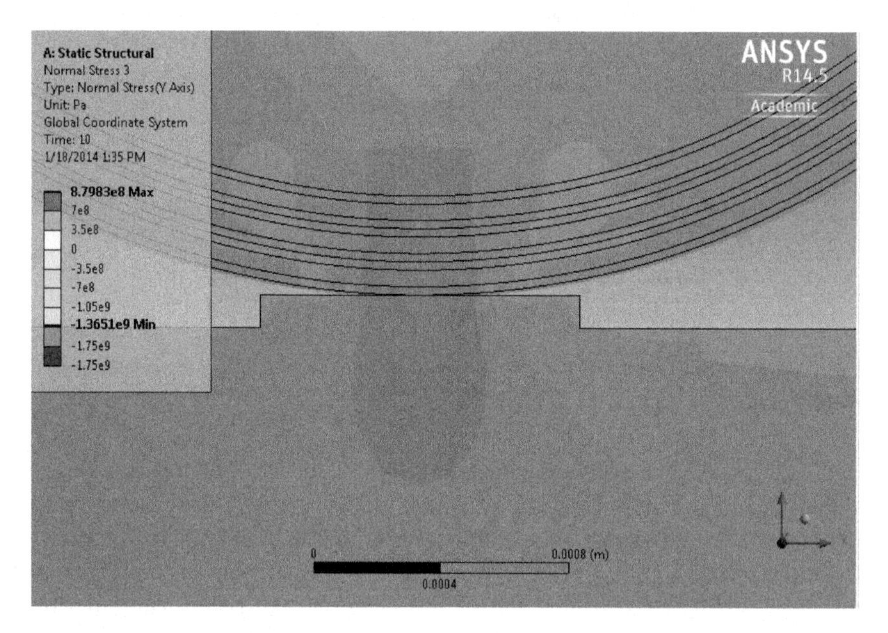

Figure 3.21. Stress contour of CORC cross-section under mechanically applied transverse compression [20].

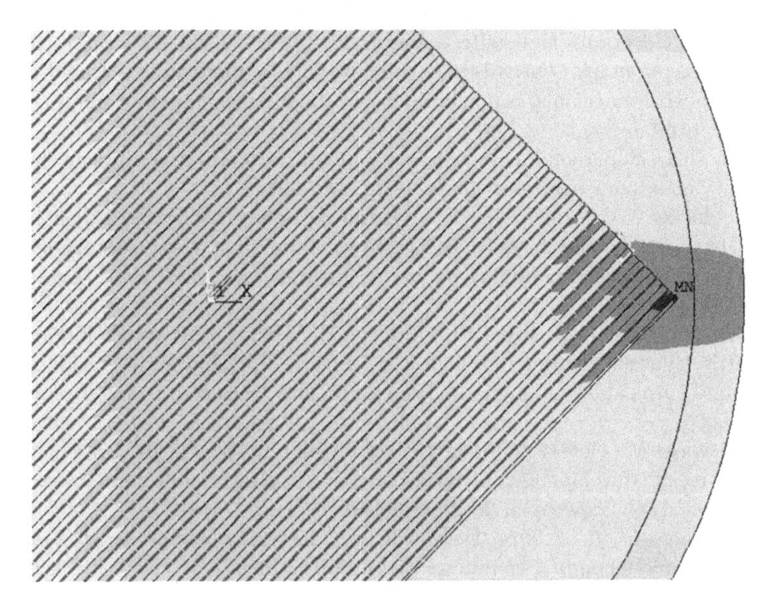

Figure 3.22. Stress contour in TSTC cross-section under internal electromagnetic load [21].

a TSTC conductor in a solder-filled copper tube, which is one support method being tested experimentally. In the model, a uniform distributed load is acting on the superconducting layer of every coated conductor in the tape stack. Depending on the rotational orientation of the stack (i.e. its position along the length of the cable), the loads from every tape can accumulate, leading to local stress concentrations and regions of high deformation, as shown in Figure 3.22. The determination of appropriate support structures for cables is actively being investigated.

References

[1] SuperPower Inc website (online). 2016. 2G HTS Wire page, http://www.superpower-inc.com/content/2g-hts-wire.

[2] SuNAM Co, Ltd website (online). 2016. 2G HTS Wire Catalog, http://www.i-sunam.com.

[3] SuperOx website (online). 2016. 2G HTS Tape page, http://www.superox.ru/en/products/42-2G-HTS-tape/.

[4] Fujikura Europe Ltd website (online). 2016. 2G High Temperature Superconductor page, http://www.fujikura.co.uk/products/energy-and-environment/2g-ybco-high-tempurature-superconductors/.

[5] Bruker Corporation website (online). 2016. YBCO 2G HTS Superconductors page, https://www.bruker.com/products/superconductors-and-metal-composite-materials/superconductors/ybco-2g-hts-superconductors/over view.html.

[6] American Superconductor website (online). 2016. HTS Wire page, http://www.amsc.com/solutions-products/hts_wire.html.

[7] Goldacker W, Grilli F, Pardo E, Kario A, Schlachter S and Vojenciak M. 2014. Roebel cables from REBCO coated conductors: a one-century-old concept for the superconductivity of the future. *Supercond. Sci. Technol.* 27, 093001.

[8] Van der Laan DC, Noyes PD, Miller GE, Weijers HW and Willering GP. 2013. Characterization of a high-temperature superconducting conductor on round core cables in magnetic fields up to 20 T. *Supercond. Sci. Technol.* 26, 045005.

[9] Takayasu M, Chiesa L, Allen NC and Minervini JV. 2016. Present status and recent developments of the twisted stacked-tape cable conductor. *IEEE Trans. Appl. Supercond.* 26(2), 6400210.

[10] Celentano G. *et al.* 2014. Design of an industrially feasible twisted-stacked HTS cable-inconduit conductor for fusion application. *IEEE Trans. Appl. Supercond.* 24, 4601805.

[11] Bykovsky N, Uglietti D, Wesche R and Bruzzone P. 2015. Design optimization of round strands made by twisted stacks of HTS tapes. *IEEE Trans. Appl. Supercond.* 26, 4201207.

[12] Glasson N. *et al.* 2011. Development of a 1 MVA 3-phase superconducting transformer using YBCO Roebel cable. *IEEE Trans. Appl. Supercond.* 21, 1393–1396.

[13] Yagotintsev K. *et al.* 2020. AC loss and contact resistance in REBCO CORC®, Roebel, and stacked tape cables. *Supercond. Sci. Technol.* 33, 085009.

[14] Allen NC, Chiesa L and Takayasu M. 2016. Structural modeling of HTS tapes and cables. *Cryogenics.* 80, 405–418.

[15] Allen NC, Chiesa L and Takayasu M. 2014. Combined tension-torsion effects on 2G REBCO tapes for twisted stacked-tape cabling. *IEEE Trans. Appl. Supercond.* 25, 4800805.

[16] Allen NC, Chiesa L and Takayasu M. 2015. Numerical and experimental investigation of the electromechanical behavior of REBCO tapes. *IOP Conf. Ser. Mater. Sci. Eng.* 102, 012025.

[17] Ilin K. *et al.* 2015. Experiments and FE modeling of stress-strain state in REBCO tape under tensile, torsional and transverse load. *Supercond. Sci. Technol.* 28, 055006.

[18] Nijhuis A. *et al.* 2015. Transverse, axial and torsional strain in REBCO tapes; experiments and modeling. *Chats-Applied Superconductivity Workshop,* Bologna, Italy.

[19] Barth C, Weiss KP, Vojenciak M and Schlachter S. 2012. Electro-mechanical analysis of Roebel cables with different geometries. *Supercond. Sci. Technol.* 25, 025007.

[20] Weiss KP, Goldacker W and Nannini M, 2011. Finite element analysis of torsion experiments on HTSC Tapes. *IEEE Trans. Appl. Supercond.* 21, 3.
[21] De Marzi G. *et al.* 2015. Bending tests of HTS cable-in-conduit conductors for high-field magnets applications. *IEEE Trans. Appl. Supercond.* 26, 4801607.
[22] Van der Laan DC. *et al.* 2015. Development of HTS conductor on round core (CORC) cables for fusion application. *HTS4Fusion Workshop,* Pieve Santo Stefano, Italy.
[23] Allen NC, Pierro F, Chiesa L and Takaysu M. 2016. Structural finite element evaluation of twisted stacked-tape cables for high-field magnets. *IEEE Trans. Appl. Supercond.* 27, 6900205.
[24] Zhang Y. *et al.* 2015. Stress-strain relationship, critical strain (stress) and irreversible strain (stress) of IBAD-MOCVD-based 2G HTS wires under uniaxial tension. *IEEE Trans. Appl. Supercond.* 26, 8400406.
[25] Dedicatoria M, Shin H, Ha H, Oh S and Moon S. 2010. Electro-mechanical property evaluation of REBCO coated conductor tape with stainless steel substrate. *Supercond. Cryo.* 12, 4.
[26] Reed R and Horiuchi T. 1983. Austenitic steels at low temperatures: The properties of austenitic stainless steel at cryogenic temperatures, *Springer.*
[27] Cheggour N. *et al.* 2003. Transverse compressive stress effect in Y-Ba-Cu-Ocoatings on biaxially textured Ni and Ni-W substrates. *IEEE Trans. Appl. Supercond.* 13, 2, 3530–3533.
[28] Clickner CC. *et al.* 2006. Mechanical properties of pure Ni and Ni-alloy substrate materials for Y–Ba–Cu–O coated superconductors. *Cryogenics* 46, 6, 432–438.
[29] Huang H and Spaepen F. 2000. Tensile testing of free-standing Cu, Ag and Al thin films and Ag/Cu multilayers. *Acta. Mater.* 48, 3261–3269.
[30] Verdier M, Huang H, Spaepen F, Embury J and Kung H. 2006. Microstructure, indentation and work hardening of Cu/Ag multilayers. *Phil. Mag.* 86, 5009–5016.
[31] Brechna H, Hartwig G and Schauer W. 1971. Cryogenic properties of metallic and non-metallic materials utilized in low temperature and superconducting magnets *Presented at the 8th International Conference High-Energy Accelerators,* Inst. fur Experim. Kernphysik, C71-09-20.
[32] Reed R and Mikesell R. 1967. Low temperature mechanical properties of copper and selected copper alloys. *National Bureau of Standards,* Boulder, CO.
[33] Goretta K, Cluff J, Joo J, Lanagan M, Singh J and Vasanthamohan N. 1995. Mechanical properties of high-temperature superconducting wires. *Proc. of 4th World Congress on Supercond.* NASA Conference Publication 3290, 633–638.
[34] Raynes AS, Freiman SW, Gayle FW and Kaiser DL. 1991. Fracture toughness of YBa2Cu3O6+δ single crystals: Anisotropy and twinning effects. *J. Appl. Phys.* 70, 5254.
[35] Ledbetter H and Ming L. 1991. Monocrystal elastic constants of orthotropic YBCO: An estimate. *J. Mater. Res.* 6, 2253–2255.
[36] Takayasu M, Chiesa L, Bromberg L and Minervini J. 2012. HTS twisted stacked-tape cable conductor. *Supercond. Sci. Technol.* 25, 014011.

[37] Takayasu M and Chiesa L. 2015. Analytical investigation in bending characteristic of twisted-stacked-tape cable conductor. *IOP Conf. Ser. Mater. Sci. Eng.* 102, 012023.
[38] Barth C, Mondonico G and Senatore C. 2015. Electro-mechanical properties of REBCO coated conductors from various industrial manufacturers at 77 K, self-field and 4.2 K, 19T. *Supercond. Sci. Technol.* 28, 045011.
[39] Weiss KP. *et al.* 2016. Mechanical and thermal properties of central former material for high current superconducting cables. *IEEE Trans. Appl. Supercond.* 26(4), 8800604.
[40] Basaran C and Jiang J. 1998. Measuring intrinsic elastic modulus of Pb/Sn solder alloys. *Mech. Mater.* 34, 349–362.
[41] Adams PJ. 1986. Thermal Fatigue of Solder Joints in Micro-Electronic Devices *M.Sc. Thesis Department of Mechanical Engineering*, MIT, Cambridge, MA.

Chapter 4

Thermal-Hydraulics of Superconducting Magnets

Roberto Bonifetto, Laura Savoldi and Roberto Zanino
Dipartimento Energia "Galileo Ferraris",
Politecnico di Torino, 10129, Torino, Italy

This chapter discusses the fundamental thermal-hydraulic (TH) aspects involved in the cooling of superconducting magnets, with special reference to the issues related to forced-flow cooling.

The aim of an electromagnet is to produce a magnetic field with given characteristics, such as intensity and shape, at each time instant of the operational transient at hand.

If the electromagnets need to be operated while reducing as much as possible the Joule losses, then the ohmic conductor must be replaced by a superconductor. The energy consumption from the electromagnet power supply will thus be minimized, as the Joule losses will be reduced to a sum of:

- the losses localized in the copper (Cu) joints;

- the Joule losses in the superconductor, where the electric field E can be computed as a function of the transport current I, as shown in Equation (4.1):

$$E = E_c(I/I_c)^n \tag{4.1}$$

with E_c set conventionally to $10\,\mu\text{V/m}$ for low-temperature superconductors (LTS), while the critical current I_c and the n value are usually obtained from conductor tests [1]. In particular, the critical current depends on several parameters, such as the magnetic field, temperature, and mechanical strain. Despite the fact that the electric field (and consequently the voltage) is never exactly zero, as shown in Equation (4.1), when $I \ll I_c$ (i.e. under normal operation), E is in fact negligibly small, with n being typically (for LTS) $>\sim 5$. However, when $I \sim I_c$ or $I \sim I_c$ (e.g. during a quench), the Joule losses are no longer negligible.

Besides these losses, there is another type of Joule loss, the so-called alternating current (AC) losses, caused by magnetic field variations (in turn, due to current variations in the magnet itself or in other magnets whose magnetic field affects the magnet at hand).

All electromagnets need to be cooled. For the resistive electromagnets, the power P to be removed equals that generated by the Joule losses in the ohmic conductor, namely

$$P = \Delta V \times I, \tag{4.2}$$

with ΔV being the voltage across the magnet. The superconductors show superconducting properties only if kept at cryogenic temperatures (below their so-called critical temperature), namely below 80–100 K (high critical temperature superconductors, HTS) or even below ~20 K (low critical temperature superconductors, LTS); the superconductors then must be first cooled down to their operating temperature (below their critical temperature), which is lower than the ambient temperature. As a consequence, the superconducting coils are subject to a static heat load coming from the ambient by means of radiative, conductive, and/or convective heat transfer. Once they are cooled below their critical temperature, the superconducting coils can be operated, but they must be continuously cooled to compensate (besides the power sources mentioned above, namely Joule and AC losses) for the static heat load also. Moreover, fusion magnets also experience heat generation due to particle flux coming from nuclear reactions taking place in the machine.

The cooling of superconducting magnets presents some peculiarities related to the operating conditions, which are very demanding in terms of temperature margin to keep the magnet far from the temperature at which the loss of superconductivity (quench) occurs, transport current requirements, external magnetic field, and mechanical strain (in turn, induced by the high current and magnetic field). The superconductor performance depends on all these parameters, thus impacting on the cooling design because of the temperature level to be maintained and the relatively small space available for the coolant.

Starting from the different possible applications of superconducting magnets, in Section 4.1, their structure is investigated, following a top-down approach from the magnet to its components, down to the conductors. The latter allows a discussion of the main issues and possible options for their cooling, which is described in Section 4.2, together with the refrigerator and other subsystems providing the cooling power.

Then, the modeling of the superconducting magnet cooling is addressed: After a detailed analysis of the time and space scales involved (Section 4.3), the main hydraulic (Section 4.4) and TH (Section 4.5) features are described in order to introduce their modeling peculiarities. After that, following a bottom-up approach, the aspects related to the heat transfer inside the entire magnet are dealt with in Section 4.6.

For the sake of completeness, and to give a flavor of the importance of TH modeling of superconducting magnets, the most relevant TH transients of interest are listed and described in Section 4.7.

Finally, a list of existing state-of-the-art models and of available experimental facilities with superconducting magnets is reported in Section 4.8.

4.1. Applications of Superconducting Magnets and Related Topologies/Geometries

The target of producing a rated magnetic field when and where required could be achieved by winding a proper number of ampere-turns of a single superconducting strand, which would be difficult from a technological point of view, or by winding a proper number of ampere-turns of a bundle of strands collected into a conductor, which is the adopted solution nowadays. The conductor winding results in a coil having a shape capable of producing the needed magnetic field. Moving on from the simplest shape producing a controlled magnetic field, i.e. the solenoid, the electromagnets can have several different shapes depending on the final application. The cooling strategy adopted for the different magnets and applications, to be chosen among those described in Section 4.2, depends on several factors, among which are:

- size of the magnet;
- type of superconducting material, namely HTS or LTS;
- operating conditions (temperature, magnetic field, transport current density resulting in Lorentz forces, i.e. mechanical strain);
- heat load on the magnet;
- target temperature margin from current sharing (in turn depending on the operating conditions).

Magnets for nuclear fusion applications have to face the worst conditions of almost all the above-mentioned factors: large dimensions, stringent operating conditions, large heat loads, and small temperature margin since the design phase. As a result, their TH conditions are also challenging: Most of this chapter will be thus devoted to those magnets.

4.1.1. *Magnetically confined nuclear fusion experiments*

In the past 30 years, a lot of experience in superconducting magnets has been gained thanks to their application in the field of magnetically confined fusion experiments, which aim at demonstrating the possibility of exploiting nuclear fusion reactions for energy production.

The adoption of superconducting coils is crucial for that kind of application in order to reduce the energy consumption of the magnet system for both tokamaks [2] and stellarators [3], the two most important configurations of magnetically confined fusion reactors. In these machines, the magnetic system is one of the most complex and most expensive components, aimed at controlling the plasma inside a vacuum chamber.

ITER, the world's largest experimental reactor under construction at Cadarache (France) [4] as a result of an international collaboration between China, the European Union, India, Japan, South Korea, Russia, and the United States, will have a fully superconducting magnet system, see Figure 4.1.

Depending on the position and aim of the four different magnet subsystems, namely:

- the central solenoid (CS),
- the toroidal field (TF) coils,
- the poloidal field (PF) coils, and
- the correction coils (CC),

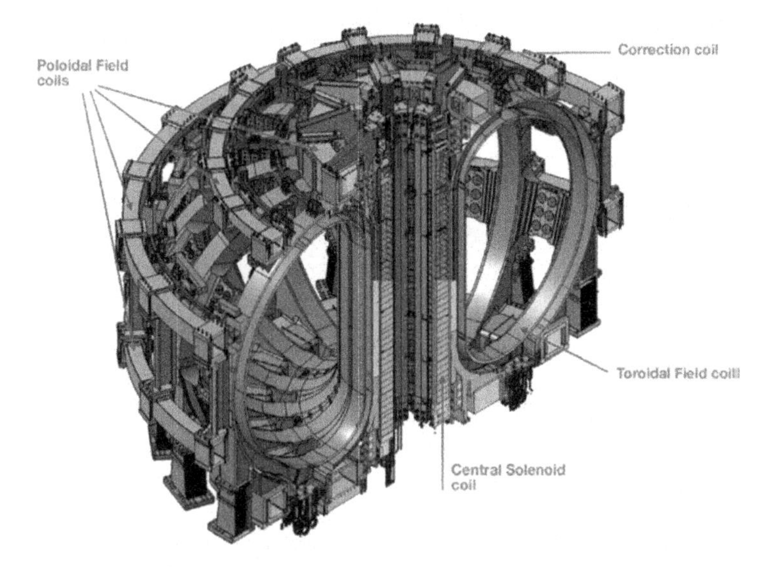

Figure 4.1. Sketch of the four superconducting magnet subsystems constituting the ITER magnet system.

Source: Reproduced from Ref. [5].

the shape of the coils varies from an almost ideal solenoid (CS, see Figure 4.2(a)), to a "D-shape" coil (TF coils, see Figure 4.2(b)) to accommodate the plasma chamber in the bore of the magnet, to a ring-shape (PF coils, see Figure 4.2(c)), and to a rectangular/trapezoidal shape (CC, see Figure 4.2(d)). All of these magnets are of an impressive size (Figure 4.2): the ITER magnet system will indeed be the largest and most integrated superconducting magnet system ever built [6]. To cool such big magnets, existing engineering solutions have been improved and new technological solutions have been adopted.

If stellarators are considered, e.g. the W7-X being operated at Greifswald (Germany) [7], some of the coils have a fully three-dimensional (3D) shape, far different from a classical solenoid, see the W7-X nonplanar coil in Figure 4.3 as an example.

All the ITER magnets, as well as most of the superconducting magnets of existing and future tokamaks (EAST [9], KSTAR [10], JT-60SA [11], EU DEMO [12]), stellarators (W7-X [13], LHD [14]), or test facilities for fusion magnets (ITER CS model coil - CSMC [15]), rely on the cable-in-conduit conductor (CICC) concept [16] for effective cooling of the superconducting coil, see Figure 4.4(a) highlighting the CICC main components and Section 4.2.

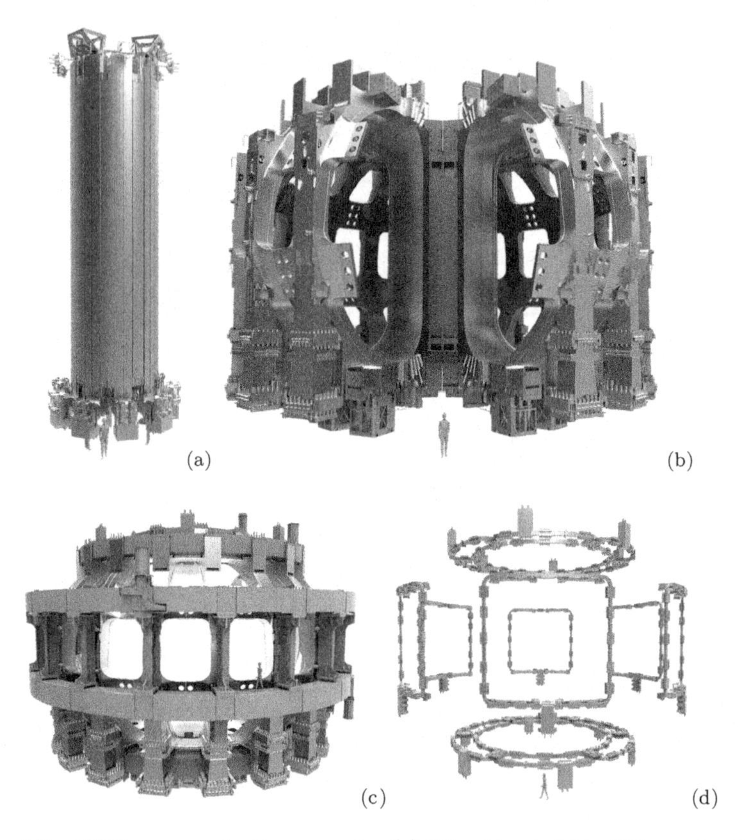

Figure 4.2. ITER superconducting magnets: (a) CS, with a height of 18 m and a diameter of 4 m; (b) TF coils (in gray), 17 m high and 9 m wide; (c) PF coils (in yellow), up to 24 m in diameter; (d) CC, up to 8 m wide.

Source: Reproduced from Ref. [6].

The CICC consists of a **cable**, in turn constituted by many superconducting strands twisted together, inserted into a conduit, called a **jacket** (see Figure 4.5). Each strand inside the cable contains superconducting filaments embedded into a Cu matrix. In the conduit, the coolant is forced to flow through one or more paths, see Figure 4.4:

- the **bundle**, i.e. the free-flow area remaining in the small spaces left free among the strands;
- a variable number of channels, usually called "**holes**" and delimitated by a metal helix, acting as pressure relief paths, in particular to mitigate the pressure rise during a quench [19].

Figure 4.3. W7-X non-planar coil.

Source: Reproduced from Ref. [8].

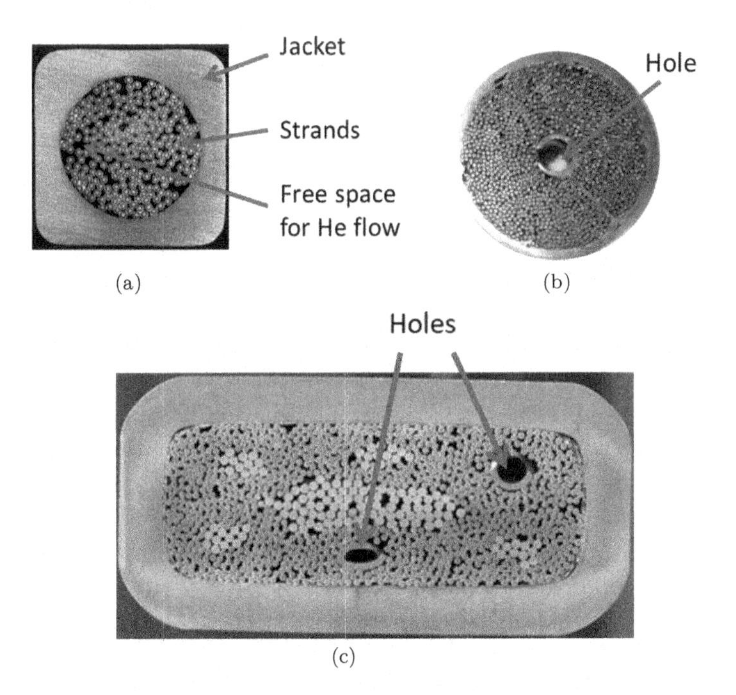

Figure 4.4. Cross-section of CICCs in: (a) Wendelstein 7-X stellarator (no holes), (b) ITER TF (one hole), (c) EU-DEMO TF (two holes) conductor (not to scale). Some of the main components of the CICC are highlighted in (a).

Source: Reproduced from (a) Ref. [17], (b) Ref. [5], and (c) Ref. [18].

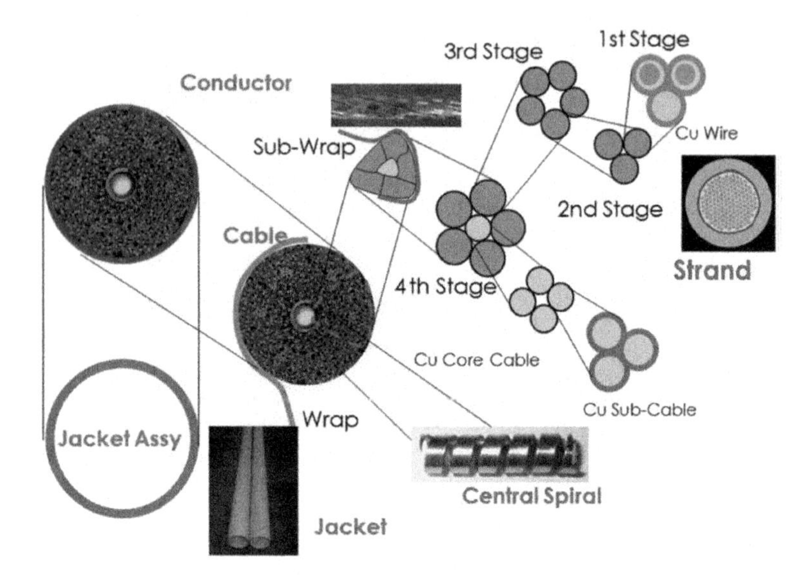

Figure 4.5. Components of the CICC.

Source: Reproduced from Ref. [20].

Different designs of the CICC, see Figure 4.6, have been produced depending on the application and the performance required.

The CICCs find great success in large magnet applications in view of their numerous advantages:

- Because the cable is made by thin, as opposed to large bulky, wires, its manufacturing is easier, as well as its handling and bending, and moreover, the hysteresis losses caused by variable magnetic fields are reduced.
- The large wetted perimeter of the strands guarantees good heat transfer with the coolant and efficient heat removal during the operation of the machine (stability).
- The stainless-steel conduit protects the strands from a mechanical point of view, withstanding large Lorentz electromagnetic forces acting on the cable.

However, there are also some disadvantages: Since the coolant and conduit walls occupy a part of the cross-section of the cable, the amount of superconductor that can be put into the cable is limited, reducing the maximum transport current and therefore the magnetic field that can be generated.

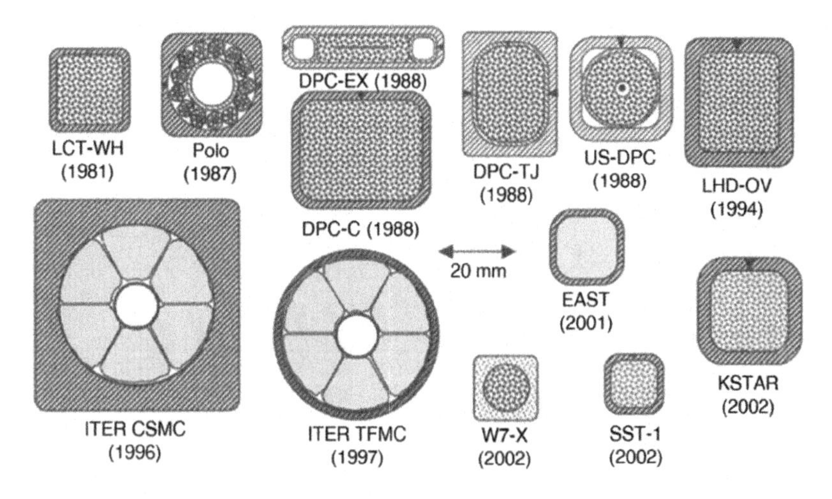

Figure 4.6. Different CICC layouts (to scale) developed in the past four decades. *Source*: Reproduced from Ref. [21].

During the coil-winding process, the conductor can be wound in several ways, depending mainly (but not necessarily only) on:

- manufacturing constraints, namely the length of the conductor (the so-called unit length);
- operational constraints (location of the expected maximum thermal and magnetic loads and, consequently, the minimum temperature margin in the coil);
- TH constraints:
  - the maximum allowed pressure drop across each cooling path,
  - the capability to remove the heat deposited by the heat source and the distance of the expected location of the minimum temperature margin from the inlet;
- space constraints, determining the location of the coolant inlets and the electrical joints.

The typical classification of the coil winding is as follows:

- Pancake winding: the different turns of the conductor are at a different radial distance from the coil axis, see for example the ITER CS case [22] in Figure 4.7(a).

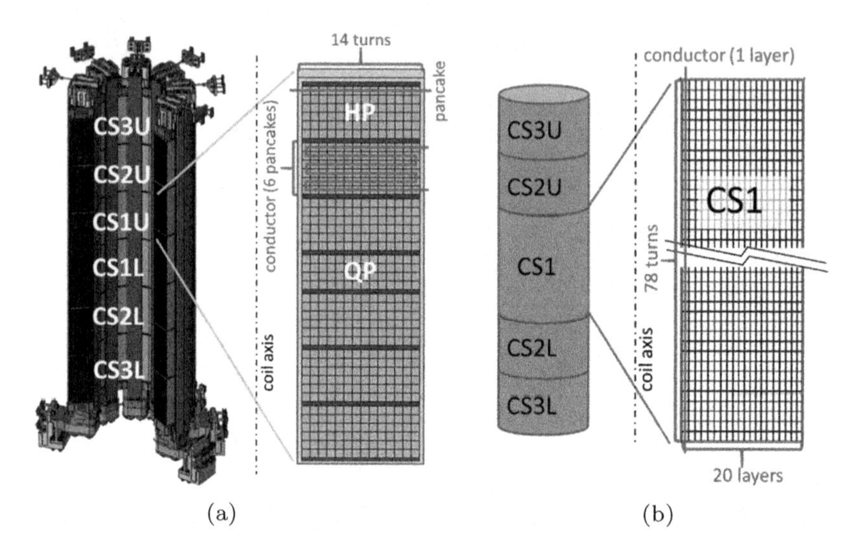

Figure 4.7. Example of (a) pancake-wound coil (ITER CS) and (b) layer-wound coil (EU DEMO CS proposed by SPC and ENEA). The numbering of the different CS modules is also reported (HP: hexa-pancake, QP: quad-pancake).

Source: (a) Partly reproduced from Ref. [22].

- Layer winding: the different turns of the conductor are at the same radial distance from the coil axis, see for example the EU-DEMO CS (proposed by SPC and ENEA) case [23] in Figure 4.7(b).

If the same conductor unit length is used to wind more than one pancake or layer, independently of the fact that they are cooled in parallel or in series, then the coil is wound in a multi-pancake (or multi-layer) way. The ITER CS is, for example, constituted by six hexa-pancakes (a single unit length used to wind six pancakes) and one quad-pancake, see Figure 4.7(a), while the ITER TF coils are wound in double-pancakes.

On the other hand, it may happen that a single conductor length is not sufficient to wind a layer or pancake. Then, for a given winding layout, namely pancake or layer wound, it is possible to wind the coil:

- with a single conductor (one-in-hand) or
- with more than one conductor (two-in-hand, three-in-hand, etc.).

Figure 4.8 shows the final coil layout for three different (layer-wound) cases. Using n conductors to wind a single layer or pancake allows us to reduce the required conductor unit length by a factor of n, with a positive

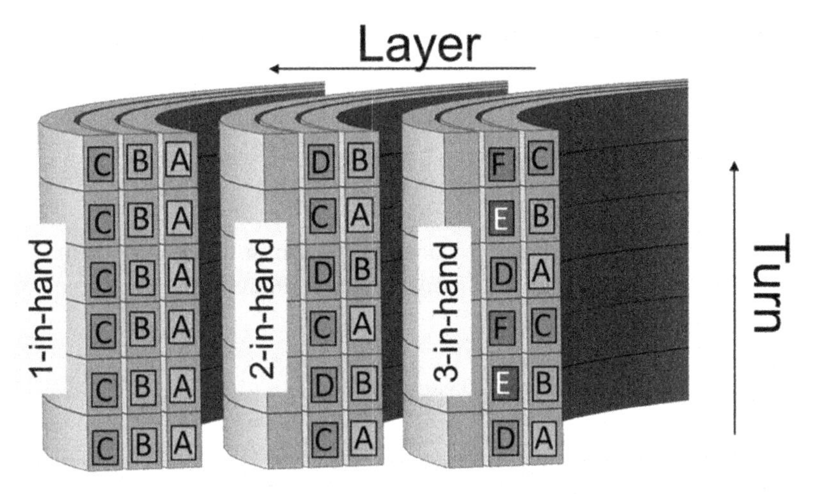

Figure 4.8. Schemes of one-, two-, and three-in-hand (layer) winding. The same conductor is indicated with the same letter and color.

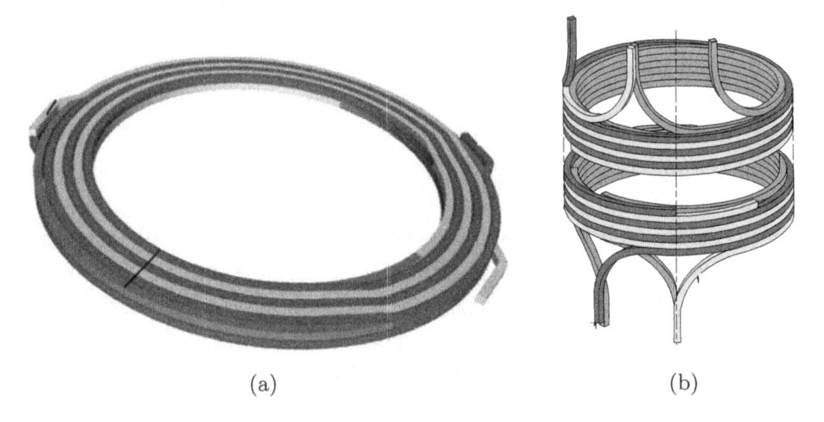

(a) (b)

Figure 4.9. Example of (a) two-in-hand pancake winding for the ITER PF coils and (b) two-in-hand layer winding for the ITER CSMC.

Source: Reproduced from (a) Ref. [25] and (b) Ref. [27].

impact on the conductor manufacturing and on the pumping power during operation.

The EU DEMO PF coils (ENEA and SPC design proposal [24]) are one-, two-, and three-in-hand layer wound depending on the coil. ITER PF coils are two-in-hand pancake wound [25], see Figure 4.9(a), and ITER CSMC is two-in-hand layer wound [26], see Figure 4.9(b).

4.1.2. *Particle accelerators*

Accelerators for high-energy physics require magnets to guide the particle beams and confine them in a relatively small and well-defined volume in the vacuum pipe. In particular, in a particle accelerator, the majority of the magnets belong to one of these two types [28]:

- Dipoles, see Figure 4.10(a): to bend the beam in its circular orbit.
- Quadrupoles, see Figure 4.10(b): to reduce (focus) the transverse beam size at the collision point.

The typical requirements for accelerator magnets are [29]:

- high current density in the targeted field range (up to more than 15 T, depending on the energy of the particles accelerated);
- precise and stable geometry (down to 2 μm tolerance);
- capability to bear mechanical stress and strain (up to 150 MPa pressure).

The CICCs are unsuitable for this type of magnet because the free area for the coolant flow and the conduit walls occupy a large part of the cross-section, reducing the fraction available for the cable; as a result, the number of superconducting wires that can be put into the cable is limited, reducing the maximum transport current and therefore the magnetic field that can be generated. For this reason, for applications needing very large magnetic fields, such as accelerators, monolithic conductors or Rutherford cables are used. The latter are composed of Cu/NbTi twisted wires wound around a core that can be, for example, a thin shim (see Figure 4.11(a)) or a round tube or cylinder (see Figure 4.11(b)). The rectangular geometry of the cable provides a high packing factor and results in a precisely controlled dimension necessary to wind coils with tight tolerances; this type of cables is also flexible enough to accommodate various geometries [28].

The conductors are insulated [34] and usually wound into a racetrack shape (see Figure 4.12) in single or double layers, with the length of the coil up to ~10 m. Such coils are then cooled by conduction (see Section 4.2) with a coolant in a bath or flowing through suitable channels in the magnet structure, see the cooling channels visible in Figure 4.10. In order to avoid any movement due to Lorentz forces that would lead to heat generation, possibly leading to a quench of the conductor, a system of collars is used to resist the electromagnetic force and to apply an initial prestress to the coil. The coil in the collar is similar in shape to a Roman arch, see each of

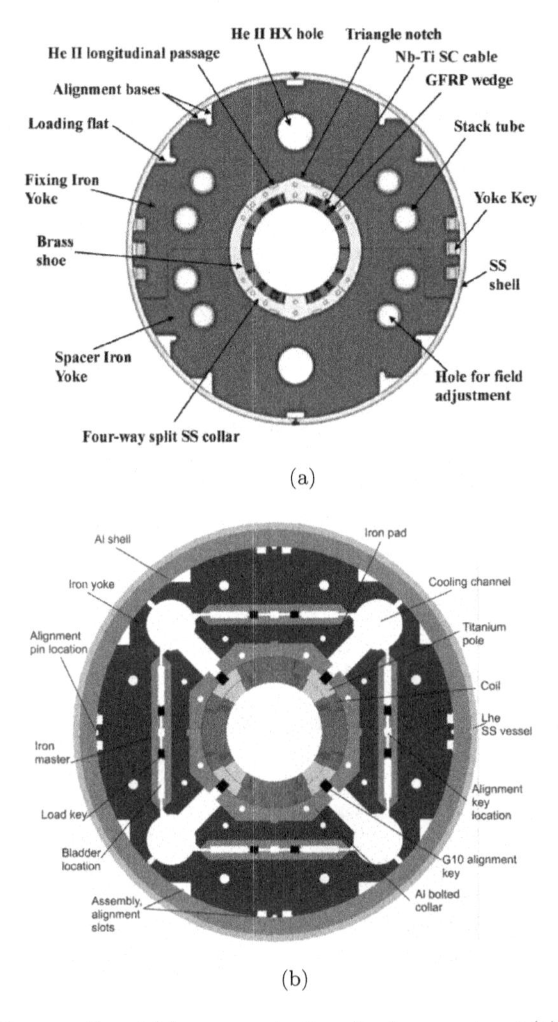

Figure 4.10. Cross-section of (a) superconducting dipole magnet and (b) superconducting quadrupole magnet for the High-Luminosity (HL) Large Hadron Collider (LHC) under design at CERN [30].

Source: Reproduced from Ref. [28].

the four red coils in the two collars in Figure 4.12, and the pre-compression is obtained by inserting an oversized wedge into the coil pole [28].

In some cases, the iron yoke, the collars, and the coil can all be inserted in an outer shell that provides helium (He) leak tightness so that the entire magnet is kept at the superconductor operating temperature. This concept

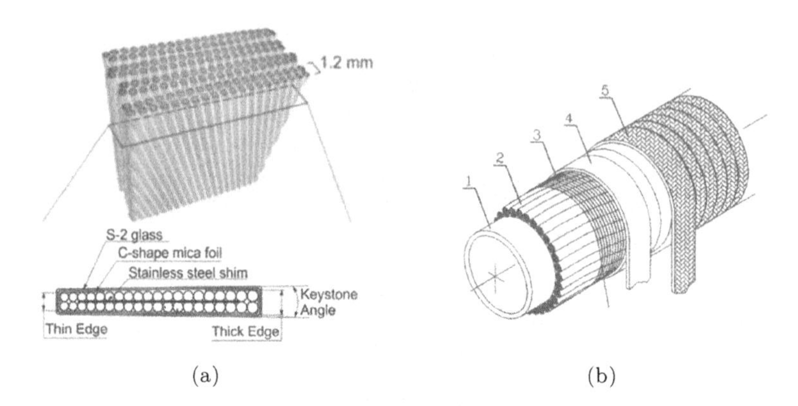

Figure 4.11. View of (a) Rutherford cable for the LHC dipoles and (b) scheme of the superconducting cable for the Nuclotron [32] dipole: (1) Cu–Ni tube, (2) superconducting strands, (3) Ni–Cr wire, (4) Kapton tape, and (5) fiberglass tape.

Source: Reproduced from (a) Ref. [31] and (b) Ref. [33].

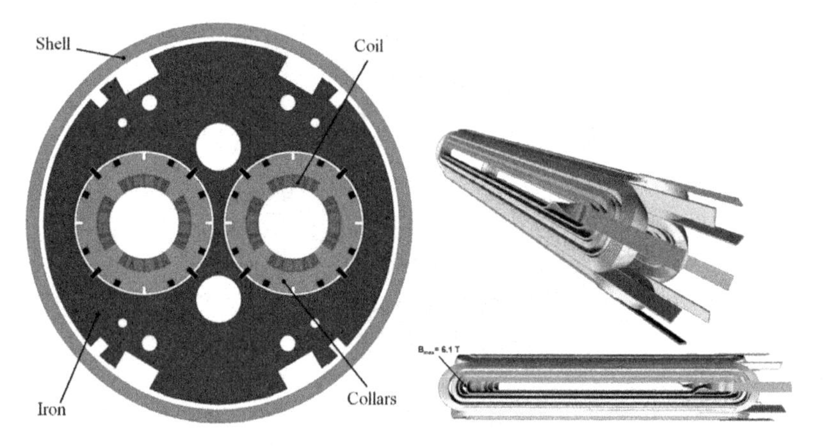

Figure 4.12. Cross-section of the two-in-one quadrupole magnet (left) for the HL-LHC (constituted by two quadrupole magnets similar to that in Figure 4.10(a)) and 3D view of coil winding (right).

Source: Reproduced from Ref. [35].

greatly simplifies the alignment and geometry of the coil but has a much larger cold mass [28], requiring more time for cool down (CD), for example.

Advanced designs include the "canted cosine-theta" coil, composed of two nested solenoids with oppositely tilted windings powered such that the solenoid components (along the solenoid axis) cancel out, leaving a nearly

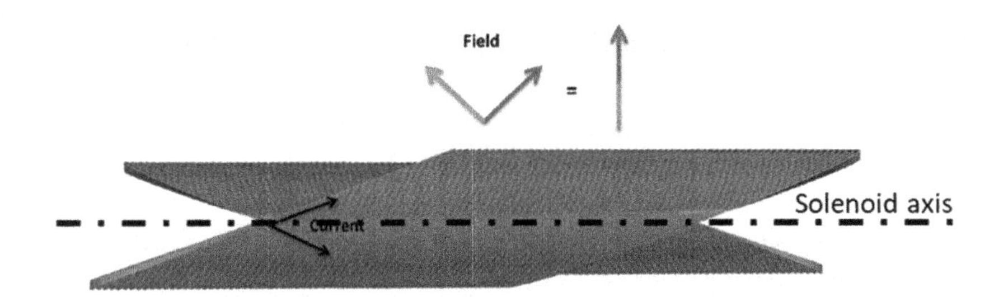

Figure 4.13. Schematic representation of a canted cosine-theta coil.
Source: Reproduced from Ref. [28].

perfect dipole field [28] in the poloidal direction (orthogonal to the solenoid axis), see Figure 4.13.

In some cases, high magnetic field ramp rates are required for particular applications in accelerators. The ideal operating mode for superconducting magnets is steady state, and fast-cycled accelerator magnets require special attention to be paid to AC losses and cooling, as the required operating margin must be maintained by removing the heat generated during ramping. These magnet coils are often wound using a cable whose strands are in turn wound on a core constituted by a pipe in which gaseous or two-phase coolants are circulated (see Figure 4.11(b)) or are realized starting from suitable CICCs [36].

4.1.3. *Others*

Besides the application to nuclear fusion devices based on magnetic confinement concepts and particle accelerators, superconducting magnets are also applied to a variety of other fields, of which an overview is given in the following sections.

4.1.3.1. *Gyrotrons*

The gyrotron is a source of microwaves used as auxiliary heating for the magnetically confined plasma in nuclear fusion installations. This technology accelerates the electrons through a high DC voltage and magnetic field. The superconducting magnet of the gyrotron has to provide the magnetic field to fire the electron beam from the electron gun into the resonator. These magnets, wound in a solenoidal shape, are usually small (a few tens of centimeters in diameter and height) so that the cooling can be provided by a liquid He (LHe) bath or they can be conduction cooled, see Section 4.2.

4.1.3.2. *Medical*

Magnetic resonance imaging (MRI), a powerful medical diagnostic tool, is the most prevalent commercial application of superconductivity. A high image quality requires a field of at least 1.5 T, a high field homogeneity, and temporal stability in a large volume.

The superconducting magnets currently used by the MRI industry have horizontal annular cryostats with circular patient bores with a diameter of ~1 m; the coils used are typically wound using NbTi wires, and they are bath cooled or cryogen-free conduction cooled [37], see Section 4.2.

In the case of nuclear magnetic resonance (NMR), the magnets' characteristics are similar, but the magnetic field is much higher, up to more than 10 T, so that high-temperature superconducting materials are adopted [38].

A relatively recent application of superconducting accelerator magnets is the development of compact, lightweight gantries for particle beam cancer therapy. Current gantry designs rely on heavy, resistive magnets. The requirements for these magnets are a large bore, the capability to combine bending and focusing, and a quickly changing field. The topology of those magnets is similar to that of the accelerator magnets described above [28] in Section 4.1.2.

4.1.3.3. *Power grid*

As far as the applications to electric utility grids are concerned, superconducting magnets can be employed, in particular, for energy storage. However superconducting cables are also used (not in the form of magnets) in power transmission, generators (e.g. generators for wind turbines [39]), motors [40], synchronous condensers, transformers, and fault-current limiters [41]. Some of these applications are described briefly in the following:

Power transmission cables: In the case of power transmission cables, the superconducting cable is not wound in a coil to generate a magnetic field but is used only to carry a given current in a limited space, minimizing the Joule losses (excluding the AC losses and the joints). Even if this application is not a magnet, it is mentioned here as forced-flow cooling (see Section 4.2), and the topology of the conductor (similar to a CICC) makes this application of superconducting cables interesting from a TH point of view. The interest in superconducting cables for power transmission has increased a lot after the discovery of HTS: The use of HTS takes advantage of the use of a cheap fluid for the cooling (liquid nitrogen, LN2, or gaseous He, GHe) [42].

According to the type of dielectric adopted, two main types of superconducting power cable concept have been developed [42]:

(1) Warm dielectric design, which is based on a flexible support with HTS tapes in one or several layers, surrounded by a thermal insulation based on two concentric flexible stainless-steel corrugated tubes with vacuum (and superinsulation) in between (see Figure 4.14(a)); the dielectric

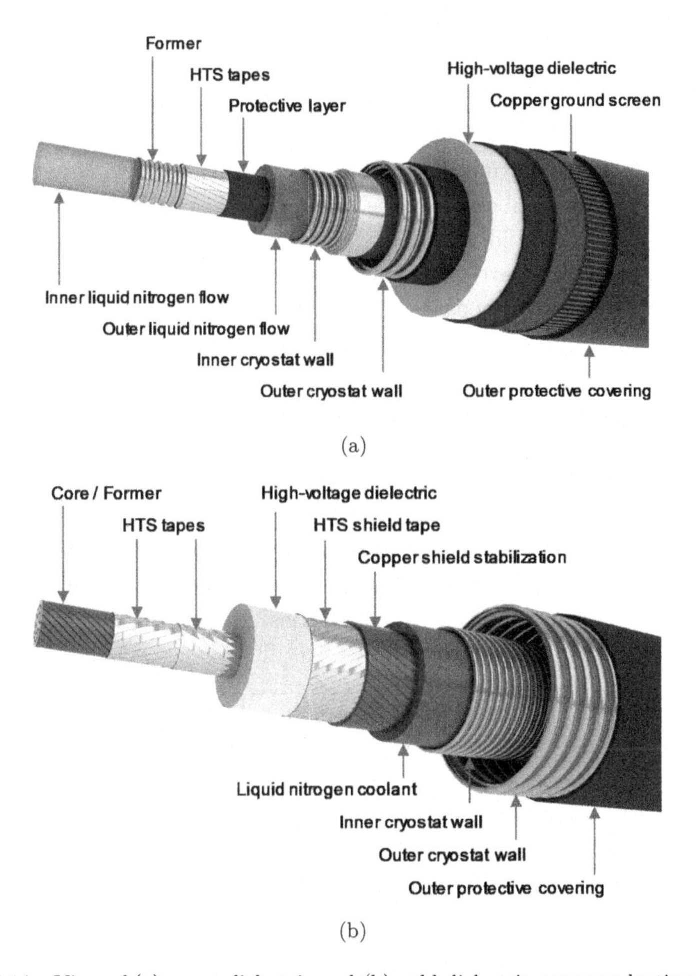

Figure 4.14. View of (a) warm dielectric and (b) cold dielectric superconducting cable. *Source*: Reproduced from Ref. [43].

insulation and the cable screen are the outermost layers at room temperature. The coolant flows into two separate regions, namely inside the flexible support and the innermost steel tube.
(2) Cold dielectric design, having a high-voltage insulation constituted by a layered dielectric impregnated with liquid N, which is then also part of the dielectric itself (see Figure 4.14(b)); three of these cable phases can be put into individual or a single cryogenic envelope, which is again based on two concentric flexible stainless-steel corrugated tubes with vacuum (and superinsulation) in between.

Energy storage: Superconducting magnetic energy storage (SMES), where the energy is stored in the magnetic field produced by an electric current flowing with no resistance in the superconducting coils, is an attractive energy storage option thanks to an overall electrical efficiency larger than 97% [44] and the possibility to withstand a very large, virtually unlimited number of cycles. The two most used configurations of SMES are the solenoid and toroidal configurations. SMES could be an important part of smart grids since they are suitable for power management purposes to compensate for electric power fluctuations on a wide spectrum of time scales, ranging from seconds to several minutes [45], in view of their high power density in the range 10–105 kW/kg [44], much higher than batteries. This is due to their capability to withstand fast current variations, down to 0.1–1 s, so that the entire energy stored (~0.01 MJ/kg [44]) can be released in that short time, resulting in a high power transfer to the utility. However, they have a low energy density (directly proportional to the square of the current, the square of the number of turns, and the coil radius but inversely proportional to the large coil mass), much lower than batteries. This is mainly driven by two factors:

(1) the critical current in the superconductors, which decreases as B increases, reducing the maximum allowed operating current;
(2) the mechanical strength (often dominating limiting factor): when either the maximum hoop stress due to Lorentz force reaches the yield stress [46] or the stress amplitude reaches the fatigue limit of the structural materials [47], the operational limit is reached.

The technology adopted so far for SMES portends the use of LTS, such as NbTi, with a peak B of ~5–6 T, cooled by stagnant He at 4.2 K [48], by forced-flow supercritical He (SHe) [44], or, rarely, by conduction (cryogen free) [44], see Section 4.2. HTS have also begun to be considered in the design of SMES [49] in view of their potential to generate a much larger magnetic field than LTS at 4.2 K or to achieve a similar magnetic field but with a much higher temperature [50] that would allow us to decrease the operation costs of the cryogenic system.

4.2. Superconducting Magnet Cooling Methods

Several cooling strategies are available to keep the superconducting magnets operating temperature at the desired value so that the superconducting state is maintained. The cooling method depends on the magnet geometry/topology and on its application.

4.2.1. *Cooling fluids*

Several cooling fluids are available depending on the superconductor operating conditions (mainly the temperature). Besides the most used (He and N_2) that are listed and for which some more details are provided in the following, other fluids are potentially available, such as hydrogen (H_2) and neon (Ne). Both of them could be used for HTS cooling in the ~20–30 K temperature range in order to reduce the operating cost of the refrigerator.

Different fluids are presented and compared here based on their thermodynamic properties.

Fluid properties are usually given as a function of couple of thermodynamic independent variables, e.g. pressure and temperature or pressure and enthalpy. The most distinguished sets of properties are those of the National Institute of Standards and Technology (NIST), a non-regulatory agency of the United States Department of Commerce [51]. An alternative open-source fluid property library is CoolProp [52].

Figure 4.15(a) shows the latent heat of vaporization (h_v) at 1 bar, compared also with the enthalpy variation (dh) when the fluid temperature increases by 1 K (namely, the specific heat at constant pressure) at different temperatures. A similar comparison but per unit volume is shown in Figure 4.15(b). It is shown that for a given allocated flow area, the highest specific heat is provided by He, Ne, and N_2 at 5, 30, and 70 K, respectively, which are the reference temperatures for the cooling of LTS (5 K) and HTS depending on their application.

4.2.1.1. *Helium*

He is traditionally used in LTS cooling in view of its very low critical (5.2 K) and boiling (4.2 K at 1 bar) temperatures, besides its low critical pressure

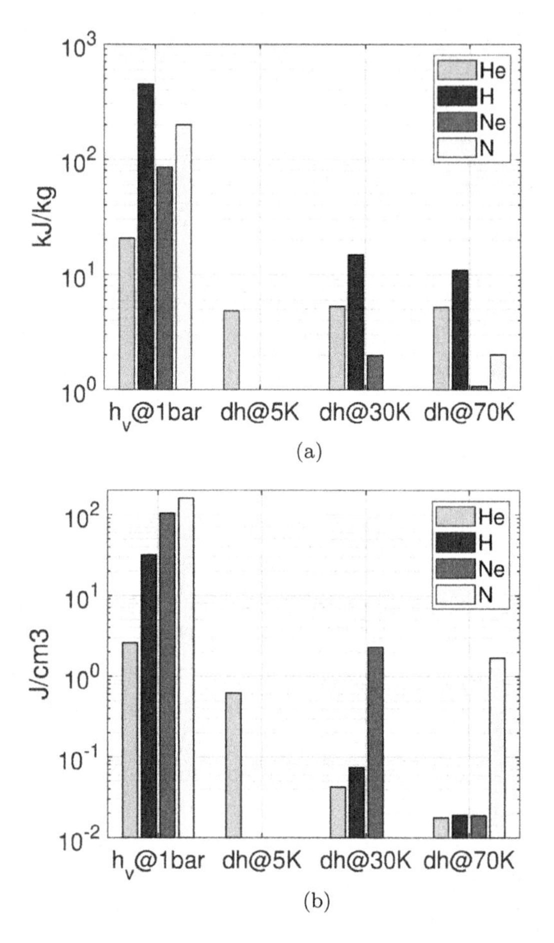

Figure 4.15. Comparison of the latent heat of vaporization at 1 bar and the enthalpy variation when the fluid temperature increases by 1 K at different temperatures (a) per unit mass and (b) per unit volume among different cryogenic fluids [51].

(2.28 bar). It can be used in different thermodynamic states. Figure 4.15 shows that its latent heat at 1 bar is the lowest, smaller than, for example, that of H_2, but it is the only fluid that allows the temperature to reach values lower than 5 K at a pressure of ∼5 bar, as the triple point of H_2, Ne, and N_2 is at 14, 24.6, and 63.2 K, respectively.

However, He production around the world is reducing and its cost is sharply increasing so that in the near future, other cooling solutions may become competitive [53].

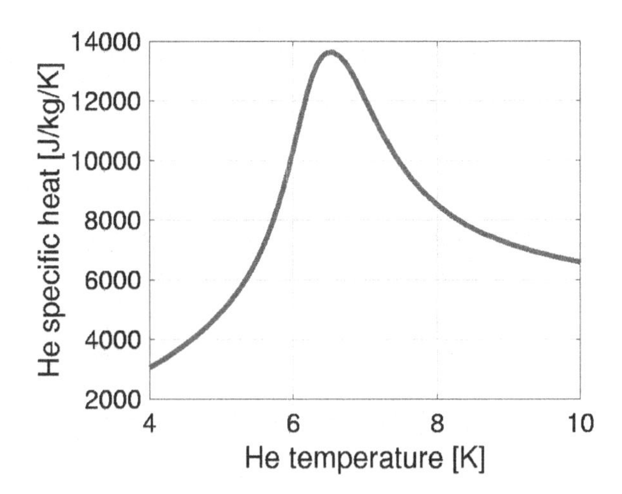

Figure 4.16. Specific heat at constant pressure of He as a function of temperature at a pressure of 5 bar [51].

Supercritical He: SHe, i.e. He at a pressure higher than its critical pressure (0.228 MPa) or at a temperature higher than its critical temperature (5.2 K), is the most commonly used coolant for superconducting magnets for fusion applications. He is an inert gas, and SHe has a peak value of specific heat at a temperature of ∼6–7 K at a pressure of 0.5 MPa (see Figure 4.16) so that it acts as an excellent coolant under the operating conditions favorable for the LTS used in the magnets that are currently the state-of-the-art technology in magnetically confined nuclear fusion, with typical critical temperatures between 10 and 20 K.

Superfluid He: In some special applications, superfluid He is used as cooling fluid. Superfluidity is a different phase state (superfluid He is also known as He-II) that is reached at a temperature below 2.17 K (He-4 at ambient pressure) and characterized by a null viscosity and a very high thermal conductivity. He-II is used as a cooling fluid in some accelerator magnets and for the superconducting magnets of the Tore Supra [54] (now WEST [55]), the French tokamak in operation at Cadarache since 1988 (WEST has started its operation in 2016).

Two-phase He: High-speed, two-phase He flow is used as a cooling fluid in the central channel of the POLO conductor [56], see Figure 4.17. The bundle region contains stagnant, pressurized SHe, which transfers heat by conduction to the central channel [19].

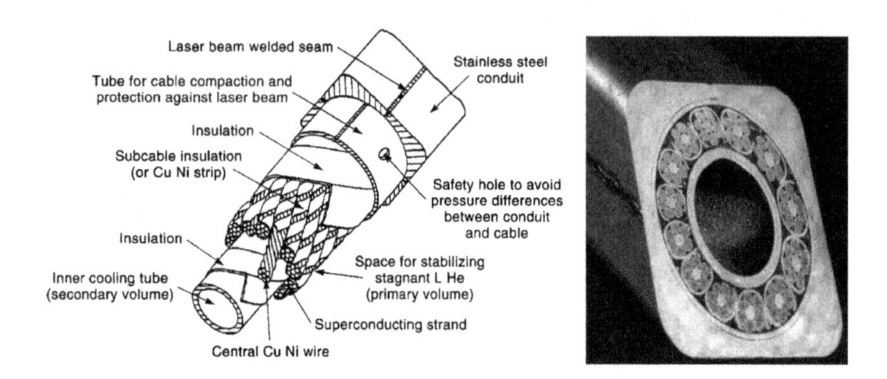

Figure 4.17. The POLO conductor, with two separate He chambers.
Source: Reproduced from Ref. [19].

4.2.1.2. *Hydrogen*

H_2 has a saturation temperature at ambient pressure of ~20 K so that it is still interesting for the cooling of HTS at that temperature level. It is the most advantageous cooling medium, exhibiting the highest latent heat and specific heat per unit mass, see Figure 4.15(a). This leads to a faster thermal mass CD and a lower boil-off [57], but it is highly flammable — this constraint limits its application, requiring caution in its handling and operation.

4.2.1.3. *Neon*

In view of its higher density, Ne shows the highest specific heat per unit volume at ~30 K (see Figure 4.15(b)), making it an interesting alternative to H_2. Indeed, its saturation temperature at 1 bar is ~27 K. However, because of its scarcity, its cost is higher than that of H_2, while the price of He is increasing (see above).

4.2.1.4. *Nitrogen*

N is a cheap and environmental friendly fluid. It is abundant in the atmosphere and easy to be used. Its saturation temperature at ambient pressure is 77 K, and with its melting temperature being equal to 63 K, it is not suitable for cooling LTS. The interest in N_2 as a coolant has thus grown after the discovery of HTS.

Indeed, the use of N_2 as a coolant is also limited for HTS to the cases where the operating conditions (in terms of the required current density and background magnetic field) are not too stringent, otherwise the operating temperature of the magnet must be reduced to liquid H_2/Ne or LHe one. As a result, N_2 turns out to be a promising coolant for power transmission lines based on HTS, rather than for superconducting magnets [42]. As shown in Figure 4.15(b), its specific heat per unit volume is excellent both in single- and two-phase flow.

4.2.2. *Cooling options*

Besides the cooling fluids, different cooling options are available for superconducting cables. They are listed in this section.

4.2.2.1. *Forced flow*

Forced-flow cooling is typically adopted when the heat transfer coefficient (HTC) must be higher than in natural convection (e.g. for magnets immersed in a pool) to guarantee a given temperature margin on a large critical portion of the magnet (e.g. in most fusion devices) and/or the heat load on the magnet is high. This may happen due to, for example:

- AC losses induced by magnetic field variations in fusion [58] and some accelerator [28] magnets,
- static heat load from the environment, and
- nuclear heat load from the plasma in fusion magnets [59, 60].

SHe is circulated in the CICC used in the fusion magnets' inside cooling loops described in Section 4.2.3.2, i.e. the circuit directly cooling the magnets and hydraulically decoupled from the refrigerator. The latter, see Section 4.2.3.1, still uses He as the process fluid but in the form of gaseous or liquid He.

Forced-flow, high-speed, two-phase He flow is used as the cooling fluid in the central channel of the POLO conductor [56], see Figure 4.17. On the other hand, the bundle region contains stagnant, pressurized SHe, which exchanges heat only by conduction to the central channel [19]. In principle, if the central channel wall is solid, the coolant used in it can be different from the working fluid cooling the bundle.

Forced-flow N_2 is usually adopted for the cooling of HTS for power transmission.

4.2.2.2. *Conduction*

Conduction cooling is based on the thermal contact between the refrigerator cold head and the magnet cryostat assembly [61]. The use of metallic and high-conductive materials in the bulky coil guarantees the cooling of the entire magnet. This cooling option with dry magnets is suitable only for small coils or in the case where the coil is operated in a steady state (with small or no magnetic field variation) and with very small heat loads from the ambient, as in accelerators and medical applications.

4.2.2.3. *Pool*

In the very first applications, LHe in a pool bath was the only option to cool a superconducting magnet. LHe can be easily stocked in large amounts independently of the size of the cryoplant.

In a pool-cooled magnet, LHe is in direct contact with the metallic conductor so that the large enthalpy inventory of the bath is exploited to obtain a high thermal stability. In order to allow the contact between LHe and each conductor, the winding pack (WP) must ensure gaps for the He that also has the function of electrical insulation. The gaps are, in some cases, guaranteed by a number of spacers between conductors, limiting the mechanical stiffness and allowing dangerous deformations of the WP under operating loads, thus inducing mechanical instabilities.

The requirements of mechanical stiffness and reliability of the high-voltage electrical insulation have ruled out the pool cooling option for fusion devices of the present and future generations [62].

Tore Supra [54] (now WEST [55]) and LHD [14, 63] are currently the two fusion devices running with magnets cooled in He-II pools. In the Tore Supra case, the choice of operating at 1.8 K was dictated by the need to enhance the magnetic field of the tokamak without moving away from the NbTi technology [62].

4.2.3. *Cryoplant description*

The cryoplant is responsible for the cooling of the superconducting magnets to the required operating (cryogenic) temperature. This function is accomplished by:

- the refrigerator, producing the cooling power by means of a process fluid, and
- the cryodistribution system, delivering the cold fluid to the client and collecting from it the warm fluid.

4.2.3.1. *Refrigerator*

The refrigerator is a complex component whose aim is to produce the desired cooling power needed to maintain the operating temperature. Its cost is directly proportional to the peak cooling power: To reduce the investment cost, it is thus required to operate it at a constant (averaged) power and to reduce the peak power.

At present, most of the superconducting magnets are based on LTS and cooled by He. The He refrigerator is thus described in this section.

The thermodynamic cycle featured by most of the He refrigerators for fusion purposes (e.g. the one operated at the ITER CSMC facility in Naka, Japan [64]), whose scheme is shown in Figur 4.18, is a conventional Collins cycle, typically with LN2 precooling (HX1 and HX2), two or three isentropic turbo-expansion stages (T1 and T2) connected in series (which distinguish the Collins from the single expansion typical of the Claude cycle [65]), and one or more Joule–Thomson (JT) isenthalpic expansions [64]. He refrigerators for other applications, such as to cool accelerator magnets, follow the same thermodynamic cycle and have a similar layout.

The compressor unit is usually composed of an inter- and post-cooled (IC and PC, respectively) two-stage warm compression system (screw compressors are typically adopted). The compressor unit is equipped with a GHe storage tank (B1) at ambient temperature for the loop pressure control (as well as with gas purifiers and oil separators), see the dashed lines denoting the He charging and discharging lines in the compressor unit shown in Figure 4.18.

The refrigerator is equipped with a lot of temperature, pressure, and flow sensors to control its operation. The GHe exiting the compressor unit is driven through a series of heat exchangers (HXs), constituting the pre-cooling stage (HX1-3 in Figure 4.18), some of which featuring LN2 pre-cooling; in most of the cases (e.g. in the ITER CSMC [64], EAST [66], and Tore Supra [67] refrigerators), the LN2 is stored in a proper tank and vented to the atmosphere after being used for pre-cooling. Then, the He enters the cooling stage, composed of a series of HXs (HX4-6 in Figure 4.18) in parallel to the expansion stages, i.e. the turbo-expanders T1 and T2. Finally, the He is driven to the after-cooling stages (HX7-8 in Figure 4.18), where the expansion into the JT valve(s) takes place.

The He finally reaches the client to cool, which is different depending on the operation phase:

- During CD, the refrigerator is connected directly to the magnet, acting as a client.
- During normal (cold) operation, the client of the refrigerator is instead a saturated LHe bath, thermally coupled by means of one or more HXs to the loop cooling the magnet.

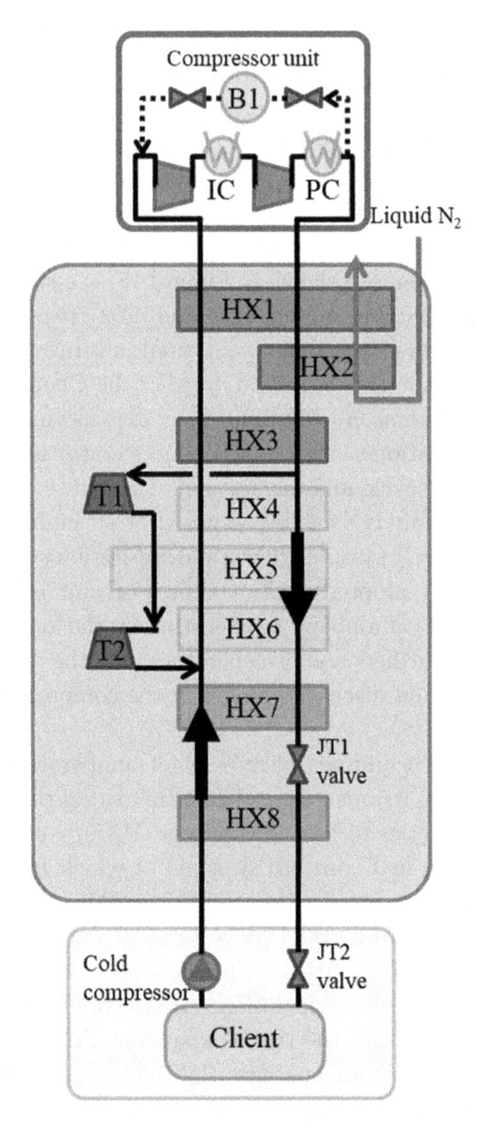

Figure 4.18. Scheme of a He refrigerator for superconducting magnets in fusion devices.

Only in small applications, where the coolant mass flow rate is small, the forced-flow cooling can be obtained directly from the refrigerator also during the cold operation: thanks to the direct connection to the refrigerator, where a two-stage compression usually results in a high pressure level (larger than 10 bar), series cooling of the magnets is adopted because a large pressure drop is allowed, and the mass flow rate can be kept small as required [62].

Besides the magnets, there are several other different clients for the refrigerator in a facility containing superconducting magnets, e.g. the thermal shields, current leads, diagnostics, and, in the case of a tokamak, cryopumps and the pellet injection system. The operating temperature of the different utilities can be different (the thermal shields are usually operated at ∼80 K): the different flows of the process fluid in the refrigerator will thus be properly split in order to optimize the power consumption.

4.2.3.2. *SHe loop*

In fusion devices, the magnets are cooled by SHe flowing in a pressurized (∼0.6 MPa) loop hydraulically separated from the refrigerator cycle, see Figure 4.19. This loop is composed of:

- a cold circulator (driving the SHe flow in the magnets);
- cryolines, connecting the cold box, which contains the LHe bath, to the magnets;

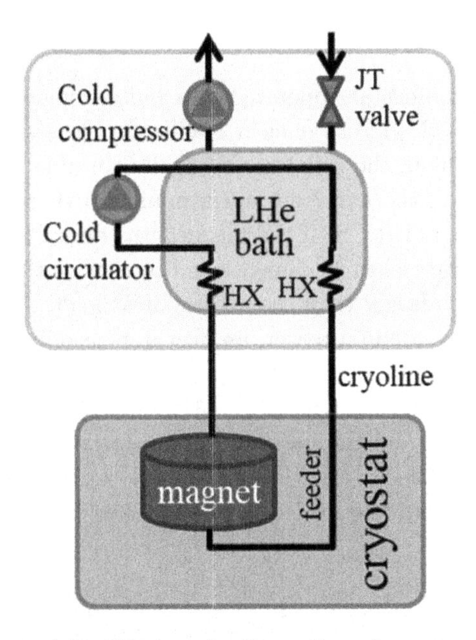

Figure 4.19. Scheme of the SHe loop for the cooling of superconducting magnets in fusion devices.

- feeders, i.e. the pipelines entering the cryostat (containing the magnets) to supply the cooling fluid directly to the coil cooling the channel inlets;
- the HX(s) providing the thermal coupling between the SHe loop and the LHe bath, i.e. between the magnet cooling loop and the refrigerator; more than one HX can be present, and in that case, they are normally one upstream of the cold circulator (to remove from the SHe loop the heat from the magnets before pumping, thus reducing the required pumping power) and one downstream of the cold circulator (to remove the power introduced by the cold circulator, after the pumping process);
- control valves with suitable controllers (typically proportional–integral, PI, or proportional–integral–derivative, PID), operated on the basis of process values measured by a set of temperature, pressure, and mass flow sensors along the loop itself.

To limit the pumping power of the cold circulator, the pressure drop across the coils must be kept low, typically within 0.1 MPa. The conductor hydraulic characteristics (pressure drop during operation and maximum pressure at quench) are indeed one of the major drivers for the design of fusion conductors [62].

4.2.3.3. *Interfaces*

Current fusion machines are operated in a pulsed mode so that the heat loads on the magnets (due to nuclear load or AC losses induced by magnetic field variations in the coils) are cyclically variable: During a period, there will be a peak thermal load and a minimum thermal load, e.g. see Figure 4.20 for the ITER CS. If the magnet was directly connected to the refrigerator, the latter is to be dimensioned to remove the peak heat load, which may be much larger than the average heat load. As the refrigerator cost is driven by its nominal power, an over-dimensioning of the refrigerator power will result in a higher investment cost. For this reason, the role played by the SHe loop and the LHe bath is crucial: The active controls acting on the former and the passive thermal inertia of the latter aim at smoothing the variable thermal loads removed from the superconducting magnets. In particular, the load can be smoothed by storing temporarily the energy:

- in the bulky metal structures supporting the coil (e.g. in the TF coils case), by means of a magnet (or HXs) bypass valve in the SHe loop or reducing the cold circulator speed;

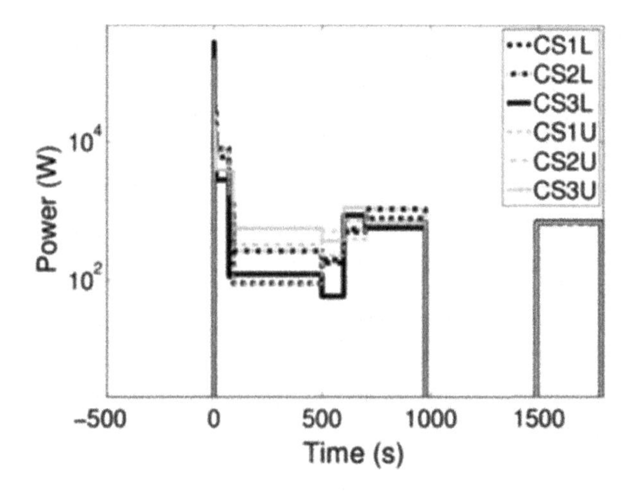

Figure 4.20. Evolution of heat load (AC losses) in each module of the ITER CS during a 15 MA standard plasma operating scenario. Note that the power scale is logarithmic. *Source*: Reproduced from Ref. [58].

- in the LHe bath, reducing the GHe mass flow rate sent back to the refrigerator by means of a suitable control valve at the LHe bath outlet or varying the cold compressor speed.

Experimental facilities [68] and numerical tools [69–75] have been developed to design suitable heat load smoothing strategies for the existing and future fusion reactors.

Magnets operated in the steady-state mode (as the accelerator ones) do not need specific load-smoothing strategies, so the control of the coolant distribution system is simpler.

4.2.4. *Solid properties*

The thermophysical properties of solids are usually given as functions of temperature only, excluding few exceptions. To deal with the cooling of superconducting magnets, the most relevant properties are density (which can be considered constant if no mechanical calculations are being performed), specific heat, thermal conductivity, and electrical resistivity. The last one is needed to compute the power generation in the conductor when the superconductor loses its superconducting state and the current is transported by Cu and other metallic components, such as the jacket.

4.2.4.1. *Metals*

As far as metals are concerned, attention should be paid to Cu. It is used as stabilizer in the cable, as all the superconducting strands are, for a given fraction, made of Cu, and in some cases, additional pure Cu strands are added to the superconducting ones. The role of Cu in the TH of the CICC is very important: It provides a low impedance path for the current transport during quench transients. The correct definition of its electrical resistivity is thus required. It must be noted that the electrical resistivity of Cu, as well as its thermal conductivity, depends not only on the temperature but also on the magnetic field and residual resistivity ratio (RRR, the ratio of the electrical resistivity at 300 K and at 0 K), the latter being higher for purer Cu. Reliable thermophysical properties of Cu can be obtained from the NIST database [76].

Other relevant metals and alloys are those adopted for the jacket, including stainless steel, titanium, Incoloy®, and aluminum.

4.2.4.2. *Superconductor*

For superconductors, besides the typical solid thermophysical properties (some of which may also depend on the magnetic field, current-sharing temperature, and critical temperature), there is also another set of properties defining its critical current density, i.e. the so-called scaling. Scaling depends on the superconducting material and its manufacturing technique, while some parameters of scaling are defined for particular types of strands and manufacturers. This set of information must be available when building a model of a superconducting cable in order to reproduce its critical surface. The latter delimits the volume of the magnetic field, current density, and temperature parameter space in which the material behaves as a superconductor. A typical, numerical definition of this surface returns the critical current density for a given magnetic field and temperature [77]. For some superconducting materials (e.g. Nb_3Sn), the critical current density is also a function of the mechanical strain [78].

4.2.4.3. *Insulations*

Insulation material properties are usually given only as a function of temperature. However, they are affected by a nonnegligible uncertainty because of the way the insulation is applied, i.e. in several superposed layers of different materials, usually impregnated by resins.

The thermophysical properties are typically given for the pure, bulky material. As a result, if the insulation is wrapped around the CICC ("turn insulation") or the coil ("ground insulation") with a given overlapping, its multi-layer nature usually alters the resulting properties and the insulation layer does not act as if it were a single layer of the same thickness. This is of paramount importance when computing the heat transfer between neighboring CICCs across the turn insulation: an error in the definition of the thermal conductivity will result in a different heat flux. Typically, the multi-layer structure reduces the thermal conductivity of the insulation layer by a factor of ~10 with respect to the nominal value referred to the bulky material [79–82].

4.3. Modeling

In view of the complex nature of large magnets, such as those used in nuclear fusion facilities, their TH analysis (both from the conceptual and from the modeling point of view) must deal with multiscale phenomena.

A multiscale approach was proposed for the TH modeling of fusion magnets [83], but it is useful also in the description of the TH phenomena. TH is indeed fundamental especially for forced-flow-cooled superconducting magnets. In the following sections (from 4.4 to 4.7), this multiscale nature of the TH of large superconducting magnets will be highlighted while dealing with the physics and the corresponding modeling (and issues).

The need for adequate models is motivated by three needs:

- Interpretation of the results obtained from existing test facilities; there are no diagnostics inside the conductors/magnets, thus the "local" conductor performance in dedicated tests must be reliably deduced from "global" (inlet, outlet) measurements (temperature, mass flow rate, pressure, voltage, current, etc.) by means of suitable computational tools.
- Prediction of the magnet operation during the design phase; superconducting magnets must be kept sufficiently below the current-sharing temperature, so the capability to reproduce/predict TH transients is needed to assess the design choices.
- Design of operational scenarios, including suitable controls for the magnets during and after the commissioning of the machine they are integrated within.

4.3.1. *Space scales*

The TH-relevant spatial scale range is as follows (see Figure 4.21) [83]:

- Macroscale: 10–100 m (magnet size/CICC length).
- Mesoscale: 10^{-2}–1 m (CICC transverse sizes/WP).

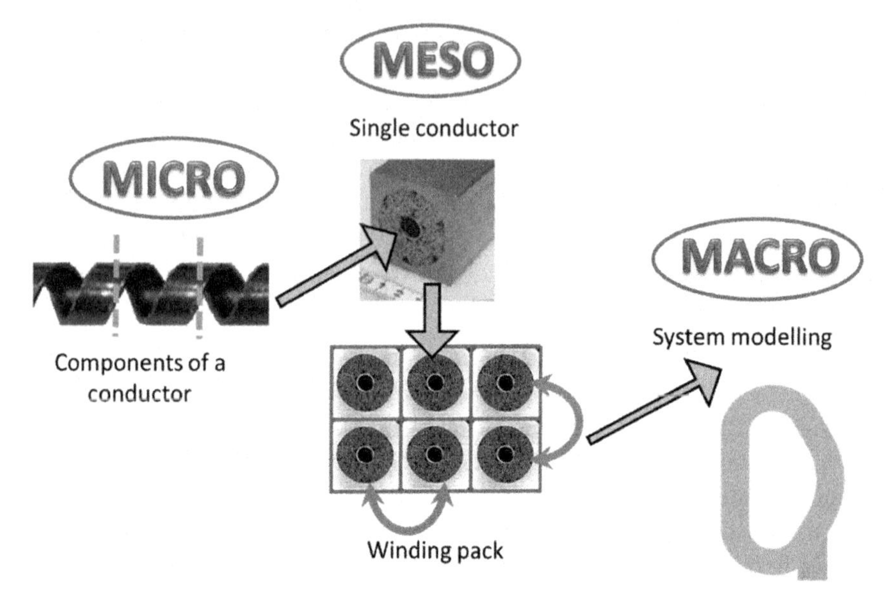

Figure 4.21. Three relevant scales for the TH modeling of the ITER superconducting magnets.

Source: Reproduced from Ref. [83].

- Microscale: 10^{-3}–10^{-2} m (strand diameter and insulation thickness/transverse size of a CICC).

From the modeling point of view [83], on the macroscale, the entire magnet (coil + structures + cryogenic circuit) is considered. The structures include all the metallic, bulky casing that surround the coil to provide mechanical stability. In fusion magnets, the largest structures are those surrounding the TF coils, providing mechanical support against gravity to the CS and PF coils also.

In order to successfully model the entire, macroscale system, the treatment of the mesoscale dimension is required: The individual CICCs are modeled, usually as one-dimensional (1D) objects along the conductor axis, weakly thermally coupled inside a winding and with the casing (if applicable), as well as the cooling pipes inside the structures (sometimes needed to provide additional cooling to the structures), also approximated as 1D objects.

The 1D mesoscale models need constitutive relations for the heat, momentum, and mass transfer inside the CICC, namely friction factors and

HTCs: They may, in turn, be derived by analyzing a limited portion of the CICC at the microscale level by means of detailed 2D or 3D computational thermal-fluid-dynamics (CFD) models.

4.3.2. *Time scales*

Similar to the spatial scales, the relevant TH time scales also cover several orders of magnitude. The different time scales can be identified on the basis of the transient under investigation and the heat transfer mechanism considered.

The typical transients of superconducting magnets for fusion applications, which will be described in Section 4.7, involve the following time scales:

- Normal operating conditions:
    - Cool down: The duration of the transient is typically from a few days to one month, depending on the size of the magnet to be cooled from room temperature to cryogenic temperature.
    - Plasma operation mode: The normal operation of a tokamak identifies the plasma pulse; in this case, the time scales depend on the plasma duration (driving the nuclear load on the TF coils) or on the duration of the most steep current/magnetic field ramps, inducing the highest heat loads in the pulsed coils. On average, the typical time scales range from 10 to 1000 s.
- Off-normal operating conditions:
    - Quench: The thermal stability of the superconducting cable is lost at the 10^{-3} s time scale, while a quench propagates for a few seconds. The time needed to recovery the initial (cold) operating conditions then depends on the cold mass and can be up to several hours.
    - Fast current discharge: It depends on the coil inductance and the maximum voltage that the insulation can withstand. Typically, relevant time scales are in the range 1–10 s, while as in the case of a quench, the time needed to recover the initial operating conditions can be up to several hours.
    - Severe accidents, such as loss of flow or loss of coolant: Their time scales are in the range of a few seconds, but in the case of a small loss of flow/coolant, an accidental transient can be much longer.

The different heat transfer mechanisms that could be involved in forced-flow cooling act at different time scales. In particular:

- Advection of the coolant along the channel: It depends on the coolant speed and channel length and, usually, is in the range 100–1000 s for a CICC of fusion magnet in cold operation; it reduces to 10–100 s if the pipes for the cooling of the structures are considered.
- Convection: The excellent thermal coupling between the coolant and the solids in a CICC for fusion magnets, especially in the bundle region (thanks to the large wetted perimeter), reduces the time scale of this heat transfer mechanism well below 1 s, usually to the order of 1–10 ms.
- Conduction: In this case, the thermophysical properties of the materials considered determine the time scale of the thermal coupling, which at the cold operating temperature of 4.5 K ranges from ~100 s (heat conduction across 10 cm in the bulky stainless-steel structures) to ~1 s (heat conduction across 1 cm of the ground insulation, wrapped around the coil) and to 0.1–1 s (heat conduction across a few mm of the insulation separating the conductors in the coil).

4.4. Forced-Flow CICC Superconductor Hydraulics

Most of the interest in the hydraulic analysis of a CICC is focused on the assessment of its pressure drop per unit length because, as already discussed above, together with the maximum pressure during a quench, it is the major driver of the design of fusion conductors. On the other hand, the most interesting TH aspects concern the capability to properly cool the superconducting material (strands in a CICC) in order to maintain the superconducting state. Moreover, in the case of dual-channel CICCs, the complexity of the analysis is increased because of the presence of multiple regions for the coolant flow, see Figure 4.22, typically coupled from both the hydraulic and the TH points of view.

4.4.1. *Multiple flow regions*

SHe coolant flows both in the annular region, where the cable bundle is present, and in the central channel, see Figure 4.23.

As the CICC axial–transverse size ratio of a CICC is typically ~1000 in a fusion coil, 1D (axial) models are customarily used for a CICC. Compressible Euler-like flow of one fluid component for each region is computed, solving the three coupled equations of the conservation of mass (Equation 4.3),

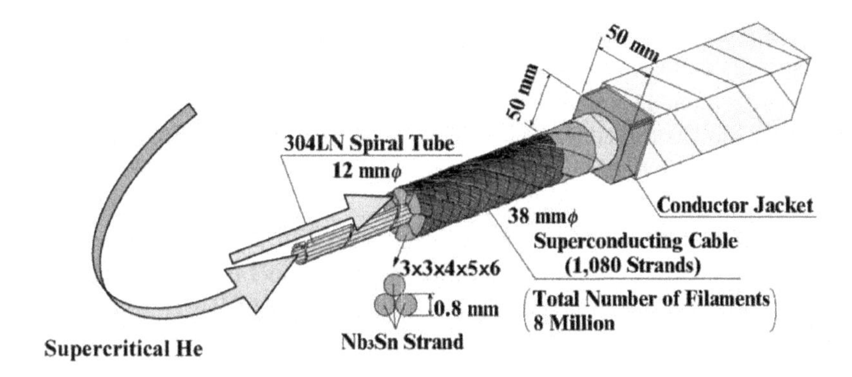

Figure 4.22. Sketch of the CICC used for the ITER central solenoid model coil (CSMC) as a representative of a typical CICC. Note the dual-channel structure.

Source: Reproduced from Ref. [84].

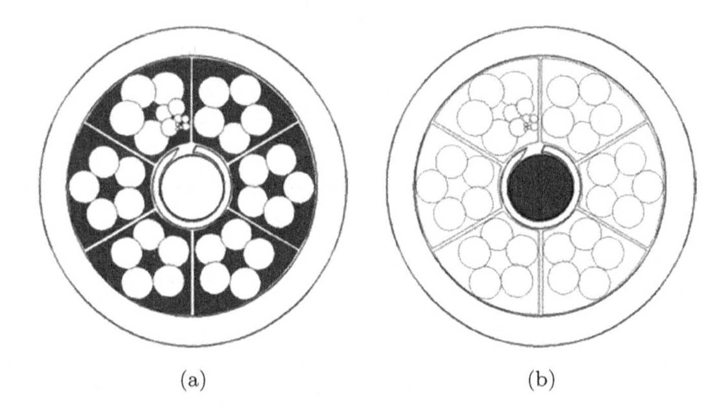

Figure 4.23. (a) Bundle and (b) hole flow regions in a CICC.

momentum (Equation 4.4), and energy (Equation 4.5), usually rewritten to make explicit the non-conservative variables velocity, pressure, and temperature:

$$\left\{ \frac{\partial \rho}{\partial t} + \frac{\partial \rho v}{\partial x} = \Lambda_\rho \right. \tag{4.3}$$

$$\left\{ \frac{\partial \rho v}{\partial t} + \frac{\partial \rho v^2}{\partial x} + \frac{\partial p}{\partial x} = \Lambda_v \right. \tag{4.4}$$

$$\left\{ \frac{\partial \rho e}{\partial t} + \frac{\partial \rho e v}{\partial x} + \frac{\partial p v}{\partial x} + \frac{\partial q_c}{\partial x} = \Lambda_e \right. \tag{4.5}$$

In this set of equations, ρ, v, p, and e are the fluid density, velocity, pressure, and specific (sum of internal and kinetic) energy. Λ indicates the sources of mass (Λ_ρ), momentum (Λ_v), and energy (Λ_e) in the respective equations. The variable t is the time, x the space, and q_c the thermal flux exchanged within the fluid by conduction (it is usually neglected as the thermal conductivity of SHe is very low).

The 1D, single-channel fluid model reported in Equations (4.3–4.5) is based on the assumption that each hydraulic channel can be considered a single conductor, independent of the winding strategy followed to build the magnet.

If, besides the bundle (B), one (or more) central channel (hole, H) is also present, a second set of these equations is solved for the second coolant region.

The heat transfer between the two fluid regions also account for additional terms described in Section 4.5.3.

As far as solids are concerned, heat conduction along at least two solid components (strands and jacket, see Figure 4.24) is solved, assuming a uniform temperature across the respective cross-sections. This is a limiting assumption, as transverse temperature gradients between different cable components may arise (and have indeed been measured [84]) even at the sub-petal level and can be particularly relevant in the case of coupled electromagnetic TH transients:

$$A\rho C\frac{\partial T}{\partial t} - \frac{\partial}{\partial x}\left(k\frac{\partial T}{\partial x}\right) = Q, \tag{4.6}$$

where C is the specific heat, k the thermal conductivity, and Q the heat source, including the volumetric heat generation and the heat transfer with

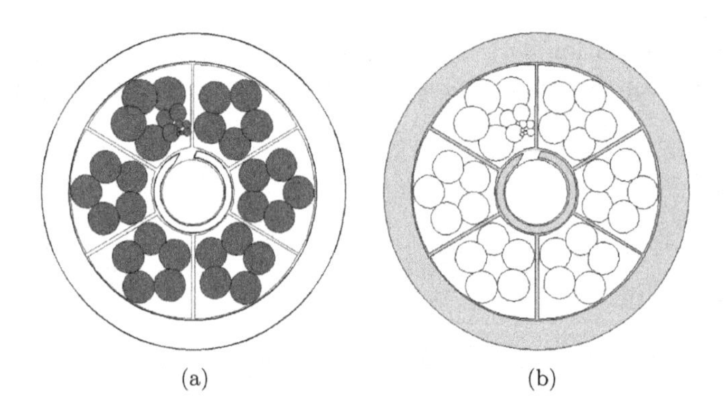

Figure 4.24. (a) Strands and (b) jacket solid cross-sections in a CICC.

other solid regions (including the structures) and fluid(s). For the latter, the heat transfer with solids is accounted for in the source of the energy equation.

Note that, as shown in Figure 4.24(b), the cross-section of the wall of the central channel and that of the sub-cable wrappings (metallic sheets used to partially wrap each of the last cabling stage, usually called "petal") are also considered part of the solid cross-section in order to remove them from the free-flow area that otherwise would be overestimated.

4.4.1.1. *Bundle*

The bundle flow region of a CICC, see Figure 4.23(a), is characterized by a very large wetted perimeter of the strands, up to more than 1 m, providing an excellent thermal coupling between the cable, constituted by small-diameter strands, and the coolant. However, the hydraulic diameter of the bundle region is very small, typically $\sim 10^{-4}$ m. As a result, the pressure drop is very high, and therefore, the mass flow rate is limited. Consequently, the coolant speed in the bundle region is also small, normally 10^{-2}–10^{-1} m/s during cold operation. The small velocity has a negative impact on the stability, as the HTC increases with speed. For these reasons, the use of a channel ("hole") with low hydraulic impedance is a common practice.

4.4.1.2. *Hole*

The hole flow region of a CICC, see Figure 4.13(b), is included in several designs (see Figure 4.4(b) and 4.4(c) and Figure 4.6) as a pressure release channel to mitigate the pressure rise during a quench. In channels with larger hydraulic diameter (from 5 mm in the EU DEMO TF coil conductor proposed by ENEA [18] to 10 mm of the ITER PF coils [85]), the coolant speed is large (up to 1 m/s) and the average residence time of the coolant in the winding is shortened [19].

The walls of the channel are typically constituted by a spiral (or a spring as in some of the CSMC conductors [26]).

4.4.1.3. *Coupling between bundle and hole*

In the most general case, mass, momentum, and energy exchanges are allowed between the two fluid regions across the wall separating them, see also Section 4.5.3.

The two regions are coupled by means of mass transfer, which is accounted for in the mass conservation equation (Equation (4.3)) as a

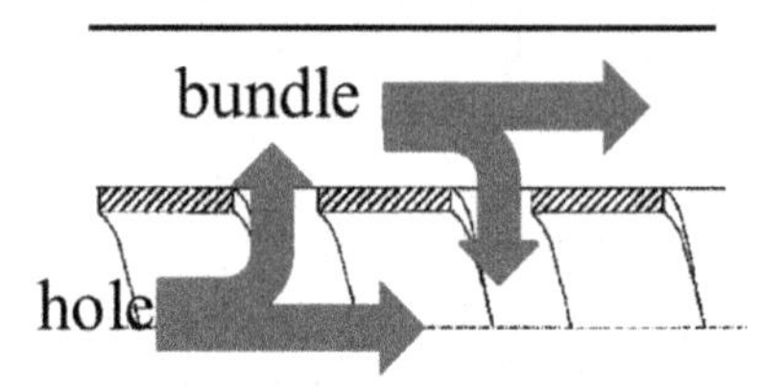

Figure 4.25. Scheme of the coolant mass transfer across the perforated part of the spiral.

source/sink proportional to the pressure difference between the two regions and to an equivalent localized pressure drop coefficient (K), in turn related to the area available for the cross flow:

$$\Lambda_\rho = \frac{K(p_B - p_H)}{lA} \tag{4.7}$$

where, l and A are the length and cross-section of the fluid region, respectively. In other words, a valve-like model is typically adopted in many codes [86] to compute the Λ_ρ term in Equation (4.3) used to model the radial fluid transport in the CICC, see Figure 4.25.

The mass transfer also causes a momentum and energy transfer, which is accounted for in the source in the momentum and energy equations proportionally to the Λ_ρ term.

4.4.2. *Friction factors*

The momentum transfer between the coolant and solids is very important as it determines, for a given mass flow rate, the pressure drop along a CICC and thus the related cost of pumping.

The momentum source Λ_v includes the friction term F,

$$F = 2f\rho\frac{v^2}{D_h}, \tag{4.8}$$

with f being the friction factor coefficient and D_h the hydraulic diameter.

For analysis, as the two regions are modeled separately as two parallel channels for simplicity (while microscale analyses reveal that there is mass transfer with zero net flux between the two regions), separate friction factors are needed for the bundle region and the hole so that at least two sets of independent measurements are needed. Once both friction factors are known, it is possible to compute both the pressure drop and the flow repartition between the different regions of a CICC. Note that even for CICCs with a single hole, the two-channel model is a simplification of the reality, as several parallel flow channels are actually present:

- a "third" channel may naturally appear during coil operation because of Lorentz forces and finite cable stiffness;
- "triangles" appear on the CICC cross-section at the contact between two petals and the (inner jacket) cable wrap.

For the bundle region, the correlation based on the Darcy–Forchheimer equation for flow in porous media [84, 87, 88] is recommended as the primary correlation to be used in TH simulations:

$$f = \frac{D_h^2 \varphi}{2K} \frac{1}{Re} + \frac{D_h \varphi^2}{2} \frac{C_F}{\sqrt{K}}, \tag{4.9}$$

where φ is the bundle void fraction, Re is the Reynolds number, C_F is the drag coefficient that characterizes a specific porous medium, and K is the permeability. The drag coefficient can be defined as

$$\frac{C_F}{\sqrt{K}} = \frac{2.42}{\varphi^{5.80}} \, [\mathrm{m}^{-1}], \tag{4.10}$$

while for the permeability, a formulation is available:

$$K = 19.6 \times 10^{-9} \frac{\varphi^3}{(1-\varphi)^2} \, [\mathrm{m}^2], \tag{4.11}$$

where it should be noted, however, that this dependence on only porosity cannot be too realistic since it is expected that K should also depend on the tortuosity of the flow path, and therefore, in the case of a CICC, on the different cabling twist pitches. A friction factor correlation taking into account the tortuosity of the flow path, unfortunately, does not exist yet (only the dependence on the cabling pattern — braided vs. non-braided conductors — has been investigated so far in Ref. [89]).

Alternative correlations include the one given by Katheder [90], which is of the type

$$f = \frac{1}{\varphi^{0.72}} \left(\frac{A}{Re^{0.88}} + B \right), \tag{4.12}$$

or others that can be found in Ref. [84].

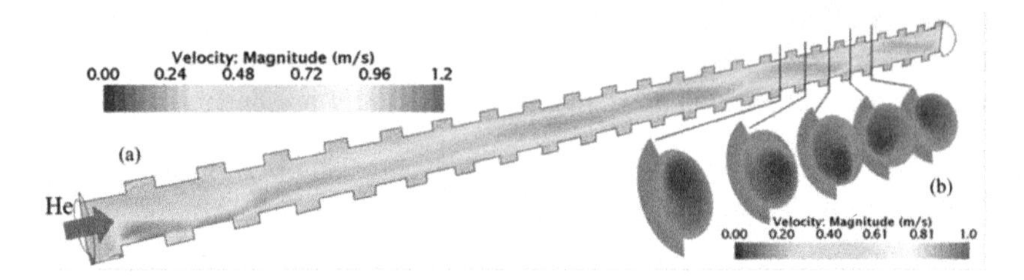

Figure 4.26. (a) Velocity field in a longitudinal cross-section of a hole. The inlet is located on the left. (b) Velocity field in five cross-sections separated by one pitch from each other.

Source: Reproduced from Ref. [18].

For the pressure-relief channel, the friction factor correlation depends, of course, on the geometry (cross-section) of the channel. If (as usual) an helix is adopted, then the correlations from Refs. [91–93] can be used.

Advanced CFD methods are used to address the problem of friction in the central channel at the microscale level, in particular to predict the effect of gap size [92, 94, 95], aiming at developing suitable friction factor correlations to be adopted for the 1D modeling of He flow.

Recent analyses also show that a periodic flow field is established in a spiral-walled channel with a pitch longer than a single-spiral pitch [18], see Figure 4.26. As a consequence, most of the previous correlations obtained from numerical experiments considering a single-spiral pitch with periodic boundary conditions at the inlet and outlet should be carefully reconsidered.

4.5. Forced-Flow CICC Thermal-Hydraulics

The major heat transfer processes *across* a CICC section are summarized in Figure 4.27, emphasizing the different conductor components involved. They are accounted for in the form $h \times \Delta T$ in the source of the energy equation of the fluid region(s) (Λ_e term in Equation (4.5)) and of the heat conduction equation of the solids (Q term in Equation (4.2)), with h being the HTC.

The heat transfer significantly affects, together with friction, the most relevant time scales of TH transients in a CICC [84].

The other most important heat transfer mechanism is advection, dominating the heat transfer *along* the CICC axis; it is described by one of the 1D equations on which the typical models are based, namely the energy conservation equation (Equation 4.5).

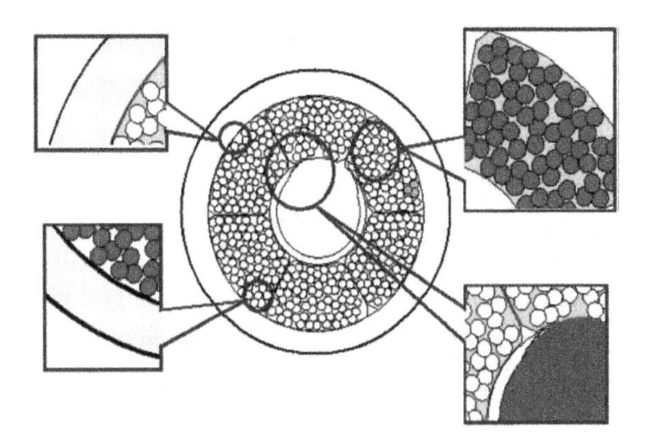

Figure 4.27. Different heat transfer channels in the ITER CICC: strand–He (upper right), bundle He — hole He (lower right), strand–jacket (lower left), bundle He — jacket (upper left). A four-temperature model (jacket uniform, all strands common and uniform, bundle He uniform, and hole He uniform) is implicitly assumed in this representation.

Source: Reproduced from Ref. [84].

4.5.1. *Heat transfer coolant–solids*

For the heat transfer between the coolant and solids, the series of

- the minimum between the transient [96] and steady-state boundary layer thermal resistance
- and Kapitza resistance [97]

can be typically adopted [84]. The steady-state heat transfer is derived from correlations for the Nusselt number $Nu = f(Re, Pr)$. The Dittus–Boelter correlation is typically used, even if it was derived for pipes, for any type of ducts and bundles and to a good approximation both for the uniform surface temperature and heat flux conditions:

$$Nu_{turbulent} = 0.023Re^{0.8}Pr^{y}, \tag{4.13}$$

where Pr is the Prandtl number and y in the exponent of Pr should be taken as 0.4 if the wall is hotter than the fluid and 0.3 if the wall is colder than the fluid. Equation (4.13) has been experimentally confirmed in the range $0.6 < Pr < 160$ and $Re > 10,000$ for small to moderate temperature differences and properties evaluated at the average wall/bulk temperature.

The resulting values of the HTC, h, are of several 10^2 W/(m^{2}K). The same recipe is adopted for the calculation of the HTC between any solid (jacket, strands, and hole wall) and the coolant.

It is recommended not to rely on heat–momentum transfer analogies (the so-called Colburn analogies), which, as far as momentum transfer is concerned, are only valid if there is no form drag [98] — i.e. in the case where the surface transferring heat through a thermal boundary layer is also the one originating friction through the dynamic boundary layer. In the particular case of the ITER CICC central channel, the non-applicability of the Colburn analogy was also clearly confirmed by detailed CFD simulations [99].

The HTC between the jacket and the coolant is relevant for the following issues [84]:

- analysis of stability tests;
- interpretation of the jacket temperature sensor signals;
- access to the jacket heat capacity during the later quench phase.

4.5.2. *Heat transfer between different solids*

The heat transfer between the strands and the jacket provides [84]:

- a channel for direct strand cooling in parallel to the direct cooling from the fluid (but the strand-to-jacket contact perimeter is two orders of magnitude lower than the strand wetted perimeter);
- a channel for accessing the jacket heat capacity in the later quench phase, as above 30 K, it is about an order of magnitude larger than the coolant one. The time scale of this thermal coupling may thus be relevant for quench propagation and hot-spot temperature;
- the only direct channel for strand heating if the jacket is heated first by, for example, a heater or thermal contact with the jacket of a neighboring warmer CICC.

Historically, the HTC between the strands and the jacket was assumed to be constant and equal to 500 W/(m^{2}K) [100]. The contact area between the strands and the jacket strongly depends on the conductor design and the applied Lorentz forces.

4.5.3. *Heat transfer between different coolant regions*

The thermal coupling between the hole and the bundle contributes to determining the time scale on which the central channel coolant heat capacity becomes accessible for cooling the cable. This is not relevant (being too slow) for stability purposes, but it is important for the coil re-cooling after transient events.

The most important heat transfer paths [86] between the coolant in the hole and in the bundle are [101]:

- advection due to the mass transfer between one region and another;
- thermal conduction across the spiral wall.

The former is already accounted for as a source in the 1D energy conservation equation for the fluid (Equation (4.5)). The thermal coupling between the hole and the bundle across the spiral wall is modeled while accounting for two parallel paths (in addition to the advection across the spiral perforated fraction of the wall), namely:

- the series of hole–spiral boundary layer + spiral wall + spiral–bundle boundary layer (weighted with the unperforated fraction of spiral) in parallel to
- the series of hole–wall boundary layer + wall–bundle boundary layer, which qualitatively accounts for the heat transfer across the spiral gap, weighted on the perforated fraction of the spiral.

The latter component is needed because several CFD analyses (at the microscale level) performed in order to better investigate the heat transfer between the hole and the bundle [99, 102] showed that a lot of turbulence and stagnation regions alternate among the small gaps of the spiral. This enhancement in the heat transfer across the perforated fraction of the spiral wall is thus multiplied by a suitable factor that needs to be calibrated (usually empirically) for each conductor, $H_{\text{no-wall}}$.

As a result,

$$h_{\text{global}} = H_{\text{no-wall}} \times h_{\text{open}} \times \mathit{perfor} + h_{\text{closed}} \times (1 - \mathit{perfor}) \tag{4.14}$$

with *perfor* being the fraction of the perforated area,

$$h_{\text{open}} = \frac{1}{\frac{1}{h_B} + \frac{1}{h_C}}, \tag{4.15}$$

and

$$h_{\text{closed}} = \frac{1}{\frac{D_{\text{ave}}}{D_{\text{out}} h_B} + \frac{D_{\text{ave}}}{2k_{\text{steel}}} \ln\left(\frac{D_{\text{out}}}{D_{\text{in}}}\right) + \frac{D_{\text{ave}}}{D_{\text{in}} h_C}} \tag{4.16}$$

where D_{ave} is the average spiral diameter, $(D_{\text{in}} + D_{\text{out}})/2$ and h_B and h_C are the HTCs between the coolant in the bundle and the wall of the channel and between the coolant in the channel and the wall of the channel, respectively. In the case of h_{open}, the perturbation introduced in h_B and h_C by the presence of a permeable wall is neglected.

The thermal conductivity k_{steel} is evaluated at the tube temperature, considered equal to the average of the coolant temperatures in the two hydraulic channels.

4.6. Heat Transfer Mechanisms in the Magnet

From the heat and mass transfer phenomena described in Sections 4.4 and 4.5, the attention is now directed toward the heat transfer in the meso- and macroscale, i.e. at the coil WP and casing (i.e. at the magnet) level. In this case, excluding the case when there is an active cooling of the structures by means of a fluid, the heat is transferred by conduction.

4.6.1. *Heat transfer within the winding*

When there are no radial plates to mechanically sustain the conductors, in view of their packing in the coil winding, neighboring CICCs are obviously in thermal contact with each other; heat is transferred by conduction across the (electrical) insulation layer separating the turns and the layers or pancakes, while the heat capacity of the insulation is typically negligible. For this reason, and as most of the models dealing with CICC-wound magnets are 1D along the conductor axis, a possibility to account for the transversal thermal conduction across the turn insulation is to consider the insulation (including the turn insulation and the inter-layer or inter-pancake insulation) as a pure thermal resistance based on the electrical analogy [103]. This allows us to compute the heat transfer between the jacket of two neighboring CICCs, see Figure 4.28.

Note that this approximation does not account for dynamics, as the thermal capacity is neglected in the electrical analogy (in a typical CICC for fusion applications, the insulation thermal capacity is approximately two orders of magnitude smaller than that of the jacket), but the contribution of the insulation thermal capacity can be added to that of the jacket.

The (minimum) thermal resistance ($R_{\text{th}} = 1/\text{HTC}$) between different turns/layers/pancakes can be evaluated as

$$R_{\text{th}} = \frac{\delta}{k}, \tag{4.17}$$

where k is the thermal conductivity of the insulation material between turns/layers/pancakes and δ is the thickness of the insulation layer. In the

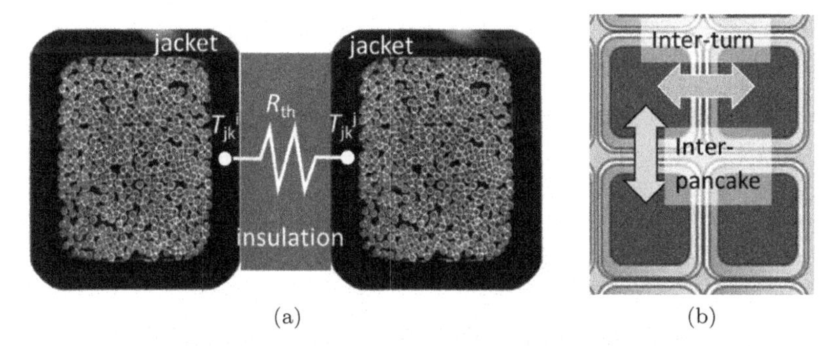

(a) (b)

Figure 4.28. Sketch of (a) inter-turn/inter-pancake conductor insulation and (b) inter-turn and inter-pancake (or inter-layer) thermal coupling paths in the WP models.

case of a multi-layer insulation, the total thermal resistance should be computed as a series of thermal resistances of each layer. These recipes provide a lower bound for the total thermal resistance since the contact resistances are neglected, see Section 4.2.4. The resulting heat flux between the jacket (at temperature $T_{\mathrm{jk}}(x)$) of the two neighboring turns/layers/pancakes i and j is then

$$q''(x) = \frac{(T_{\mathrm{jk}}(x_i) - T_{\mathrm{jk}}(x_j))}{R_{\mathrm{th}}}. \tag{4.18}$$

Both the insulation material (in particular, its thermal conductivity) and the value of δ depend on the WP design. The thermal conductivity is evaluated at the mean temperature of $T_i(x)$ and $T_j(x)$. This is the recipe adopted in the multi-conductor Mithrandir code [103].

A different approach, more expensive from a computational point of view, is to model the 2D (or 3D) insulation layers, discretizing numerically their thickness, so that the dynamics are also taken into account. This is done, for example, with the 4C code [104] when, as in the case of the ITER TF coils, the turns/pancakes are separated by thick stainless-steel radial plates [105].

The inter-turn/inter-layer/inter-pancake thermal coupling has been proven to play a nonnegligible role in Refs. [82, 106, 107]. The effect of the thermal coupling within the WP is evident for several transients scanning a wide range of time scales:

- during slow CD [108];
- during normal operation, for example, of the EU DEMO TF coils WP [109];
- during fast quench transients [106, 110].

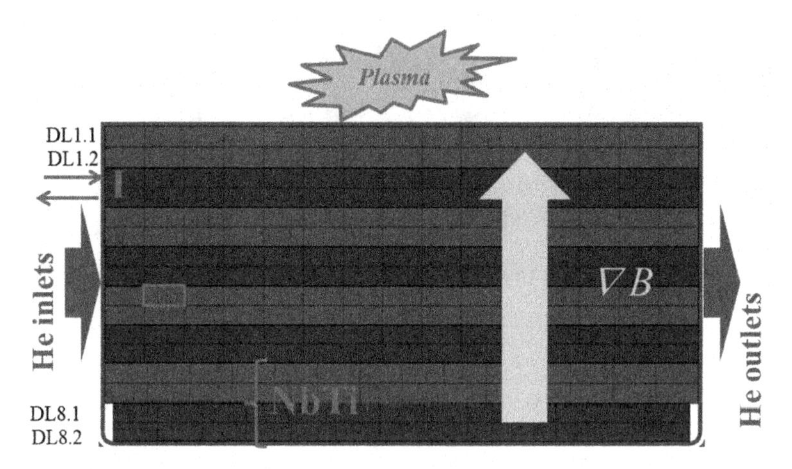

Figure 4.29. Schematic cross-section of the DEMO TF coil WP proposed by ENEA [109]. The red rectangle highlights a single CICC turn. The convention adopted for the numbering of the double-layers (DLs) is also shown, as well as the He flow direction (blue arrows) and DL current (red arrows) inlets/outlets.

In the latter case, accounting for the inter-turn/inter-layer/inter-pancake heat transfer completely changes the picture of the transient evolution. As an example, the results of the quench propagation analysis of the EU DEMO TF coils WP (2015 ENEA proposal [109], see Figure 4.29) performed with the 4C code [104] is reported.

The normal zone length shown in Figure 4.30(a) is shown to be strongly dependent on the thermal coupling between neighboring turns: The sudden slope changes are indeed caused by the normal zone initiation in the adjacent turns of the same layer, Figure 4.30(b), caused by the inter-turn heat transfer rather than by the heat advection due to He expulsion.

The same is true if the inter-layer heat transfer within the WP is considered, see Figure 4.31. Note that neglecting the thermal coupling within the WP is not necessarily conservative: Besides the maximum hot-spot temperature, the total amount of He expulsed from the coil and heated up must also be considered, as it must then be re-cooled to the operating temperature with a nonnegligible cost in terms of refrigeration power and time.

4.6.2. *Heat transfer within the magnet structures*

The thermal conduction within the bulky stainless-steel structures typical of the TF coils of fusion magnets is usually slow (see the time scales reported in

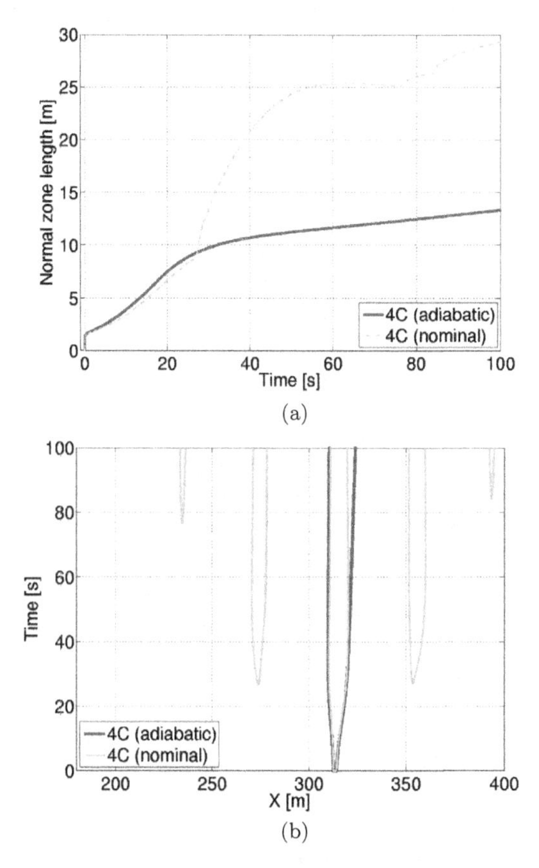

Figure 4.30. Evolution of the computed (a) normal zone length and (b) quench front propagation in DL 6.2 for the DEMO TF coil WP proposed by ENEA [109].

Section 4.3.2) so that the approximation of modeling 3D structures as a set of 2D cross-sections thermally coupled with the CICC holds, which is commonly adopted by many numerical tools, among which is the 4C code [104]. The thermal coupling between neighboring cross-sections is indeed provided mainly by the coolant advection in CICCs.

However, the heat transfer within the structures is important because they are subject to heat loads and sometimes actively cooled by additional cooling paths so that their role in the TH of the magnet could become very important in establishing a steady-state temperature map in the WP, as well as acting as an additional heat source/sink during slow transients.

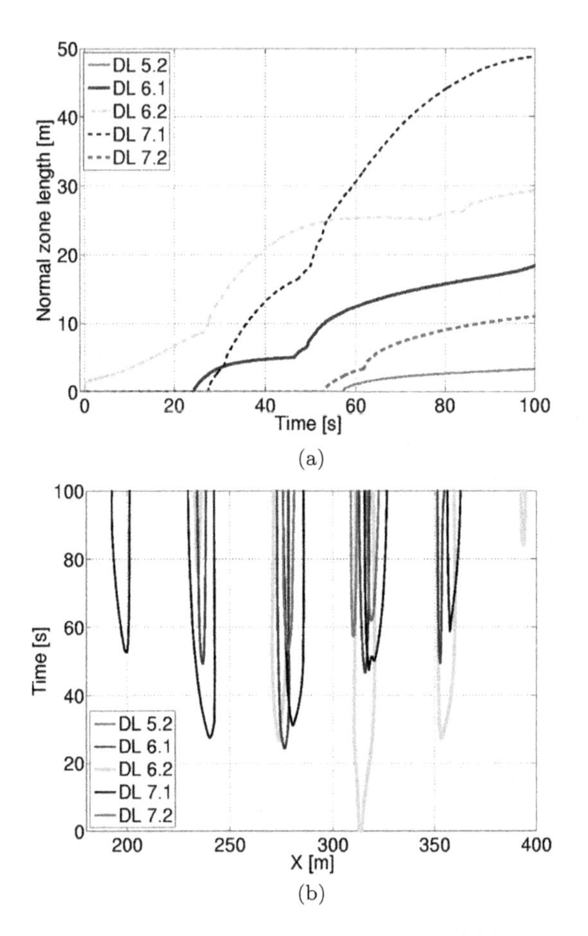

Figure 4.31. Evolution of the (a) normal zone length and (b) quench front propagation in DLs 5.2–7.2 in the nominal scenario for the DEMO TF coil WP proposed by ENEA [109].

During normal (cold) operation, the massive stainless-steel structures of the TF coils adopted in fusion reactors are needed to withstand the huge thermo-mechanical forces acting on the WP. However, the external surface of the structures is facing the cryostat, typically at ∼80 K, and thus exposed to a radiative thermal load. Moreover, the huge mass of the TF magnets must be supported against gravitational load so that dedicated supports (the so-called gravity support, see Figure 4.32) are installed. Since they are in contact with the basement of the building housing the magnets, they act as a heat bridge to the cold magnets.

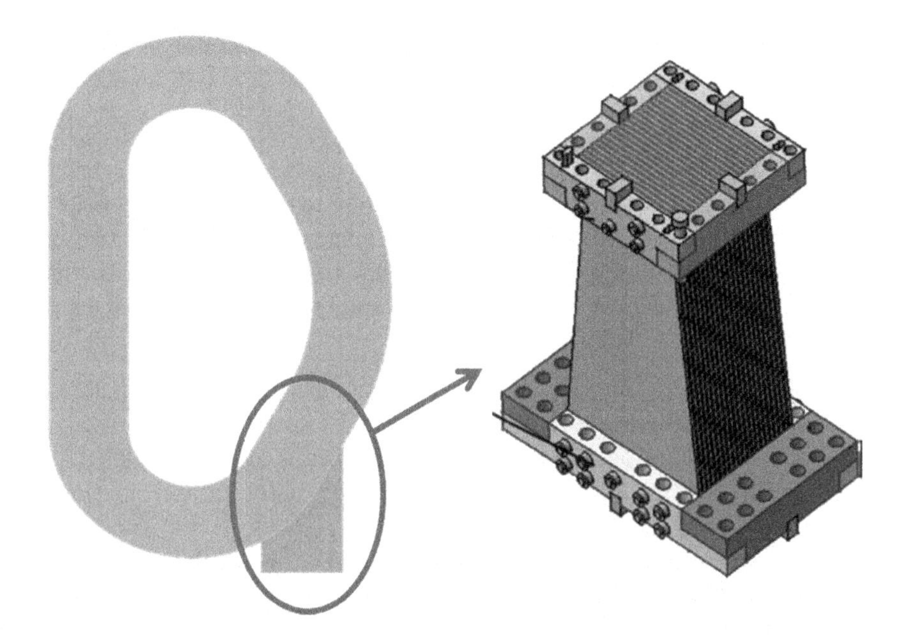

Figure 4.32. View of the gravity support of the ITER TF magnets.
Source: Reproduced from Ref. [111].

Finally, during fast magnetic field variations (induced, for example, by a fast current discharge from the coil), eddy currents are generated in the bulky stainless-steel structures so that Joule heat load is deposited within them.

All these thermal loads may affect the coil operation temperature and thus erode the temperature margin during operation, possibly leading to a quench. For this reason, on the one hand, suitable cooling of the structures must be guaranteed, and on the other hand, a proper model of the thermal conduction in the structures (and of their cooling paths) is required for a reliable TH simulation of a fusion magnet.

4.6.2.1. *Cooling of the coil casing*

The cooling of the stainless-steel structures surrounding the WP in large fusion magnets is provided:

- by the thermal contact with the actively cooled WP;
- in some cases, by the additional installation of dedicated cooling pipes inside the structures (as in the case of the ITER TF coils [105])

or attached to their surface (as in the W7-X stellarator nonplanar coils [8]).

The key issues in the design of these channels are:

- a proper design of their spatial location in order to take care of all possible hot spots during the different transients the magnet may be subjected to;
- the definition of a proper strategy for their thermal connection to the magnet; the problem is still open [112], despite several options that have been attempted so far:
  - welding/brazing/impregnation into suitably machined grooves [105] or on the magnet surface [113];
  - wrapping Cu stripes around the pipe and the magnet surface [8];
  - forcing the pipe surrounded by a Cu mesh into suitably machine grooves [108];
  - coolant flow in the grooves machined and covered by a leak-tight welded cover [81], providing the best thermal coupling but also introducing severe leak-tightness issues.

4.6.3. *Heat transfer between structures and winding*

The amount of heat transferred between the WP and the casing strongly depends on the casing cooling channel (CCC) design and the thermal properties of the ground insulation surrounding the WP, as well as on the coupling between the channels cooling the structures and the structures themselves.

In the specific case of the TF coils of a fusion reactor, it is important to note that when the coil is charged, Lorentz forces tend to push the WP away from the plasma. As a result, the WP normally detaches from the structures so that almost no heat transfer between the two is present during normal operation on the plasma-facing edge of the WP.

4.6.3.1. *Issues in the ground insulation modeling*

As mentioned above, the model of the ground insulation is crucial for a proper calculation of the heat exchanged between the structures and the WP. The impact of a variation in the thermal conductivity of the ground insulation on, for example, the temperature margin in the ITER TF coils is shown in Figure 4.33 to be as large as 0.3 K, which is ~50% of the minimum requirement of 0.7 K.

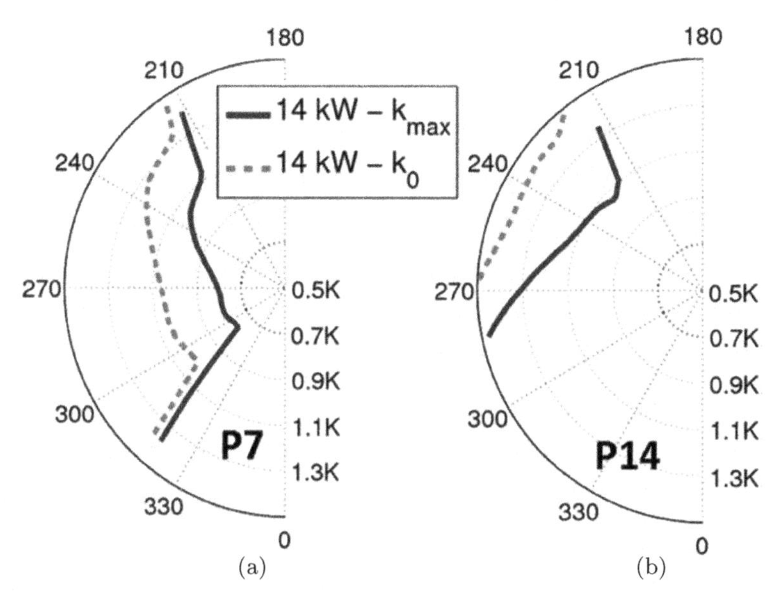

Figure 4.33. Polar distribution of the temperature margin on the inboard (straight) leg of the ITER TF coils computed during the standard operation scenario (with the baseline nuclear heat load of 14 kW on the entire TF magnet system) for (a) a central pancake and (b) a side pancake, with maximum (solid) and minimum (dashed) thermal conductivities of the winding ground insulation. The inboard equatorial location is at 270°.

Source: Reproduced from Ref. [60].

4.7. Relevant TH Transients

This section is devoted to describing the most interesting transients in the superconducting magnets for fusion reactors. The aim is to introduce the relevant transient drivers and constraints on which the magnet design is based.

4.7.1. *Cool down*

In order to reach its superconducting properties, a superconducting coil must be cooled from room temperature to (at least) its critical temperature. This necessary (and ordinary) transient is very long (up to one month, depending on the cold mass of the magnet to be cooled) because of:

- the limited cooling power of the refrigerator, which is directly hydraulically coupled to the magnet and drives the transient evolution;
- the thermo-mechanical constraints.

The latter are dictated by the fact that large thermal gradients in the magnet may result in large secondary mechanical stresses, arising from the different contraction (shrinking) of the (different) materials during the cooling (and, of course, different expansions during the warming-up). The commonly adopted, well-established rule of thumb consists of limiting the decrease in the inlet temperature of the coolant when the maximum temperature difference across the magnet is ~40–50 K. As in the real world it is not possible to install temperature sensors anywhere in the magnet, the use of reliable TH simulation tools for the design of a suitable CD strategy is of paramount importance to avoid excessive temperature gradients in the magnet that may lead to the breaking of cable strands or insulation layers during a (routine) CD transient.

As far as the LTS magnets' CD is concerned, it can be divided into two main phases [108], see Figure 4.34:

- Phase 1 from ~300 K to ~80 K (LN2 temperature): In this phase, the turbines of the refrigerator (see Section 4.2.3.1) are switched off and the cooling and after-cooling stages are bypassed. The decrease rate of the coil inlet temperature is automatically controlled at the desired value (typically ~ -1 K/h), mixing the GHe mass flow cooled by LN2 with the GHe mass flow at ambient temperature; the two mass flow rates are regulated by the opening of suitable control valves, see Figure 4.34(a).
- Phase 2 from ~80 K to ~4 K: In this phase, the turbines are turned on, letting a portion of the HP He (exiting the LN2 HXs) expand down to LP in order to cool the primary He flow bypassing the turbines through a series of regenerative HXs, namely HX3–HX6 (see Figure 4.34(b)). In this phase, the inlet temperature decrease depends on the refrigerator working point, and usually, it is slower than in phase 1 so that no issues arise in the temperature gradients.

The inlet (and outlet) temperature evolution in a large-scale superconducting magnet for fusion (e.g. the ITER CSMC [15]) is reported in Figure 4.35, where the different slopes of the temperature decrease between phase 1 and phase 2 are evident.

4.7.2. *Normal operation*

For a superconducting magnet applied to fusion reactors, the normal operation is typically the cold operation during plasma pulses. The timeline of

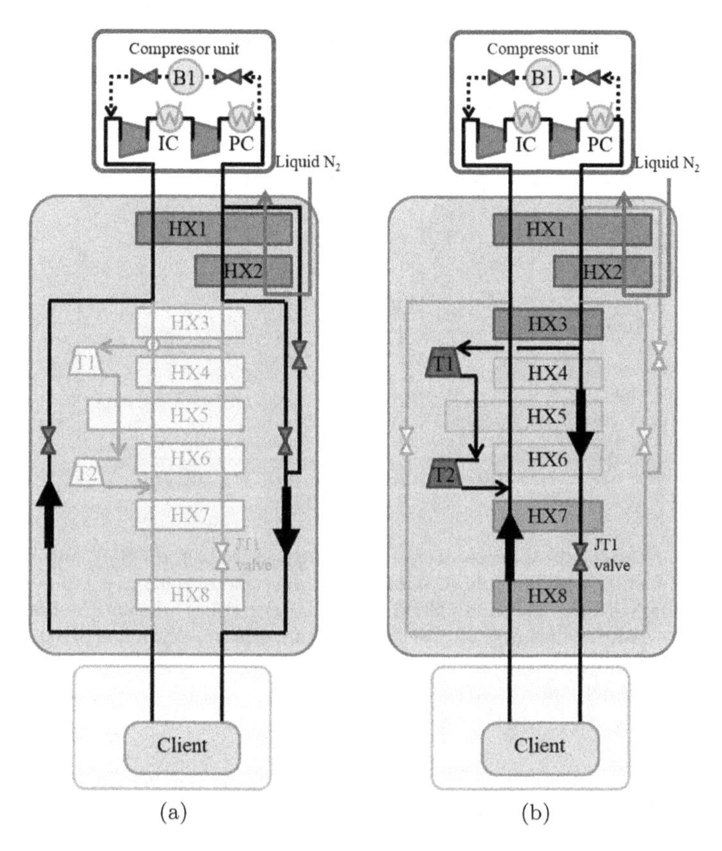

Figure 4.34. Schematic illustration of the operation of a typical refrigerator for fusion magnets cooling during a CD transient: (a) phase 1 and (b) phase 2.

such a periodic pulse for the ITER 15 MA plasma current scenario (the reference one) is reported in Figure 4.36.

The heat loads reported in Figure 4.36 refer to a TF magnet operated in direct current (DC), which mainly suffers from the static heat load on the outer surface of the structures and the nuclear heat load from the plasma. In the case of AC-operated fusion magnets, such as the CS and PF coils, most of the heat load comes from eddy current/AC losses induced by the changing magnetic field, see for example the ITER CS heat load reported in Figure 4.20. The timeline of the AC-operated magnets is thus the same, with higher AC losses, no nuclear heat load (shielded by the bulky structures of the TF coils), and reduced static heat load, especially for the CS that only faces the cold inner leg surface of the TF magnets.

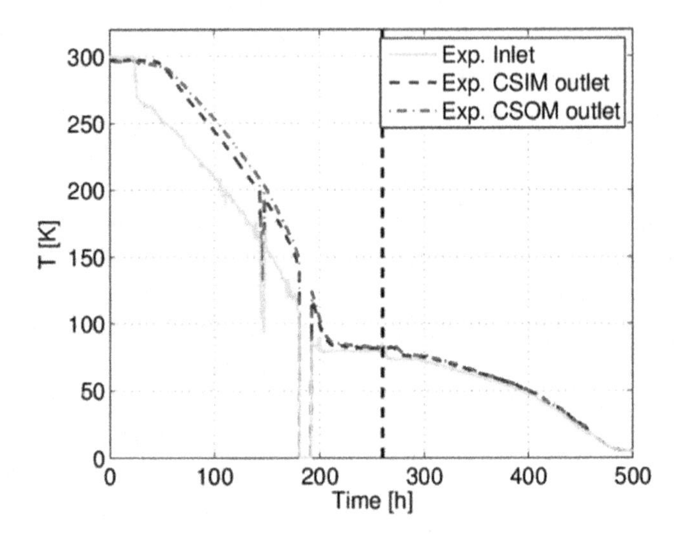

Figure 4.35. Measured evolution of the inlet temperature of the CSMC inner module (CSIM) and CSMC outer module (CSOM) (green line) and of the outlet temperature of the CSIM (dashed blue line) and CSOM (dash-dotted red line) during the first CSMC CD in the 2015 test campaign [114]. The temperature dips at ~150 h and ~190 h are due to temporary data acquisition problems. The vertical dashed line marks the separation between phase 1 and phase 2 of the CD.

Source: Reproduced from Ref. [108].

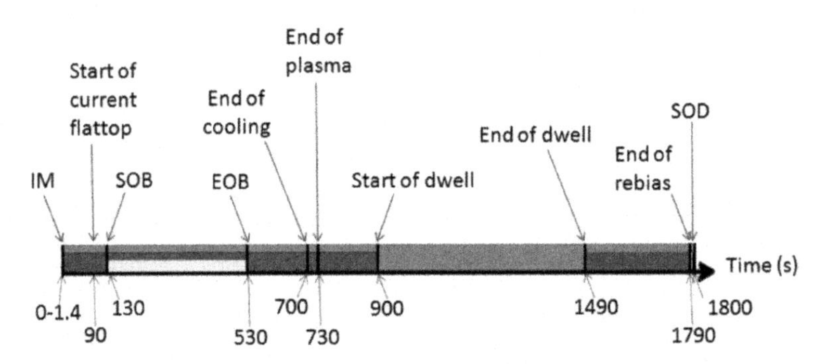

Figure 4.36. Timeline of the standard 15 MA scenario foreseen in the ITER (IM = initial magnetization, SOB = start of burn, EOB = end of burn, SOD = start of discharge). The radiative (static) load is applied during the whole 1800 s (blue line), the AC losses/eddy current load (CS, PF, and TF) according to the red line, and the nuclear load (TF) between the SOB and the EOB (yellow line).

Source: Reproduced from Ref. [59].

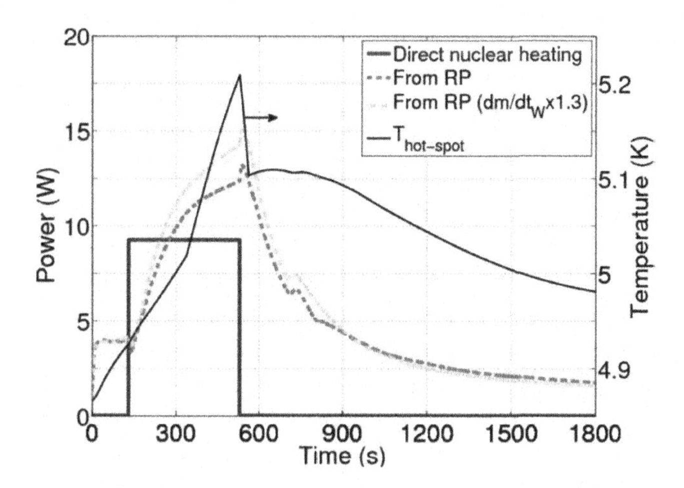

Figure 4.37. Time evolution of the computed heat load (W) on the first turn of a central pancake of the ITER TF coils directly from nuclear heating (solid) and indirectly from the radial plates (RP) through the jacket during one plasma scenario with nominal (dashed) and increased (dash-dotted) mass flow rates (dm/dt) in the WP. The computed hot-spot temperature in the central pancake is also reported on the right axis.

Source: Reproduced from Ref. [59].

The evolution of some relevant TH parameters of the ITER TF coils computed by 4C code during the same standard pulse is shown in Figure 4.37. Besides the nuclear heat load deposited during the plasma burn (from SOB to EOB, see Figure 4.36), the heat load exerted on the WP by conduction across the turn insulation is increased if the He mass flow rate is increased in the CICCs of the WP by ~30%. This is due to the fact that the heat sink in the WP is stronger. The hot-spot temperature increases during the plasma burn and is then reduced during the dwell time, which is useful to re-cool the conductor and recover the initial operating conditions.

4.7.3. *Off-normal operation*

The off-normal operating conditions for a superconducting magnet used for magnetically confined fusion reactors are all those conditions that may lead to transients capable of seriously damaging the magnet itself.

4.7.3.1. *Stability and quench*

The analysis of the stability of a CICC involves the very fast time scales on which the heat transfer within the conductor cross-section takes place.

Numerical analyses with different levels of complexity are used to assess which (thermal) perturbation that the conductor can withstand without the generation of a propagating normal zone [115–118]. The details of the transient HTCs mentioned in Section 4.5 are of paramount importance, and the time step of the simulation must be very small in order to capture the fast superconducting-to-normal transition [118].

If the thermal stability of the conductor is lost because the thermal perturbation is so large that the available coolant (and solids) thermal capacity is not sufficient to absorb it, then a normal zone develops and propagates: This is called a *quench* (note that if the normal zone develops but is reabsorbed without propagating, a recovery takes place and the stability is not lost). A quench is a severe transient during which, if no action is taken by the magnet protection system (unprotected quench), the temperature in the cable can rapidly increase so much to damage or destroy it owing to Joule heat generation in the normal zone.

Because of its dangerous nature, this transient in CICCs has been deeply investigated both from experimental [114, 119–124] and numerical [101, 106, 110, 125–135] points of view.

The typical features of a quench transient in a CICC, with reference to the data collected during the latest ITER CS Insert (CSI) test campaign in 2015 [114], see Figure 4.38 for the sensors location, are reported here. This case, although quite complex, has been chosen as a reference because the conductor was very well instrumented in terms of TH and electrical measurements:

- The timeline of a quench test in a full-scale CS conductor (the CSI) is shown in Figure 4.39. After heating (by means of an inductive heater IH01 wrapped around the CICC), a normal zone is initiated, and a quench is detected when the integral over 1 s of the total voltage measured across the coil is > 0.1 Vs. Then, a delay is imposed (only for testing purposes) before discharging the current from the CSI and CSMC to protect them.
- The increase in the resistive voltage measured across the entire coil during the quench propagation is evident from Figure 4.40(a), until the current is discharged from the coil. The voltage measurement is usually adopted as the first quench detection in LTS fusion magnets, as it reacts immediately and can be measured easily.
- The normal zone front propagation upstream and downstream of the IH location is shown in Figure 4.40(b), where an acceleration of the propagation speed occurs due to the preheating of the cable ahead of the

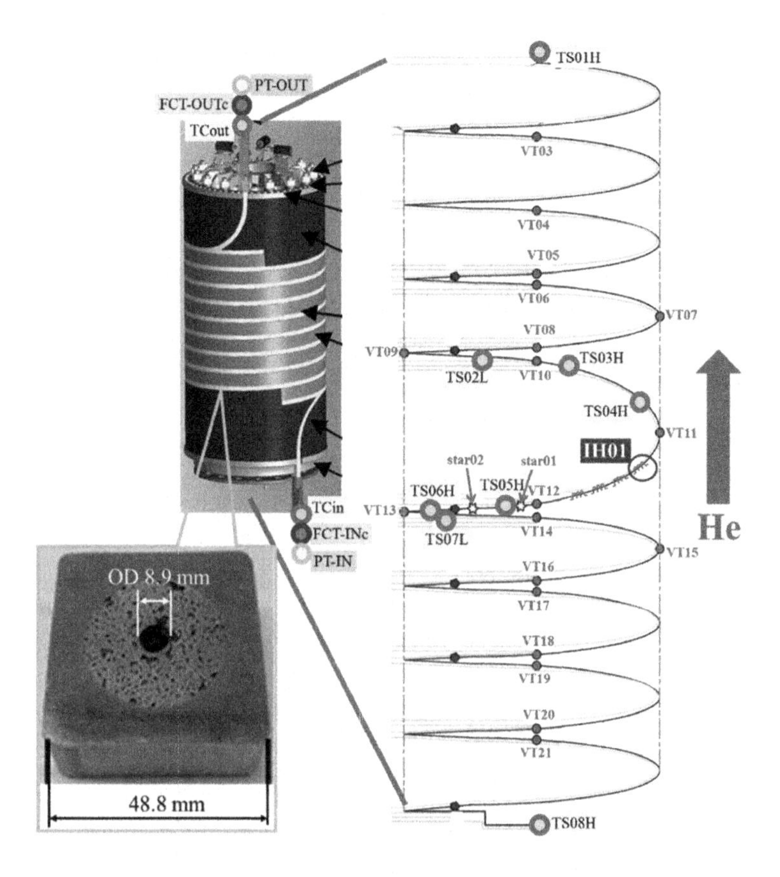

Figure 4.38. CSI conductor cross-section and sketch of the available diagnostics installed on the coil: VT-xx and star-xx voltage taps, TS-xx and TC-xx jacket and He thermometers, PT-xx pressure taps, and FCT-xx flow meters. IH01 indicates the inductive heater used as a driver of the quench tests.

Source: Reproduced from Ref. [101].

propagation direction because of warm He expulsion ("pressure-driven" quench [131]).

- The latter is evident from Figure 4.41(a), where the mass flow rate measured at the coil extremities is reported. The heating in the central part of the coil causes a violent He expulsion from the coil endings at the sound speed time scale. Thanks to its quick reaction, it is used as a secondary (redundant) quench detection system in (LTS) fusion magnets.
- The He expulsion causes a pressurization of the SHe circuit, as measured by the pressure taps and shown in Figure 4.41(b).

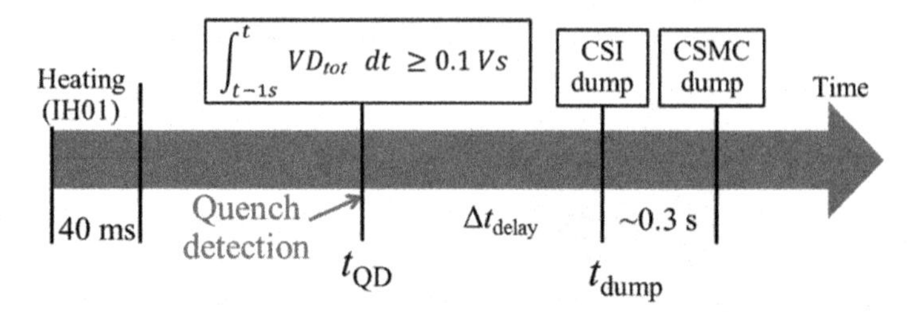

Figure 4.39. Typical quench test timeline, from the IH pulse to the CSI and CSMC current dumps. VD_{tot} is the total voltage measured across the coil, t_{QD} is the quench detection time, t_{dump} is the current discharge time, and Δt_{delay} is the hold time manually set to delay the current discharge after the quench detection.

Source: Reproduced from Ref. [101].

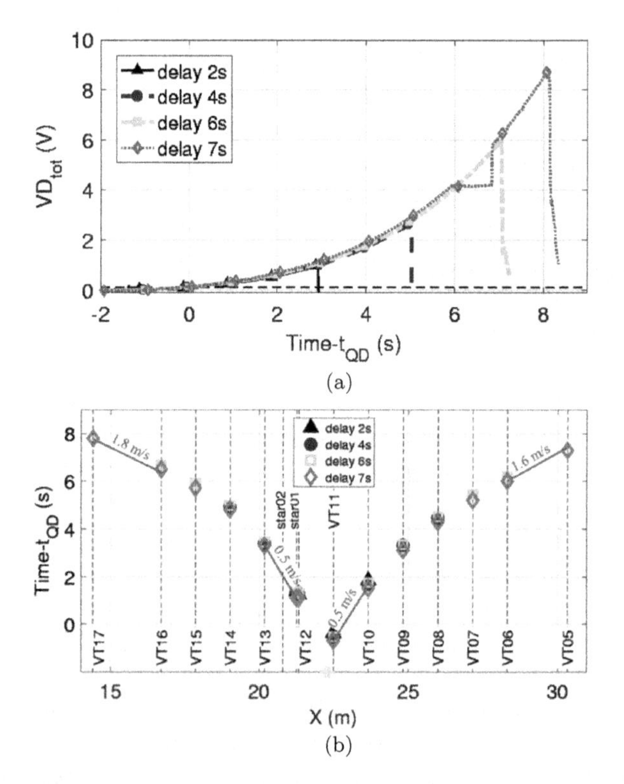

Figure 4.40. (a) Measured evolution of the total voltage (sum of the local voltages) across the CSI for all EOB quench tests. The time coordinate is shifted in order to synchronize them at the quench detection time (t_{QD}). (b) Measured propagation of the quench front for different quench tests. Initial and final quench front speeds are reported. The cyan crosses on the x-axis at ~22 m indicate the IH01 location.

Source: Reproduced from Ref. [101].

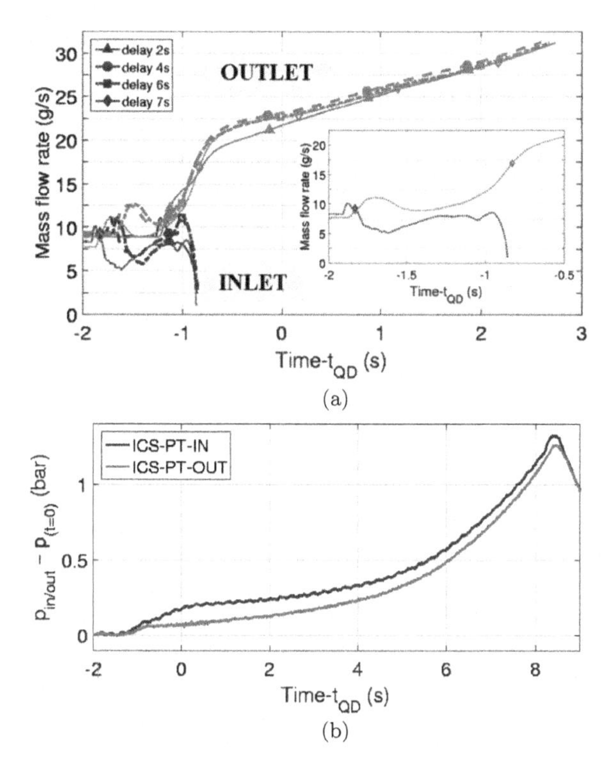

Figure 4.41. (a) Measured evolution of inlet (blue) and outlet (pink) mass flow rates in the EOB quench tests. The mass flow rates during the first phase of the transient for the 7 s delay quench are reported in the inset (the IH is operated at time: $t_{QD} \sim -2\,s$). (b) Measured inlet and outlet pressurization evolution for the 7 s delay quench.

Source: Reproduced from Ref. [101].

- As far as the temperature is concerned, the most interesting one is the hot-spot temperature of the cable, shown in Figure 4.42(a). It cannot be measured but can be estimated by local voltage measurements (the so-called "virtual sensor"), exploiting the (known) temperature dependence of the electrical resistivity of Cu [132, 135]. The conductor design must ensure that the maximum hot-spot temperature during a protected quench does not overcome a given threshold to avoid thermo-mechanical damages to the strands.
- Also, the jacket temperature is an important TH parameter during a quench, see Figure 4.42(b), as it drives the thermo-mechanical stresses in the WP.

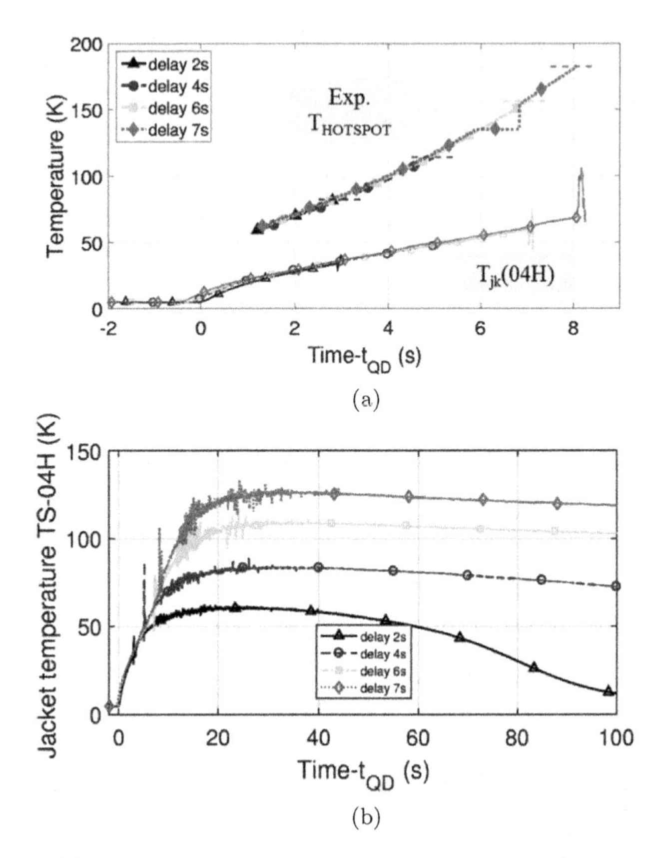

Figure 4.42. (a) Measured evolution of the jacket temperature (open markers) just downstream of the IH (TS04H location) and of the virtual sensor signal deriving the hot-spot temperature (solid markers) from the VD1112 measurement, for all the EOB quench tests. The hot-spot temperature values from the virtual sensor just before the CSI dump are also reported as horizontal dashed segments. (b) Measured evolution of the jacket temperature at TS04H location for ~100 s after the quench initiation. The spikes at ~ 5–10 s are induced by the strong magnetic field variation during the dump. *Source*: Reproduced from Ref. [101].

4.7.3.2. *Fast discharge/current ramps*

The fast magnetic field variations, usually induced by quick current ramps or exponential discharges (see Figur 4.43(a) for the fast current discharge from the ITER TF coils) forced by the magnets protection system, cause large heat deposition into the superconducting cable and in the stainless-steel structures because of AC losses, see Figure 4.43(b).

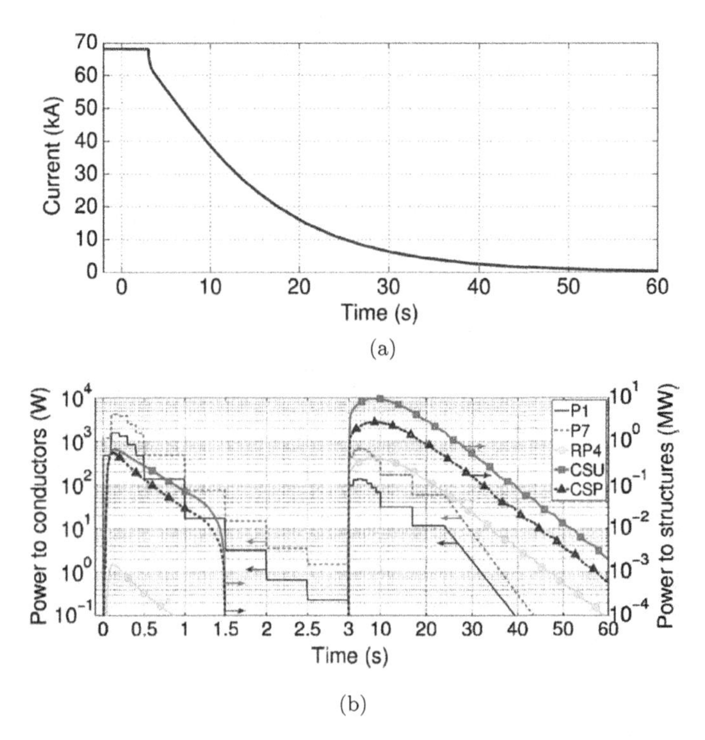

Figure 4.43. (a) ITER TF coil current discharge evolution. (b) Evolution of the power deposited in: side pancake P1 and central pancake P7 on the left axis; a central radial plate (RP4), in the outer part of the casing (CSU), and in the inner (plasma facing) part of the casing (CSP) on the right axis.

Source: Reproduced from Ref. [136].

The power deposition can lead to a fast temperature increase (see the effects on the straight leg of the ITER TF structures cross-section shown in Figure 4.44(a)) and, consequently, to a quench even if the current is being discharged from the magnet. In the case of a quench during the ITER TF fast discharge, the amount of He vented to the quench tanks, according to the simulations in Ref. [136], is larger than 150 kg, see Figure 4.44(b). That He must then be re-cooled and re-charged in the SHe loops, with a large cost in terms of cooling power and time.

An accurate numerical analysis of the fast current discharge can help in the design of the dump resistors to which the energy is discharged in order to optimize the exponential time constant (constrained also by the maximum voltage tolerated by the conductor insulation).

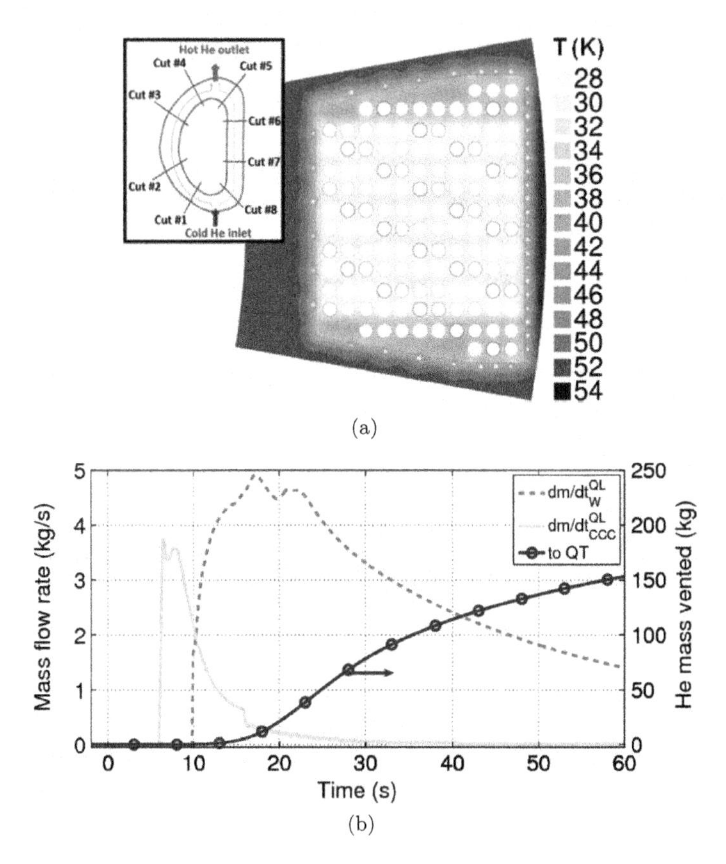

Figure 4.44. (a) Temperature distribution on cut #5, see the inset, at time $t \sim 20$ s, when the coil current is still ~20 kA. (b) Evolution of mass flow rate from winding (W) and casing cooling channels (CCC) to the quench line (QL), left axis, and of the total He mass in the quench tanks (QT), right axis.

Source: Reproduced from Ref. [136].

4.7.3.3. *Loss of flow/coolant accidents*

Two other relevant accidents for an actively cooled superconducting magnet are the loss of flow accident (LOFA) and the loss of coolant accident (LOCA).

The LOFA can be triggered by several initiating events [137], among which, for example, is a cold circulator trip. If the LOCA is unprotected and the plasma operation is not inhibited or interrupted, the LOFA in the ITER TF coil could lead to a quench, as computed, for example, in Ref. [138], see Figure 4.45.

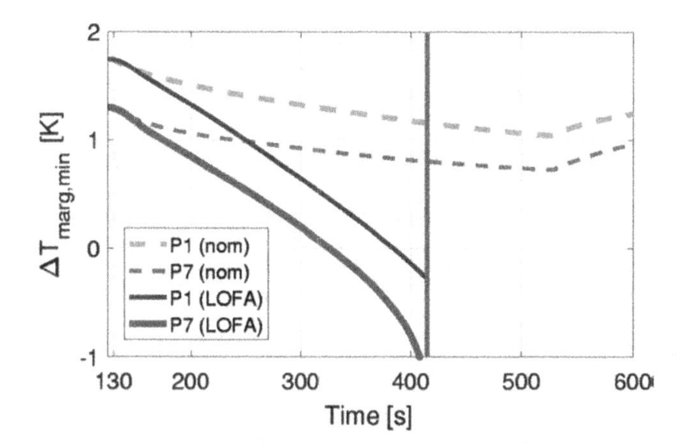

Figure 4.45. Comparison of the temperature margin (evaluated at the peak magnetic field on the conductor cross-section) evolution in a side (P1) and a central (P7) pancake in both nominal (dashed curves) and LOFA (solid curves) simulations.

Source: Reproduced from Ref. [138].

If the protection system accomplishes its functions, the protected LOFA can evolve into a quench only for some choices of the current accelerated discharge time from the coils [137]. However, the time required to re-cool the coil is estimated to be as long as 1 h.

In the case of LOCA, the consequences can be twofold:

- the active cooling of the coil is lost, similar to the LOFA but with a depressurization of the cooling circuit;
- if the break is located inside the cryostat, the loss of vacuum accident in the latter is also initiated.

Existing, validated TH codes could be used for detailed simulations of accidental transients at an early magnet design stage in order to pursue a “safety-based” design. In particular, the simulations can be carried out to assess the deterministic consequences of given events [137, 138] or to generate the database needed to perform an integrated deterministic-probabilistic safety analysis [139].

4.8. Available Models and Experimental Facilities

Even if the need for numerical modeling of superconducting magnets is evident, the effort toward verification and validation of the existing TH models has been rather limited so far, sometimes notwithstanding the existence of

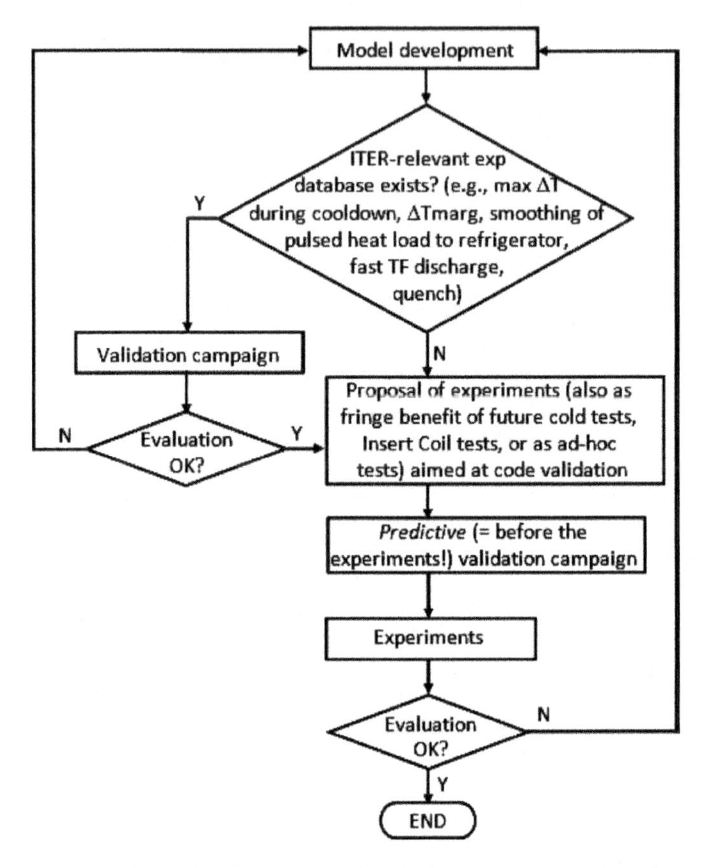

Figure 4.46. Possible roadmap to confirm the reliability of existing TH codes for fusion superconducting magnets system.

Source: Reproduced from Ref. [83].

a significant experimental database. Particular attention should also be devoted to the assessment of the predictive capabilities of the existing TH codes, according to the roadmap depicted in Figure 4.46.

In this section, both the most important existing code and the experimental facilities (in principle, available for data collection with validation purposes) are described.

4.8.1. *Thermal-hydraulic codes*

A short description of the more important tools available for the simulation of TH transients in superconducting magnets is given in this section.

4.8.1.1. *Venecia*

The Venecia code [140] is the next generation of the Vincenta code, developed for the simulation of TH transients in superconducting magnets cooled by forced-flow He.

The complex computational model is built as a combination of 1D nonlinear TH models for coolant flows and 1D or 2D models of heat diffusion in solids linked together. It is able to simulate the transient behavior of cryogenic systems and superconducting magnet systems simultaneously, reproducing the magnet geometry, cryogenic components, and nonlinear thermal properties of coolant and solids. It can cope with fast (stability and quench) and slow (normal operation, CD, warm-up) transients.

The applications include:

- magnets for nuclear fusion facilities;
- accelerators;
- MRI magnets;
- superconducting motors, generators, and SMES;
- experimental and diagnostic devices,
- superconducting cables in general.

It has undergone some validation exercises, as reported in Ref. [141].

4.8.1.2. *4C*

The 4C code has been developed at Politecnico di Torino (Torino, Italy) as a result of a suite of tools addressing the modeling of superconducting magnets for fusion applications, from a single CICC [142] to the WP [103] and finally to the entire magnet [104]. It addresses the TH transients in superconducting magnets for fusion applications, but it could also be suitable to simulate other magnets, such as those of the SMES and superconducting cables for power transmission.

The architecture of the 4C code is shown in Figure 4.47.

The coil (WP) is modeled as a set of CICCs hydraulically in parallel, as described in Sections 4.3–4.5.

The CICCs can be thermally coupled to their neighbors (M&M [103]).

As far as the structures are concerned, a poloidal discretization of a 3D transient heat conduction problem is performed using several 2D cuts on suitable cross-sections; the transient 2D heat conduction problems are solved with finite elements, accounting for all the details of the magnet cross-section (insulation layers, wedged shape, etc.).

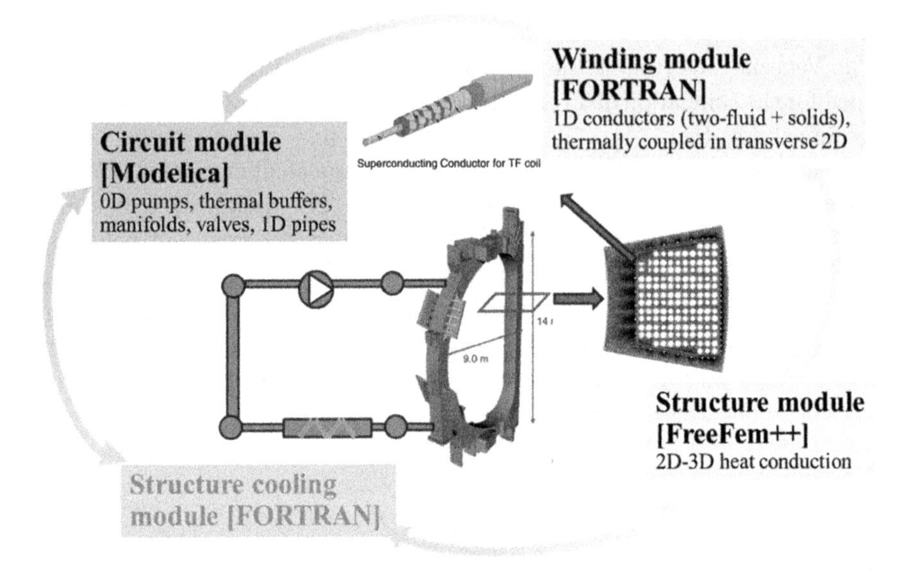

Figure 4.47. Architecture of the 4C code.

The circuit components are described in terms of differential-algebraic equations; these components can then be connected hierarchically to build arbitrarily complex system models in a way that can be represented graphically in terms of object diagrams. A new Modelica "Cryogenics" library has been developed for the modeling of cryogenic circuits using He as a working fluid [143].

After its development, the 4C code has undergone a long series of verification (including benchmarking [105, 143]) and validation exercises [8, 70, 81, 101, 108], including predictive validation [126, 144], as schematically represented in Figure 4.48, so that it can be claimed to be a reliable tool for the TH analysis of the superconducting magnet systems of existing and future fusion facilities.

4.8.1.3. *Supermagnet*

Supermagnet [145, 146] allows superconducting magnet designers to consider different configurations of superconductors in a magnet, with disparate cooling methods and power supply connections, and for each of them, several operating conditions.

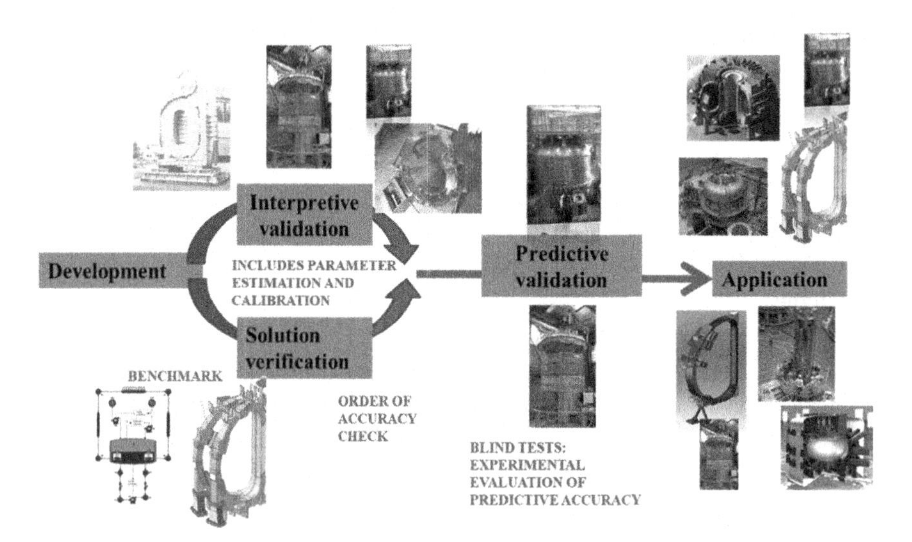

Figure 4.48. Benchmarks, verifications, validations, and applications of the 4C code so far.

The CryoSoft suite of codes THEA (Thermal, Hydraulic and Electric Analysis of Superconducting Cables), FLOWER (Hydraulic Network Simulation), POWER (Electric Network Simulation of Magnetic Systems), HEATER (Simulation of Heat Conduction), MrX (Generic Data Exchange) provides a basis for well-optimized and flexible tools for the analysis of specific issues in superconducting magnet systems. Supermagnet is the manager application that launches two or more of the above codes, schedules their communication, and terminates execution as appropriate.

The codes communicate through a data exchange mechanism that achieves the desired physical coupling and makes it possible to describe a series of processes, such as:

- effect of He expulsion during thermal transients on the proximity cryogenics;
- regulation of cryogen flow and valving conditions, subject to transient response in the superconducting cables;
- evolution of the coil current during a quench, including the effect of quench resistance, and coupling within segments of the same magnetic system;
- cooling of a coil with thermally coupled parallel channels.

4.8.1.4. *Others*

Several general-purpose commercial tools have been used to model part of a superconducting magnet system depending on the needs:

- EcosymPro [73] and Aspen HYSYS [147] for the analysis of the cooling circuits;
- ANSYS Fluent [148] for a full 3D modeling of the superconducting magnet.

Moreover, a computational tool similar to those described in Sections 4.8.1.1–4.8.1.3 has been developed by the Chinese Academy of Science [149].

4.8.2. *Conductor test facilities*

One of the most important conductor test facility is SULTAN (Villigen, Switzerland) [150]. It is devoted to the TH and electromagnetic testing of full-size, short-length (up to $\sim 2.5\,\mathrm{m}$) straight CICC samples.

It can produce over a length of $\sim 0.5\,\mathrm{m}$ a uniform magnetic field in relevant conditions for superconducting magnets of fusion devices. A huge number of samples have then been tested for all the existing (and future) nuclear fusion facilities around the world, aiming at qualifying the conductors.

Cyclic tests can be done so that the conductor performance can also be assessed after a lot of thermo-mechanical cycles.

Moreover, a fast magnetic field variation is allowed so that the AC behavior of the conductor can also be assessed.

4.8.3. *Magnets test facilities*

Concerning the facilities capable of coping with the testing of large-scale magnets, especially for fusion applications, the following must be mentioned:

- ITER CS Model Coil (CSMC) at Naka, Japan [15, 26]: It is a model coil magnet for the ITER CS, but it is also used to test single-layer, full-scale, long-length coils (the so-called insert coils) in its bore in a magnetic field up to 13 T under fully relevant conditions for their application in superconducting magnets for fusion.

- TOSKA facility at KIT, Karlsruhe [151]: It provides the cryostat, power supply, and cooling power for the test of large-scale magnets, such as the ITER Toroidal Field Model Coil [80].
- The Cold Test Facility for the JT-60SA Tokamak TF Coils at CEA Saclay, France [152]. This facility is equipped with a large cryostat, power supply, and suitable cryoplant.
- The ITER Central Solenoid Module Final Test Facility at San Diego, CA, United States [153]. This facility is also equipped with a large cryostat, power supply, and suitable cryoplant.

4.8.4. *Available experiments*

The available fusion experiments where the magnetic confinement is performed by means of superconducting magnets are listed in this section.

4.8.4.1. *Superconducting tokamaks in operation*

The superconducting tokamaks currently in operation are:

- EAST in Hefei, China [9];
- KSTAR in Daejeon, Korea [154];
- WEST (formerly, Tore Supra [54] in Cadarache, France [55].

Under (advanced) construction are the JT-60SA in Naka, Japan [11] and ITER in Cadarache, France [6].

4.8.4.2. *Superconducting stellarators in operation*

The stellarators in operation, relying on superconducting magnets, are:

- W7-X in Greifswald, Germany [7];
- Large Helical Device (LHD) in Toki, Japan [63].

References

[1] Martovetsky N. 2013. Correlation between degradation and broadness of the transition in CICC. *Supercond. Sci. Technol.* 26(10), 104001.,
[2] Kadomtsev BB, Troyon FS, Watkins ML, Rutherford PH, Yoshikawa M and Mukhovatov VS. 1990. Tokamaks. *Nucl. Fusion* 30(9), 1675–1694.
[3] Spitzer L. 1958. The stellarator concept. *Phys. Fluids* 1(4), 253–264.
[4] Ikeda K. 2010. ITER on the road to fusion energy. *Nucl. Fusion* 50(1), 014002.

[5] Mitchell N. *et al.* 2008. The ITER magnet system. *IEEE Trans. Appl. Supercond.* 18(2), 435–440.
[6] www.iter.org.
[7] Marushchenko NB. *et al.* 2016. Main results of the first experimental campaign in the stellarator W7-X. *Probl. At. Sci. Technol.* 106(6), 3–8.
[8] Bonifetto R, Kholia A, Renard B, Riße K, Savoldi Richard L and Zanino R. 2011. Modeling of W7-X superconducting coil cool-down using the 4C code. *Fusion Eng. Des.* 86(6–8), 1549–1552.
[9] Wan YX, Weng PD, Li JG, Gao DM, Wu ST and EAST Team. 2004. Progress of the EAST project in China. *20th International Atomic Energy Agency (IAEA) Fusion Energy Conference.*
[10] Oh YK. *et al.* 2009. Commissioning and initial operation of KSTAR superconducting tokamak. *Fusion Eng. Des.* 84(2–6), 344–350.
[11] Tomarchio V. *et al.* 2017. Status of the JT-60SA project: An overview on fabrication, assembly and future exploitation. *Fusion Eng. Des.* 123, 3–10.
[12] Corato V. *et al.* 2018. Progress in the design of the superconducting magnets for the EU DEMO. *Fusion Eng. Des.* 136, 1597–1604.
[13] Risse K. *et al.* 2016. Wendelstein 7-X-commissioning of the superconducting magnet system. *IEEE Trans. Appl. Supercond.* 26(4), 4202004.
[14] Yanagi N. *et al.* 1998. Development of superconductors for the Large Helical Device. *J. Nucl. Mater.* 258–263(PART 2 B), 1935–1939.
[15] Tsuji H. *et al.* 2001. Progress of the ITER central solenoid model coil programme. *Nucl. Fusion* 41(5), 645–651.
[16] Hoenig MO and Montgomery DB. 1975. Dense supercritical-helium cooled superconductors for large high field stabilized magnets. *IEEE Trans. Magn.* 11(2), 569–572.
[17] Risse K. *et al.* 2003. Fabrication of the superconducting coils for Wendelstein 7-X. *Fusion Eng. Des*, 66, 965–969.
[18] Bonifetto R. *et al.* 2018. Thermal-hydraulic test and analysis of the ENEA TF conductor sample for the EU DEMO fusion reactor. *IEEE Trans. Appl. Supercond.* 28(4), 4205909.
[19] Bruzzone P. 2001. Superconductors and joints, tests and trends for future development. *Fusion Eng. Des.* 56–57, 125–134.
[20] Mitchell N, Devred A, Libeyre P, Lim B and Savary F. 2012. The ITER magnets: Design and construction status. *IEEE Trans. Appl. Supercond.* 22(3), 4200809.
[21] Bruzzone P. 2015. Superconductivity and fusion energy — The inseparable companions. *Supercond. Sci. Technol.* 28(2), 024001.
[22] Libeyre P. *et al.* 2016. Status of design and manufacturing of the ITER Central Solenoid and Correction Coils. In *Proceedings — Symposium on Fusion Engineering*, pp. 1–8.
[23] Wesche R, Sedlak K, Bykovsky N, Bruzzone P, Zani L and Coleman M. 2016. Winding pack proposal for the TF and CS coils of European DEMO. *IEEE Trans. Appl. Supercond.* 26(3), 4200405.
[24] Zappatore A. *et al.* 2018. Performance analysis of the NbTi PF coils for the EU DEMO fusion reactor. *IEEE Trans. Appl. Supercond.* 28(4), 4901005.

[25] Lim B. *et al.* 2011. Design of the ITER PF coils. *IEEE Trans. Appl. Supercond.* 21(3 PART 2), 1918–1921.
[26] Ogata H, Ando T and Ito T. 1996. Design of the ITER central solenoid (CS) model coil. *IEEE Trans. Magn.* 32(4 PART 2), 2320–2323.
[27] Zanino R, Mitchell N and Savoldi Richard L. 2003. Analysis and interpretation of the full set (2000–2002) of TCS tests in conductor 1A of the ITER central solenoid model coil. *Cryogenics*, 43(3), 179–197.
[28] Bottura L, Gourlay SA, Yamamoto A and Zlobin AV. 2016. Superconducting magnets for particle accelerators. *IEEE Trans. Nucl. Sci.* 63(2), 751–776.
[29] Bottura L, De Rijk G, Rossi L and Todesco E. 2012. Advanced accelerator magnets for upgrading the LHC. *IEEE Trans. Appl. Supercond.* 22(3), 4002008.
[30] Apollinari G, Bruening O, Nakamoto T and Rossi L. 2017. High luminosity large Hadron Collider HL-LHC. *Cern Yellow Rep.* 1–19.
[31] Pulikowski D. *et al.* 2017. Testing mechanical behavior of Nb3Sn Rutherford Cable during coil winding. *IEEE Trans. Appl. Supercond.* 27(4), 4802105.
[32] Kekelidze V. *et al.* 2014. The NICA project at JINR dubna. In *EPJ Web of Conferences*, p. 00127.
[33] Baldin AM. *et al.* 1995. Superconducting fast cycling magnets of the nuclotron. *IEEE Trans. Appl. Supercond.* 5(2), 875–877.
[34] Devred A. *et al.* 2006. Overview and status of the Next European Dipole Joint Research Activity. In *Superconductor Science and Technology*, p. S67.
[35] Rossi L, Stavrev S and Szeberenyi A. 2014. 2Nd Periodic Hilumi Lhc Report. Grant Agreement number 284404.
[36] McIntyre P, Breitschopf J, Chavez D, Gerity J, Kellams J, Sattarov A and Tomsic M. 2018. Block-coil high-field dipoles using superconducting cable-in-conduit. *IEEE Trans. Appl. Supercond.* 28(3), 4005307.
[37] Parizh M, Lvovsky Y and Sumption M. 2016. Conductors for commercial MRI magnets beyond NbTi: Requirements and challenges. *Superconductor Science and Technology*, 30(1), 014007.
[38] Marshall WS, Bird MD, Godeke A, Larbalestier DC, Markiewicz WD and White JM. 2017. Bi-2223 test coils for high-resolution NMR magnets. *IEEE Trans. Appl. Supercond.* 27(4), 4300905.
[39] Qu R, Liu Y and Wang J. 2013. Review of superconducting generator topologies for direct-drive wind turbines. *IEEE Trans. Appl. Supercond.* 23(3), 5201108.
[40] Hirakawa M, Inadama S, Kikukawa K, Suzuki E and Nakasima H. 2003. Developments of superconducting motor with YBCO bulk magnets. *Phys. C: Supercond. Appl.*, pp. 392–396(PART 1) 773–776.
[41] Scanlan RM, Malozemoff AP and Larbalestier DC. 2004. Superconducting materials for large scale applications. In *Proceedings of the IEEE*, pp. 1639–1654.
[42] Schmidt F and Allais A. 2004. Superconducting cables for power transmission applications — a review. In *Workshop on Accelerator Magnet Superconductors Proceedings*, p. 352.

[43] Muñoz-Antón J, Marian A, Lesur F and Bruzek C-E. 2020. Dichotomic decision optimization for the design of HVDC superconducting links. *Entropy* 22, 1413.

[44] Coombs TA. 2015. High-temperature superconducting magnetic energy storage (SMES) for power grid applications. In *Superconductors in the Power Grid: Materials and Applications*, pp. 261–280.

[45] Nomura S, Shintomi T, Akita S, Nitta T, Shimada R and Meguro S. 2010. Technical and cost evaluation on SMES for electric power compensation. *IEEE Trans. Appl. Supercond.*, 20(3), 1373–1378.

[46] Moon FC. 1982. The virial theorem and scaling laws for superconducting magnet systems. *J. Appl. Phys.* 53(12), 9112–9121.

[47] Mbaruku AL and Schwartz J. 2008. Fatigue behavior of Y-Ba-Cu-O/hastelloy-C coated conductor at 77 K. *IEEE Trans. Appl. Supercond.* 18(3), 1743–1752.

[48] Nagaya S. *et al.* 2006. Field test results of the 5 MVA SMES system for bridging instantaneous voltage dips. *IEEE Trans. Appl. Supercond.*, 632–635.

[49] Tixador P. *et al.* 2007. Design and first tests of a 800 kJ HTS SMES. *IEEE Trans. Appl. Supercond.* 17(2), 1967–1972.

[50] Friedman A, Shaked N, Perel E, Sinvani M, Wolfus Y and Yeshurun Y. 1999. Superconducting magnetic energy storage device operating at liquid nitrogen temperatures. *Cryogenics (Guildf).* 39(1), 53–58.

[51] NIST. 2013. NIST Standard Reference Database 23: Reference Fluid Thermodynamic and Transport Properties (REFPROP), Version 9.1. *NIST Standard Reference Database 23: Reference Fluid Thermodynamic and Transport Properties (REFPROP), Version 9.1.*

[52] Bell IH, Wronski J, Quoilin S and Lemort V. 2014. Pure and pseudo-pure fluid thermophysical property evaluation and the open-source thermophysical property library coolprop. *Ind. Eng. Chem. Res.* 53(6), 2498–2508.

[53] Massol O and Rifaat O. 2018. Phasing out the U.S. Federal Helium Reserve: Policy insights from a world helium model. *Resour. Energy Econ.* 54, 186–211.

[54] Saoutic B. *et al.* 2011. Contribution of Tore Supra in preparation of ITER. *Nucl. Fusion* 51(9), 094014.

[55] Missirlian M. *et al.* 2014. The WEST project: Current status of the ITER-like tungsten divertor. *Fusion Eng. Des.*, 89(7–8) 1048–1053.

[56] Darweschsad M. *et al.* 1997. Development and test of the poloidal field prototype coil POLO at the Forschungszentrum Karlsruhe. *Fusion Eng. Des.* 36(2–3), 227–250.

[57] Lvovsky Y, Stautner EW and Zhang T. 2013. Novel technologies and configurations of superconducting magnets for MRI. *Supercond. Sci. Technol.* 26(9), 093001.

[58] Savoldi Richard L, Bonifetto R, Carli S, Froio A, Foussat A and Zanino R. 2014. Artificial Neural Network (ANN) modeling of the pulsed heat load during ITER CS magnet operation. *Cryogenics (Guildf).* 63, 231–240.

[59] Savoldi Richard L, Bonifetto R, Foussat A, Mitchell N, Seo K and Zanino R. 2013. Mitigation of the temperature margin reduction due to the nuclear radiation on the ITER TF coils. *IEEE Trans. Appl. Supercond.* 23(3), 4201305.
[60] Savoldi Richard L. *et al.* 2014. Analysis of the effects of the nuclear heat load on the ITER TF magnets temperature margin. *IEEE Trans. Appl. Supercond.* 24(3), 4200104.
[61] Iwasa Y, Bascuñán J, Hahn S and Park DK. 2012. Solid-cryogen cooling technique for superconducting magnets of NMR and MRI. *Physics Procedia*, 36, 1348–1353.
[62] Bruzzone P. 2010. Superconductors for fusion: Achievements, open issues, roadmap to future. *Phys. C: Supercond. Appl.*, 470(20), 1734–1739.
[63] Fujiwara M. *et al.* 1996. Large helical device (LHD) program. *J. Fusion Energy* 15(1–2), 7–153.
[64] Hamada K. *et al.* 1994. Final design of a cryogenic system for the ITER CS model coil. *Cryogenics (Guildf).* 34(SUPPL. 1), 65–68.
[65] Jha AR. 2006. *Cryogenic Technology and Applications.* Oxford: Butterworth-Heinemann.
[66] Bai H. *et al.* 2006. Cryogenics in EAST. *Fusion Eng. Des.* 81(23–24), 2597–2603.
[67] Garbil R. *et al.* 2001. The Tore Supra Cryoplant control system upgrade. *Fusion Eng. Des.* 58–59, 217–223.
[68] Hoa C. *et al.* 2012. Investigations of pulsed heat loads on a forced flow supercritical helium loop — Part A: Experimental set up. *Cryogenics (Guildf).* 52(7–9), 340–348.
[69] Vallcorba R, Hitz D, Rousset B, Lagier B and Hoa C. 2012. Investigations of pulsed heat loads on a forced flow supercritical helium loop: Part B: Simulation of the cryogenic circuit. *Cryogenics (Guildf).* 52(7–9), 349–361.
[70] Zanino R, Bonifetto R, Casella F and Savoldi Richard L. 2013. Validation of the 4C code against data from the HELIOS loop at CEA Grenoble. In *Cryogenics*, Vol. 53, pp. 25–30.
[71] Savoldi Richard L, Bonifetto R, Carli S, Grand Blanc M and Zanino R. 2013. Modeling of pulsed heat load in a cryogenic SHe loop using Artificial Neural Networks. *Cryogenics (Guildf).* 57, 173–180.
[72] Zanino R, Bonifetto R, Hoa C and Savoldi Richard L. 2013. 4C modeling of pulsed-load smoothing in the HELIOS facility using a controlled bypass valve. *Cryogenics (Guildf).* 57, 31–44.
[73] Lagier B, Hoa C and Rousset B. 2014. Validation of an EcosimPro® model for the assessment of two heat load smoothing strategies in the HELIOS experiment. *Cryogenics (Guildf).* 62, 60–70.
[74] Carli S, Bonifetto R, Savoldi L and Zanino R. 2015. Incorporating Artificial Neural Networks in the dynamic thermal-hydraulic model of a controlled cryogenic circuit. *Cryogenics (Guildf).* 70, 9–20.
[75] Froio A, Bonifetto R, Carli S, Quartararo A, Savoldi L and Zanino R. 2016. Design and optimization of Artificial Neural Networks for the modelling of

superconducting magnets operation in tokamak fusion reactors. *J. Comput. Phys.* 321, 476–491.

[76] Drexler NJ, Reed ES and Simon RP. 1992. *Properties of Copper and Copper Alloys at Cryogenic Temperatures.* Washington, DC.

[77] Bottura L. 2000. Practical fit for the critical surface of NbTi. *IEEE Trans. Appl. Supercond.* 10(1), 1054–1057.

[78] Bottura L and Bordini B. 2009. Jc(B,T,ε) Parameterization for the ITER Nb 3Sn production. *IEEE Trans. Appl. Supercond.*, 19(3), 1521–1524.

[79] Duchateau JL. *et al.* 2003. ITER toroidal field model coil test: Analysis of heat transfer from plates to conductors. *Fusion Eng. Des*, 66, 1007–1011.

[80] Ulbricht A. *et al.* 2005. The ITER toroidal field model coil project. *Fusion Eng. Des.* 73(2–4), 189–327.

[81] Zanino R, Bonifetto R, Heller R and Savoldi Richard L. 2011. Validation of the 4C thermal-hydraulic code against 25 kA safety discharge in the ITER Toroidal Field Model Coil (TFMC). *IEEE Trans. Appl. Supercond.* 21(3 PART 2), 1948–1952.

[82] Savoldi Richard L. *et al.* 2013. 4C code analysis of thermal-hydraulic transients in the KSTAR PF1 superconducting coil. In *Cryogenics*, Vol. 53, pp. 37–44.

[83] Zanino R and Savoldi Richard L. 2013. Multiscale approach and role of validation in the thermal-hydraulic modeling of the ITER superconducting magnets. *IEEE Trans. Appl. Supercond.* 23(3), 4900607.

[84] Zanino R and Savoldi Richard L. 2006. A review of thermal-hydraulic issues in ITER cable-in-conduit conductors. *Cryogenics (Guildf).* 46(7–8), 541–555.

[85] Ilyin Y. *et al.* 2010. Performance analysis of the ITER poloidal field coil conductors. *IEEE Trans. Appl. Supercond.*, 20(3), 415–418.

[86] Zanino R, De Palo S and Bottura L. 1995. A two-fluid code for the thermo-hydraulic transient analysis of CICC superconducting magnets. *J. Fusion Energy* 14(1), 25–40.

[87] Bottura L and Marinucci C. 2008. A porous medium analogy for the helium flow in CICCs. *Int. J. Heat Mass Transf.* 51(9–10), 2494–2505.

[88] Bagnasco M, Bottura L and Lewandowska M. 2010. Friction factor correlation for CICC's based on a porous media analogy. *Cryogenics (Guildf).* 50(11–12), 711–719.

[89] Lewandowska M and Bagnasco M. 2011. Modified friction factor correlation for CICC's based on a porous media analogy. *Cryogenics (Guildf).* 51(9), 541–545.

[90] Katheder H. 1994. Optimum thermohydraulic operation regime for cable in conduit superconductors (CICS). *Cryogenics (Guildf).* 34(SUPPL. 1), 595–598.

[91] Zanino R, Santagati P, Savoldi Richard L, Martinez A and Nicollet S. 2000. Friction factor correlation with application to the central cooling channel of cable-in-conduit super-conductors for fusion magnets. *IEEE Trans. Appl. Supercond.* 10, 1066–1069.

[92] Zanino L, Giors R and Savoldi Richard S. 2007. CFD modeling of ITER cable-in-conduit superconductors. Part III: correlation for the central channel friction factor. In *21th International Cryogenic Engineering Conference (ICEC21)*, p. 207.

[93] Nicollet S. *et al.* 2006. Review of singular cooling inlet and linear pressure drop for ITER coils cable in conduit conductor. In *AIP Conference Proceedings*, pp. 1757–1764.

[94] Zanino R, Giors S and Mondino R. 2006. CFD modeling of ITER cable-in-conduit superconductors. Part II. Effects of spiral geometry on the central channel pressure drop. *Fusion Eng. Des.* 81(23–24), 2605–2610.

[95] Zanino R, Giors S and Mondino R. 2006. CFD modeling of ITER cable-in-conduit superconductors. Part I: Friction in the central channel. In *AIP Conference Proceedings*, pp. 1009–1016.

[96] Koizumi N, Takahashi Y and Tsuji H. 1996. Numerical model using an implicit finite difference algorithm for stability simulation of a cable-in-conduit superconductor. *Cryogenics (Guildf).* 36(9), 649–659.

[97] Van Sciver SW. 1986. *Helium Cryogenics.* New York: Plenum.

[98] Brodkey H and Hershey R. 1988. *Transport Phenomena: A Unified Approach.* New York: McGraw-Hill.

[99] Zanino R and Giors S. 2008. CFD modeling of ITER cable-in-conduit superconductors. Part V: Combined momentum and heat transfer in rib roughened pipes. In *AIP Conference Proceedings*, pp. 1261–1268.

[100] Koizumi N, Takeuchi T and Okuno K. 2005. Development of advanced Nb3Al superconductors for a fusion demo plant. *Nucl. Fusion* 45(6), 431–438.

[101] Bonifetto R, Isono T, Martovetsky N, Savoldi L and Zanino R. 2017. Analysis of quench propagation in the ITER central solenoid insert (CSI) Coil. *IEEE Trans. Appl. Supercond.* 27(4), 4700308.

[102] Zanino R, Giors S and Savoldi Richard L. 2010. CFD model of ITER CICC. Part VI: Heat and mass transfer between cable region and central channel. *Cryogenics (Guildf).* 50(3), 158–166.

[103] Savoldi L and Zanino R. 2000. M & M: Multi-conductor Mithrandir code for the simulation of thermal-hydraulic transients in superconducting magnets. *Cryogenics (Guildf).* 40(3), 179–189.

[104] Savoldi Richard L, Casella F, Fiori B and Zanino R. 2010. The 4C code for the cryogenic circuit conductor and coil modeling in ITER. *Cryogenics (Guildf).* 50(3), 167–176.

[105] Bonifetto R, Buonora F, Savoldi Richard L and Zanino R. 2012. 4C code simulation and benchmark of ITER TF magnet cool-down from 300 K to 80 K. *IEEE Trans. Appl. Supercond.* 22(3), 4902604.

[106] Savoldi L, Bonifetto R, Zanino R and Muzzi L. 2016. Analyses of low- and high-margin quench propagation in the European DEMO TF coil winding pack. *IEEE Trans. Plasma Sci.* 44(9), 1564–1570.

[107] Zanino R. *et al.* 2014. Application of the 4C code to the thermal-hydraulic analysis of the CS superconducting magnets in EAST. *Cryogenics (Guildf).* 63, 255–262.

[108] Bonifetto R. *et al.* 2017. Analysis of the cooldown of the ITER central solenoid model coil and insert coil. *Supercond. Sci. Technol.* 30(1), 015015.

[109] Zanino R. *et al.* 2016. Development of a thermal-hydraulic model for the European DEMO TF coil. *IEEE Trans. Appl. Supercond.* 26(3), 4201606.

[110] Zanino R, Bessette D and Savoldi Richard L. 2010. Quench analysis of an ITER TF coil. *Fusion Eng. Des.* 85(5), 752–760.

[111] Lee PY. *et al.* 2010. Redesign of gravity support system for ITER construction. *Fusion Eng. Des.* 85(1), 33–38.

[112] Hoa C. *et al.* 2015. Experimental characterization of the ITER TF structure cooling in HELIOS test facility. In *IOP Conference Series: Materials Science and Engineering*, p. 012148.

[113] Cucchiaro A. *et al.* 2013. Manufacturing of the first toroidal field coil for the JT-60SA magnet system. In *2013 IEEE 25th Symposium on Fusion Engineering, SOFE 2013*, pp. 1–6.

[114] Martovetsky N. *et al.* 2016. ITER central solenoid insert test results. *IEEE Trans. Appl. Supercond.* 26(4), 4200605.

[115] Savoldi L. *et al.* 2017. Analysis of the ITER central solenoid insert (CSI) coil stability tests. *Cryogenics (Guildf).* 85, 8–14.

[116] Marinucci C, Savoldi L and Zanino R. 1999. Stability analysis of the ITER TF and CS conductors using the code Gandalf. *IEEE Trans. Appl. Supercond.* 9(2 PART 1), 612–615.

[117] Ciotti M, Nijhuis A, Ribani PL, Savoldi Richard L and Zanino R. 2006. THELMA code electromagnetic model of ITER superconducting cables and application to the ENEA stability experiment. *Supercond. Sci. Technol.* 19(10), 987–997.

[118] Savoldi Richard L, Bessette D and Zanino R. 2009. Stability analysis of the ITER PF coils. *IEEE Trans. Appl. Supercond.* 19(3), 1496–1499.

[119] Anghel A. 1998. QUELL experiment: Analysis and interpretation of the quench propagation results. *Cryogenics (Guildf).* 38(5), 459–466.

[120] Martovetsky N. *et al.* 2002. Test of the ITER central solenoid model coil and CS insert. *IEEE Trans. Appl. Supercond.*, 12(1), 600–605.

[121] Okuno K. *et al.* 2003. Test of the NbAl insert and ITER central solenoid model coil. *IEEE Trans. Appl. Supercond.*, 13(2), 1437–1440.

[122] Martovetsky N. *et al.* 2003. Test of the ITER TF insert and central solenoid model coil. *IEEE Trans. Appl. Supercond.*, 13(2), 1441–1446.

[123] Bessette D. *et al.* 2009. Test results from the PF conductor insert coil and implications for the ITER PF system. *IEEE Trans. Appl. Supercond.* 19(3), 1525–1531.

[124] Ozeki H. *et al.* 2017. Tcs measurement result of ITER toroidal field insert coil tested in 2016. *IEEE Trans. Appl. Supercond.* 28(3), 4202905.

[125] Zanino R, Bonifetto R and Savoldi Richard L. 2010. Analysis of quench propagation in the ITER poloidal field conductor insert (PFCI). *IEEE Trans. Appl. Supercond.* 20(3), 491–494.

[126] Zanino R, Bonifetto R, Brighenti A, Isono T, Ozeki H and Savoldi L. 2018. Prediction, experimental results and analysis of the ITER TF insert

coil quench propagation tests, using the 4C code. *Supercond. Sci. Technol.* 31(3), 035004.

[127] Zanino R, Bonifetto R, Savoldi L and Muzzi L. 2016. 4C code analysis of high-margin quench propagation in a DEMO TF coil. In *Proceedings — Symposium on Fusion Engineering*, Vol. 2016-May, pp. 1–5.

[128] Savoldi L. *et al.* 2017. Quench propagation in a TF coil of the EU DEMO. *Fusion Sci. Technol.* 72(3), 439–448.

[129] Lue JW, Schwenterly SW, Dresner L and Lubell MS. 1991. Quench propagation in a cable-in-conduit force-cooled superconductor-preliminary results. *IEEE Trans. Magn.* 27(2), 2072–2075.

[130] Zanino R, Bottura L and Marinucci C. 1997. A comparison between 1- and 2-fluid simulations of the QUELL conductor. *IEEE Trans. Appl. Supercond.* 7(2 Part 1), 493–496.

[131] Marinucci C, Bottura L, Vécsey G and Zanino R. 1998. The QUELL experiment as a validation tool for the numerical code Gandalf. *Cryogenics (Guildf).* 38(5), 467–477.

[132] Savoldi Richard R, Salpietro L and Zanino E. 2002. Inductively driven transients in the CS Insert Coil (II): Quench tests and analysis. *Adv. Cryog. Eng.* 47, 423–430.

[133] Takahashi Y, Yoshida K, Nabara Y, Edaya M and Mitchell N. 2006. Simulation of quench tests of the central solenoid insert coil in the ITER central solenoid model coil. *IEEE Trans. Appl. Supercond.*, 16(2), 783–786.

[134] Inaguchi T. *et al.* 2004. Quench analysis of an ITER 13T-40kA Nb3Sn coil (CS insert). *Cryogenics (Guildf).* 44(2), 121–130.

[135] Savoldi Richard L, Portone A and Zanino R. 2003. Tests and analysis of quench propagation in the ITER toroidal field conductor insert. *IEEE Trans. Appl. Supercond.*, 13(2), 1412–1415.

[136] Savoldi Richard L, Bessette D, Bonifetto R and Zanino R. 2012. Parametric analysis of the ITER TF fast discharge using the 4C code. *IEEE Trans. Appl. Supercond.* 22(3), 4704104.

[137] Savoldi L, Bonifetto R, Pedroni N and Zanino R. 2018. Analysis of a Protected Loss of Flow Accident (LOFA) in the ITER TF coil cooling circuit. *IEEE Trans. Appl. Supercond.* 28(3), 4202009.

[138] Savoldi L, Bonifetto R and Zanino R. 2017. Analysis of a Loss-of-Flow Accident (LOFA) in a tokamak superconducting toroidal field coil. In *Safety and Reliability — Theory and Applications*, pp. 67–74.

[139] Bellaera R. *et al.* 2020. Integrated deterministic and probabilistic safety assessment of a superconducting magnet cryogenic cooling circuit for nuclear fusion applications. *Reliab. Eng. Syst. Saf.* 201, 106945.

[140] "Venecia." http://www.alphysica.com/venecia.pdf.

[141] Amoskov V. *et al.* 2006. Validation of VINCENTA modelling based on an experiment with the central solenoid model coil of the international thermonuclear experimental reactor. *Plasma Devices Oper.* 14(1), 47–59.

[142] Zanino R, Bottura L, Savoldi L and Rosso C. 1998. Mithrandir +: A two-channel model for thermal-hydraulic analysis of cable-in-conduit superconductors cooled with helium I or II. *Cryogenics (Guildf).* 38(5), 525–531.

[143] Bonifetto R, Casella F, Savoldi Richard L and Zanino R. 2012. Dynamic modeling of a supercritical helium closed loop with the 4C Code. *AIP Conf. Proc.* 1434(57), 1743–1750.

[144] Zanino R, Bonifetto R, Hoa C and Savoldi Richard L. 2014. Verification of the predictive capabilities of the 4C code cryogenic circuit model. *AIP Conf. Proc.* 1573, 1586.

[145] Supermagnet. https://supermagnet.sourceforge.io/supermagnet.html.

[146] Bagnasco M, Bessette D, Bottura L, Marinucci C and Rosso C. 2010. Progress in the integrated simulation of thermal-hydraulic operation of the ITER magnet system. In *IEEE Transactions on Applied Superconductivity*, pp. 411–414.

[147] Sarkar B, Bhattacharya R, Vaghela H, Shah N, Choukekar K and Badgujar S. 2012. Adaptability of optimization concept in the context of cryogenic distribution for superconducting magnets of fusion machine. In *AIP Conference Proceedings*, pp. 1951–1958.

[148] Fluent A. 2009. Ansys fluent. *Ansys INC.*

[149] Peng N, Liu LQ, Serio L, Xiong LY and Zhang L. 2009. Thermo-hydraulic analysis of the gradual cool-down to 80 K of the ITER toroidal field coil. *Cryogenics (Guildf).* 49(8), 402–406.

[150] Fuchs AM, Blau B, Bruzzone P, Vecsey G and Vogel M. 2001. Facility status and results on ITER full-size conductor tests in SULTAN. In *IEEE Trans. Appl. Supercond.*, 11(1), 2022–2025.

[151] Komarek P and Salpietro E. 1998. The test facility for the ITER TF model coil. *Fusion Eng. Des.* 41(1–4), 213–221.

[152] Abdel Maksoud W. *et al.* 2016. Commissioning of the cold test facility for the JT-60SA Tokamak toroidal field coils. *IEEE Trans. Appl. Supercond.* 26(4), 9500306, June.

[153] Schaubel K, Langhorn A, Lloyd S, Piec Z, Salazar E and Smith J. 2017. The ITER Central Solenoid Module final test facility. *Fusion Eng. Des.* 124, 59–63.

[154] Lee GS. *et al.* 2001. Design and construction of the KSTAR tokamak. *Nucl. Fusion* 41(10), 1515–1523.

Index

Printed in the United States
by Baker & Taylor Publisher Services